Problems in the Behavioural Sciences
GENERAL EDITOR: Jeffrey Gray
EDITORIAL BOARD: Michael Gelder, Richard Gregory, Robert Hinde, Christopher Longuet-Higgins

Latent inhibition and conditioned attention theory

Problems in the Behavioural Sciences

1. Contemporary animal learning theory
   A. DICKINSON
2. Thirst
   B.J. ROLLS & E.T. ROLLS
3. Hunger
   J. LE MAGNEN
4. Motivational systems
   F. TOATES
5. The Psychology of fear and stress
   J.A. GRAY
6. Computer models of mind
   M. A. BODEN
7. Human organic memory disorders
   A.R. MAYES
8. Biology and emotion
   N. MCNAUGHTON

Latent inhibition and conditioned attention theory

R.E. Lubow
Tel-Aviv University

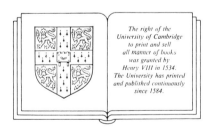

CAMBRIDGE UNIVERSITY PRESS
Cambridge
New York   Port Chester   Melbourne   Sydney

Published by the Press Syndicate of the University of Cambridge
The Pitt Building, Trumpington Street, Cambridge CB2 1RP
32 East 57th Street, New York, NY 10022, USA
10 Stamford Road, Oakleigh, Melbourne 3166, Australia

© Cambridge University Press 1989

First published 1989

Printed in Canada

Library of Congress Cataloging-in-Publication Data
Lubow, Robert E.
Latent inhibition and conditioned attention theory / by R.E. Lubow.
p. cm. – (Problems in the behavioural sciences)
Includes index.
ISBM 0-521-36307-1
1. Conditioned response. 2. Inhibition. 3. Attention.
I. Title. II. Series.
BF319.L76 1989
153.1'532 – dc19　　　　　　　　　　　　　　　　88-38530
　　　　　　　　　　　　　　　　　　　　　　　　CIP

British Library Cataloguing in Publication applied for.

ISBN 0 521 36307 1

Contents

Preface	page	ix
1 Introduction		1
2 Latent inhibition testing procedures		10
Avoidance conditioning		13
Two-way avoidance		16
One-way avoidance		18
Passive avoidance		19
Conditioned taste aversion		20
Preexposed stimuli		21
Conditioned suppression		23
Eyelid blink and nictitating membrane response conditioning		26
Other classical defensive conditioning preparations		27
Miscellaneous classical conditioning preparations		30
Conditioned defensive burying		30
Salivary conditioning		31
Autoshaping		32
Conditioning of orienting responses		37
Heart-rate conditioning		38
Conditioning of observing responses		40
Discrimination learning		44
Discrete stimulus preexposure followed by a classical conditioning test with a compound stimulus (N^+ versus NP^-) and a noxious US		45
Discrete stimulus preexposure followed by an appetitive instrumental conditioning test with a compound conditioned stimulus		47
Discrete stimulus preexposure followed by an appetitive instrumental test with simple CSs		50
Continuous stimulus preexposure followed by a simple instrumental appetitive test		52
Other preexposure–discrimination paradigms		55
Summary		56
3 Variables affecting latent inhibition		58
Similarity of preexposed stimulus and test stimulus		58

Number of stimulus preexposures	59
Duration of the preexposed stimulus	63
Intensity of the preexposed stimulus	64
Interstimulus interval	65
Retention of latent inhibition	66
Time between stimulus preexposure and acquisition	67
Interaction between duration of the preexposed stimulus and time between preexposure and conditioning	69
Time between acquisition and test	69
Role of context	74
Context change, S_oE_o versus S_oE_n	76
Independent familiarization of stimulus and context (S_o, E_o)	79
Context preexposure that continues after stimulus preexposure	80
Context exposure prior to stimulus–context preexposure	81
Second-stimulus effects	82
S_1–S_2	83
Hall-Pearce effect	87
Some comments on primacy	89
S_2–S_1 effects	89
S_1S_2: preexposing a compound stimulus	92
4 Organismic variables affecting latent inhibition	**96**
Age	96
Sex and early handling interaction	101
Species	101
Conditioning in a mollusc	101
Conditioning in the honeybee: appetitive and aversive	102
Conditioning in the goldfish	103
Pigeon	106
Subhuman mammals	106
Humans	107
Summary	107
5 Associative learning tests of the effects of stimulus preexposure in children and adults	**108**
Children	108
Lubow, Rifkin, and Alek (1976): Effects of stimulus preexposure and context	109
Kaniel and Lubow (1986): Age and social class affect latent inhibition	109
Lubow, Caspy, and Schnur (1982): Latent	

inhibition is stimulus-specific; older children require masking	110
Lucas (1984): Some contradictory findings	111
Adults	112
Classical conditioning of the eyeblink response	113
Ivanov-Smolensky conditioning	115
Classical conditioning of the GSR	116
Instrumental learning	121
Other studies	128
6 Neural substrates of latent inhibition	**130**
Septo-hippocampal system	130
Noradrenergic manipulations	132
Cholinergic manipulations	133
Serotonergic manipulations	134
Dopaminergic manipulations	134
Opiate manipulations	138
Summary	138
7 Theories and explanations of latent inhibition in animals	**141**
Conditioned inhibition	142
Competing responses	145
Habituation theories	147
Habituation of the alpha response	148
Latent inhibition and long-term habituation	150
Latent inhibition and general habituation theories	151
Habituation of the orienting response	154
Selective attention	160
Information processing theories of latent inhibition	162
Rescorla-Wagner model	164
Mackintosh's theory of attention	165
Pearce-Hall model	169
Frey-Sears model	174
Wagner's priming theory of conditioning and habituation	175
Context change	177
Context extinction	177
Priming tests	178
Other considerations	179
Evaluation of prediction theories	180
Three new hypotheses	182
Retrieval failure	182
Reminder treatments	183
Effects of delay between acquisition and test	184

Trace conditioning hypothesis	186
Inhibition-of-delay hypotheses	188

8 Conditioned attention theory of latent inhibition — 190

Predictions related to conditioning of inattention to S_1	196
Predictions related to modulation of attention to S_1 by S_1–S_2 pairings	200
Conditioned attention theory and conventional conditioning	205
Some modifications and elaborations	208
Stimulus duration considerations	209
Context effects considerations	213
Parallel processes in stimulus preexposure stage and acquisition–test stage (normal conditioning)	215

9 Conditioned attention theory as applied to latent inhibition in humans — 218

Attentional decrement overrides	223
Which attention?	223
External attentional decrement override	226
Internal attentional decrement override	228
Stimulus preexposure engages two different encoding processes	230
Property extraction	232
Stimulus relationship processing	233
More about context	235
Final comments on conditioned attention theory	237

10 Some applications of conditioned attention theory: learned helplessness and schizophrenia — 239

Learned helplessness	239
Schizophrenia, latent inhibition, and attention	247
Effects of amphetamine and haloperidol on latent inhibition in rats	250
Schizophrenia and latent inhibition	251
The Lubow et al. (1987) study	252
The Maudsley studies	257
Some speculations	260
Notes	265
References	273
Author index	311
Subject index	321

Preface

It is difficult to determine exactly when I began this book, as such. However, the research on latent inhibition was initiated in the late 1950s while I was still a graduate student at Cornell University. The next memorable date for measurement was the publication of an article reviewing the latent inhibition literature in the *Psychological Bulletin* (1973), which appeared in about half of its original length. The conditioned attention theory of latent inhibition was developed later in a series of articles in the *Journal of Experimental Psychology: Animal Behavior Processes* in the mid-1970s, and then presented more fully in a chapter in *Progress in Learning and Motivation* edited by G.H. Bower (1981). The idea of developing this material into a book emerged during a sabbatical year at Yale University, 1977–1978, while much of the writing itself was postponed until another sabbatical, 1981–1982, as an Israel-Canada Fellow at Concordia Unversity. To all of these institutions, and to the individuals who were responsible for inviting me, I express my sincerest gratitude, particularly to Allan Wagner and Eugene Rothman.

In addition, throughout the years I have been fortunate to have my research supported by a number of organizations: The National Institutes of Health were particularly encouraging, especially at the early stages of my career with a Career Development Award. Other support has come from the Scottish Rite Schizophrenia Research Program and, in Israel, from the Charles Smith Psychobiology Fund, the Israel National Academy of Science, Israel Research Trustees Foundation (Ford), and Tel-Aviv Unversity. Without the support of these agencies, the research culminating in this book could not have been accomplished.

A number of people have also substantially contributed to this volume. In particular, my warm appreciation goes to Paul Schnur and Ina Weiner, who not only read early drafts of various chapters but also coauthored many of the key journal articles that preceded the book. I have been most fortunate to enjoy their help and stimulation. Indeed, the chapter on the neural substrates of latent inhibition owes its existence to Dr. Weiner.

Finally, no book can be born without a dedicated typist, and I have been favored with two: Hedy Mendelson and Paula van der Werff. Thank you all.

Tel-Aviv   R.E. Lubow

1 Introduction

Modern science has developed to such a point that there are myriad research areas, dark corners as well as bright little chambers, that are penetrated only by a knowledgeable few who happen to be working in a given field – the specialists. I would not be surprised, then, to discover that most of my colleagues, whether they label themselves psychologists, psychobiologists, behavioral scientists, or neuroscientists, do not know what "latent inhibition" is, or at best confuse the term with older research areas that are indeed its distant cousins – latent learning and conditioned inhibition. It is for these readers that I begin this treatise with a definition, as simple as it may be.

"Latent inhibition" is defined by three characteristics. One is concerned with conditions for producing it, the second with the conditions for measuring the effect, and the third with the direction of the differences between groups. More specifically, latent inhibition is the detrimental effect of passive, nonreinforced preexposure of a stimulus on the subsequent ability of an organism to form new associations to that stimulus. To demonstrate latent inhibition, one must preexpose one group of subjects to the stimulus of interest, while not giving such stimulus preexposure to a control group. In the test phase, both groups must learn to form an association between that stimulus and a new event. When the stimulus-preexposed group learns the new association to that stimulus more poorly than does the control group, we say that latent inhibition has been demonstrated.

From the foregoing definition of the phenomenon, it should be clear that the term "latent inhibition" is completely devoid of any theoretical significance or surplus meaning. Indeed, the genesis of the term comes from the fact that the first study that demonstrated the effect (Lubow & Moore, 1959) was designed to provide a classical conditioning analog of "latent learning."[1] Latent learning was a research topic that received much attention during the decades of the 1930s through the 1950s, the era of the titanic theoretical battles between Clark Hull and Edward Chance Tolman. Studies of latent learning, together with studies of sensory preconditioning,[2] sought, for the most part, to differentiate between two theoretical approaches to associative learning theory. One of these theories stated that stimulus–response contiguity *and* reinforcement were necessary to form an association (Hull, 1943). The other stated that mere contiguity of stimuli was sufficient to form an association (Tolman, 1932).

The two classes of experiments, latent learning and sensory preconditioning, are similar; the only differences concern the types of stimuli to be associated and the form of training used to make the otherwise latent learning apparent. Sensory preconditioning experiments (e.g., Brogden, 1939) employed discrete, specifiable stimuli and classical conditioning; latent learning experiments (at least the early ones, e.g., Blodgett, 1929, and Tolman & Honzik, 1930) used vaguely identified "maze stimuli" and instrumental learning. The data from both types of research, when they produced positive results, frequently were used as the basis for concluding, either implicitly or categorically, that reinforcement was not necessary for learning and, further, that it was a stimulus–stimulus association that was learned. Although the first conclusion was acceptable when it referred to "latent reinforcement," the second conclusion was not deducible from the data of these experiments. In most cases where stimulus–stimulus associations were assumed, the experimenter had not been interested in obtaining response measures during the course of the preexposure period.

Our first latent inhibition studies (Lubow & Moore, 1959) were designed to look at these problems. We sought to recast latent learning with its instrumental conditioning procedures and ill-defined maze stimuli into a classical conditioning procedure with explicitly defined stimuli. In these original studies, we had expected to find a facilitative effect as a result of stimulus preexposure, thus paralleling the findings in latent learning and sensory preconditioning. However, as many now know, we failed to demonstrate such an improvement in learning, and instead found that stimulus preexposure *interfered* with subsequent associations to that stimulus.

It was at this point that the phenomenon was named "latent inhibition": "latent" because the experiments were initiated in order to demonstrate *latent* learning, which originally was used in the learning laboratories to mean that the learning was not visible until the application of a later test with appropriate conditions; "inhibition" was used as a descriptive word to refer to the subsequent learning *decrement*. The word "inhibition" was not meant to suggest any inhibitory processes, neither Pavlovian nor Hullian. The term "latent inhibition," then, merely paraphrased "latent learning" and pointed to an aspect of learning that resulted in a performance decrement.

Why should one devote a book to the topic of latent inhibition? Certainly the dry definition of the phenomenon, as described in the first paragraphs of this introduction, does not, by itself, suggest a compelling motive. The answer to the question can be found in two separate but interrelated reasons. One is concerned with the fact that the topic – decremental associability effects of stimulus preexposure – until recently had been completely neglected in learning theory. The causes of this neglect will be examined in a moment. As I have shown elsewhere (Lubow, 1973a), latent inhibition is a robust, ubiquitous phenomenon occurring across a wide variety of species and a diverse set of testing conditions. All mammals tested show the effect, and it has been demonstrated with such different procedures as conditioned taste aversion, conditioned suppression, classical conditioning of various responses, avoidance conditioning, appetitive instrumental condition, and others. Given the broad range of species and conditions, one would expect that latent inhibition would be a biologically significant phenomenon (i.e., that it would serve an important adaptive function). In fact, I shall argue that, without latent inhibition, learning, as we know it, would be a cumbersome and inefficient process. Latent inhibition allows for the stimulus selectivity that is a requirement for rapid, efficient learning. This, too, will be discussed in more detail shortly. First let us go back to the question of neglect.

Although Tolman and Hull had basic differences of opinion on the questions of what is learned and what are the necessary conditions for such learning to occur, they both conceived of answers to these questions in terms of a two-element paradigm: stimulus–stimulus in the first case, and stimulus–response in the second. The associative bias inherited from the British empiricists, such as Locke and Hume, shaped the basic learning paradigm so that it had to consist of at least two elements. Association, by definition, could not be conceived of as occurring to anything less than two separately identified, discrete elements.

The associationist paradigm suggests that the pairing of two elements, when successful, results in something *more* than the original inputs. As in Liebig's creation of an organic compound, urea, from the association of two inorganic elements, so learning was envisaged in terms of being an increase in, something more than, the sum of the two parts. Similarly for Hull and Tolman, evidence for the learning of an association was established in terms of an organism's *improved* performance as compared with some control condition.[3]

These two assumptions of associationist doctrine – first, that *two* or more elements are required, and, second, that the resultant is an *addition* of some quality – served to blind behavioral scientists, for many decades, to what is in fact the most ubiquitous of influential experiences, that of passive exposure to environmental stimuli. Indeed, when such studies of the effects of passive stimulus preexposure were undertaken, these research activities were carried out under the rubric of "habituation" and were, until recently, insulated from any ideas pertaining to learning.

Studies of latent inhibition, which began in the late 1950s (Lubow & Moore, 1959), broke with the two-element conceptualization of learning. Ironically, though, the expectations in those early studies were such that we still looked for an *improvement* in subsequent learning. It was only after two experiments produced impaired learning, and we could find no conditions within the design or execution of the experiments that allowed for an explanation as to why facilitation was not obtained, that the phenomenon was thought to be genuine and was labeled "latent inhibition." It is interesting to note that, about 10 years later, the single-element design with decremental effects became the basis for another popular learning paradigm, namely, learned helplessness (Maier, Seligman, & Solomon, 1969).

Within the larger framework of learning theory, a phenomenon such as latent inhibition should have been expected, if not required. Again, however, one can look to the influence of Hull as an obstacle to such an expectation. Hull embraced the notion of equipotentiality of stimuli, which required that all stimuli that impinge on the sensory surfaces of an organism at the time of reinforcement should gain equally in associative strength. Only the probabilistic or experimenter-manipulated distribution of stimuli from trial to trial causes a particular stimulus to gain a "majority" of the associative strength, thus becoming responsible for eliciting a conditioned response. Today, the idea of stimulus equipotentiality has been severely undermined by such phenomena as blocking, overshadowing, and examples of prepared learning such as conditioned taste aversion. Similarly, the phenomenon of latent inhibition denies the premise of stimulus equipotentiality. Latent inhibition data demonstrate that future associations of stimuli with responses or other events will be markedly impaired by the familiarity of the stimulus.

Unreinforced preexposures of stimuli, then, play an important role in the general learning process. They prepare an organism, so that at the time that it is required to make an association between the stimulus and another event, equipotentiality is *inoperative*. A disadvantage accompanies that stimulus that has not previously been associated with reinforcements. In the competition that stimuli encounter in gaining access to new associations, the novel stimulus is given favored status. Because it is the natural condition of an organism that at any given moment it is inundated by a virtual sea of stimuli, these various mechanisms for selecting those stimuli that will enter into new associations – often necessary for survival – are of considerable value. That latent inhibition plays an exceptionally important role in providing this filter function can be demonstrated by comparing it to the other phenomena with similar functions: blocking, overshadowing, and preparedness.

Let us take a simple example from the natural world. Consider the following situation. Lightning flashes, and then thunder is heard. Certainly, an association between the two must have been perceived by even the most primitive, prehistoric human, and probably after only one experience with these events. Although the possibility of one-trial learning may have had its detractors in the past, it is now fairly obvious that such phenomena not only exist but probably are common occurrences. There are numerous laboratory examples in the animal learning literature: for example, conditioned taste aversion, where a single pairing of a novel flavor with internal malaise is sufficient for the animal to avoid that flavor in the future; similarly in the conditioned suppression literature, where one pairing of a tone and a moderately intense shock is sufficient to form an association between the two, as manifested in the ability of that tone to disrupt subsequent behavior. But given temporal and/or spatial contiguity of events, why do we not form associations within every sequence of stimuli that passes our way? From the two animal examples, one might conclude that at least one of the stimuli must be motivationally significant, such as digestive upset, or pain from shock, or, in the lightning–thunder example, a noxiously loud noise. But the availability of a motivationally significant stimulus would account for only half of the problem. It may explain why a painting over a couch is *not* associated with the color of the couch. However, why does the child, sitting in the living room, when it first experiences a loud clap of thunder, not associate the thunder with the color of the couch or the picture, or any of the other stimulus attributes of these objects, or the objects themselves?

If the flash of lightning were visible, one could appeal to the Pavlovian notion of overshadowing, which promises better associability to the more intense or salient of the two or more stimuli present at the same time, at the expense of the weaker stimuli. Similarly, if the child had prior experience with lightning–thunder pairings, the phenomenon of blocking might come into play; a prior association between two events interferes with the ability of a new stimulus to enter into that association. Overshadowing can be thought of as creating an associative bias on the basis of certain stimulus characteristics such as intensity; blocking can be though of as creating an associative bias on the basis of past stimulus associations. Overshadowing and blocking are processes that enfeeble the equipotentiality of stimuli and as such provide some limits for what will be associated. But these two mechanisms cannot, by themselves, be major sources for structuring the otherwise associatively homogeneous equipotential stimulus world. In the laboratory, both overshadowing and blocking have been shown to be highly constrained. That is, their limits of action are sharply curtailed by specific procedural requirements, particularly the temporal parameters that relate to the two stimuli at the time that they are competing for the association with the third, motivationally significant, stimulus. Consequently, it is likely that these two phenomena have only a limited stimulus selection function in the natural world. In addition, for overshadowing, biological adaptiveness is not always best served by stimulus salience, especially when salience is defined in terms of stimulus intensity; that is, an intense stimulus is not always the one for which learning an association is adaptively appropriate.

Physical stimulus intensity aside, the premise of equipotentiality is challenged by the propensity for some stimuli to enter into certain associations more easily than others. One can find many examples of stimuli that have a preferential status, such as in imprinting, or again in conditioned taste aversion. The demonstrations that flavor–illness associations are more easily established than sound–illness associations, and, conversely, sound–shock associations are more easily established than flavor–shock associations (Domjan & Wilson, 1972b; Garcia & Koelling, 1966), provide good examples of such biases. Indeed, these preferred associations go a long way toward imposing a structure on the stimulus domain. However, to the extent that this tactic is successful, it also limits the adaptability of the organism by curtailing new, potentially significant associations. At the extreme, biologically determined preferential associations would produce a reflex machine perhaps only a little more flexible than the tropistic organisms of Loeb (1918).

What is required in order to have an organism that can conveniently form new associations that are both efficient and valid for increasing its chances of survival is *an automatic process that degrades the associability of familiar stimuli that have had no consequences in the past*. For many, this statement will immediately conjure up the notion of habituation and thus the question, So what's new? Indeed, there may be a relationship between latent inhibition and habituation. However, for our purposes here, it should be sufficient to remind the reader that habituation typically refers to the waning of a *specific* response *contemporaneously* with repeated stimulation of that response. Latent inhibition, on the other hand, refers to a decrement of future associability of the preexposed stimulus with any response. If habituation can be descriptively characterized as the waning of a specific response with the repetitive presentation of a given stimulus, then latent inhibition must be characterized by the fact that repetitive presentation of a given stimulus will interfere with the acquisition of *any* response to that stimulus. Thus, simultaneously, latent inhibition is stimulus-specific, but the stimulus preexposure experience affects any future association with that stimulus. Unlike habituation, latent inhibition, on logical grounds, demands some central processing. Although certain habituation phenomena *may* also require central processing, such as dishabituation, the simple response decrement to iterative stimulation may be demonstrated in spinal organisms (e.g., Thompson & Spencer, 1966) as well as in one-cell organisms (see Wyers, Peeke, & Herz, 1973, for a review). No such demonstrations have been presented for latent inhibition, nor is it likely that they exist.

In summary, the phenomenon of latent inhibition is exquisitely simple – poorer learning as a result of passive stimulus exposure. Yet, in spite of, or perhaps because of, this simplicity, latent inhibition would appear to play a major role on the stage of learning – setting the background in such a manner as to create a bias in favor of potentially important stimuli. This is accomplished by degrading the associability of those stimuli that have been registered as inconsequential by the organism in the past. Indeed, it is difficult to conceive how new learning could take place if there were not some constantly operating mechanism that worked to dampen out associative possibilities on the basis of such past experience.

In this short introduction, I have briefly described the latent inhibition phenomenon, explained its functional significance, and suggested why it may have been neglected in earlier learning theories. What, then, are the purposes of this book? There are several. Considering the significance of the topic, it is important to have a single source that summarizes the considerable empirical work that has been published. Although several authors have attempted such an integration in the past (Lubow, 1973a; Weiss & Brown, 1974), those reviews are now almost 15 years old and therefore quite incomplete. Also those original summaries were constrained by the page limitations of journals and consequently did not allow for full expression of the data and their meanings.

Second, there are now enough research reports in the literature that it is possible to derive *new* empirical generalizations by cross-experiment comparisons. This is of particular value in order to gain a fuller understanding of the latent inhibition phenomenon and , concurrently, to provide a basis for theory construction and evaluation. Furthermore, a fine-grained analysis of the designs and variables will lead to identification of those areas where additional research is most needed.

Third, the latent inhibition phenomenon, which is considered by many theorists to be a reflection of attentional processes, has become of increasing significance to neuroscientists, who see it as a convenient tool for measuring the effects of various manipulations, such as drug treatments and lesions, on attention itself. As such, latent inhibition provides one of the few animal paradigms that directly relates to our understanding of normal human cognitive processes as well as processes in which attentional deficits are implicated, as in schizophrenia (Lubow, Weiner, & Feldon, 1982). This volume, for the first time, makes available to these neuroscientists a comprehensive treatise of latent inhibition, not only the data but also the underlying assumptions and theories. It is hoped that this will serve to increase their understanding of the bases for their applications of the latent inhibition procedures and will encourage more investigators to explore this essential behavioral phenomenon.

Finally, I consider this book to be an exercise in determining the general value of our current approach to data collection and theory construction in psychology. Again, the reader is to be reminded that we are dealing with an empirical phenomenon of utmost simplicity – even less complex than classical conditioning. Given this simple, almost raw behavioral fact, and given that there are now hundreds of latent inhibition experiments, should we not expect to be able to construct a theory that is viable by at least those conservative criteria of good theories – integrative ability, internal consistency, and predictive accuracy?

To the extent that this goal will be achieved by the data reported in this volume, either from my own theorizing or via that of someone else, this book will have served a most useful function. To the extent that this cannot yet be accomplished, but the book identifies questions and issues that ultimately will allow for appropriate closure on the field, then, too, I shall consider the book a success. If neither of these possibilities obtains, there would seem to remain only two alternatives: Either I have failed, in some way, to present the data in a comprehensive and/or meaningful fashion or, indeed, the type of behavioral research that characterizes investigation of the latent inhibition effect, as well as many other areas of experimental psychology, may be in need of a major reconceptualization – a paradigm shift (Kuhn, 1962), if you will. The final judgment on these issues remains with the reader.

This book is divided into several sections. One major division separates the empirical bases of latent inhibition from theoretical considerations. The first section of the review of the empirical studies is divided into specific procedures in which latent inhibition has been investigated, as, for example, conditioned taste aversion, avoidance conditioning, and autoshaping. The reason for using "procedure" as a category of analysis was impelled by the belief, later substantiated by the data, that cross-procedural data are not always uniform and that differences in testing paradigms often obscure meaningful empirical generalizations. Furthermore, by looking for consistencies of data across methods, as well as differences, it may be possible to induce new generalizations, not only about latent inhibition but also about the procedures themselves.

In addition to providing an analysis of the different learning procedures, the empirical section examines the effects of a variety of variables on latent inhibition. This is done whenever there are sufficient numbers of studies reporting on the influence of a given variable so that some general statement in regard to its effects on latent inhibition might be made. The variables that have been scrutinized include not only typical experimental manipulations, such as number of stimulus preexposures, duration of stimulus preexposure, stimulus intensity, and the effects of context change, but also organismic variables such as age, sex, and species, and the effects of physiological manipulations such as drugs and lesions. Often the variable in question will not have received a parametric evaluation in a particular study, but there will have been sufficient numbers of studies using very similar procedures with different values of the variable so that an empirical generalization is warranted.

The chapters on theory immediately follow the data sections. Thus, those theories of latent inhibition that have been derived from animal data are presented separate from those theories that have been derived from human data. Because the two sets of theories have completely ignored each other, there is no apparent loss in using such a disquieting arrangement. In these chapters, specific theories are examined. For latent inhibition in animals, these include Wagner's priming theory (1976), Lubow's conditioned attention theory (Lubow, Weiner, & Schnur, 1981), Mackintosh's attention theory (1975), and Pearce and Hall's theory (1980). The formal characteristics of these theories are examined, as well as their success in integrating and/or predicting the results described in the empirical chapters. In particular, because the major theories of latent inhibition are based on the premise that something is learned during the stimulus preexposure period, two questions will be asked of each of the theories, questions that also marked our initial characterizations of the Hull–Tolman controversies, but are equally valid here: What is learned? How is it learned? Or, in more modern terminology, What is encoded, and how is it stored and retrieved?

To bridge the gap between the animal and human data, and between their respective theories, a chapter is devoted to an integration of these materials: A single theory is proposed that takes into account the ontogenetic, phylogenetic, and procedural similarities and differences that have been uncovered in the earlier chapters, a theory that it is hoped will provide some scientific basis for Bernard Berenson's insightful and eloquent observations on the *habit of inattention*, made some eight decades ago:

For nature is chaos, indiscriminately clamoring for attention. Even in its least chaotic state it has much more resemblance to a freakish and whirlingly fantastical "Temptation of St. Antony" by Bosch, than to compositions by Duccio. . . or to others by Raphael. . . . To save us from the contagious madness of this cosmic tarantella, instinct and intelligence have provided us with stout insensibility and inexorable habits of inattention, thanks to which we stalk through the universe tunneled in and protected on every hand, bigger than the ants and wiser than the bees. [Berenson, 1952/1980, p. 104][4]

Finally, a last chapter examines some possible applications of latent inhibition principles and conditioned attention theory. With increasing degrees of speculation, relationships between the data and theory of the earlier chapters are applied, first to learned helplessness and then to schizophrenia – "whirlingly fantastical."

2 Latent inhibition testing procedures

Originally, this book was to have been organized such that a separate chapter would be devoted to each of the major paradigms within which latent inhibition had been demonstrated. Although such a categorization proved to be convenient for the initial organizing of the hundreds of published reports, it soon became clear that there was an absence of *major* differences in the latent inhibition phenomena that could be attributed to the type of testing procedure employed. Therefore, it was decided to present the empirical evidence for latent inhibition on the basis of manipulated variables, organismic variables, and variations across experiments. It is, of course, in this latter category that testing procedures are to be found. Procedures are placed at the head of the list, for two reasons: first, to present evidence that latent inhibition is indeed a ubiquitous phenomenon, found across a very wide range of testing procedures; second, by accomplishing the preceding, to avoid, in subsequent sections, the necessity of repetitiously labeling the experimental procedures for each point that is made.

Before identifying those paradigms in which latent inhibition has been investigated, two general comments are in order, one concerning the number of stages in the procedure, and the second concerning the problem of differentiating the conditioned response (CR) to the target stimulus in the test from the unconditioned response (UR) to that stimulus.

By definition, a demonstration of latent inhibition requires, at a minimum, a two-stage procedure – stimulus preexposure and acquisition, whereby in the latter stage a measure of the effects of the preceding stimulus preexposure stage is extracted. However, some testing paradigms, such as conditioned suppression and conditioned taste aversion, routinely employ three stages, in which conditioned stimulus–unconditioned stimulus (CS–US) pairings are given prior to and independent of the subsequent test of the effectiveness of those pairings. This three-stage procedure introduces a number of complications, all of which center on the problem of whether to attribute stage 3 test differences to an effect of the strength of the CS–US association in stage 2 or to a *direct* effect of the stimulus preexposure on test responding. The problem can be illustrated when one examines the effects of stimulus preexposure on conditioned taste aversion. As indicated, this procedure requires three stages to demonstrate latent inhibition (stimulus preexposure, conditioning, and testing), and the major assumption is that

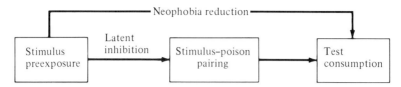

Figure 1. Two independent routes for preexposure effects in the taste aversion paradigm.

increased test consumption by the preexposed group as opposed to the nonpreexposed group reflects weaker conditioning in the second stage. However, it is also well established that flavor preexposure, by itself, will result in subsequently increased consumption of the flavored item (i.e., reduced neophobia) (e.g., Domjan, 1976; Franchina & Gilley, 1986). How, then, can one determine whether stimulus preexposure affects consumption directly or affects consumption by reducing the *conditioned* flavor–poison association? The two processes, which can be labeled as neophobia reduction and latent inhibition (Figure 1), both operate in the same direction – to increase consumption of the test fluid. It follows, then, that suitable care must be taken in all three-stage procedures to differentiate the latent inhibition effect from other effects that might produce the same results.

The situation described in the preceding paragraph is not unrelated to the problem of differentiating the CR to the target stimulus from the UR to that stimulus. The best example can be taken from the three-stage conditioned lick-suppression procedure for demonstrating latent inhibition, as employed by Carlton and Vogel (1967, Exp. 1). Two groups were differentiated, during the preexposure stage, by the fact that one received a number of tone presentations and one did not. On the following day all subjects were given one tone–shock pairing. The third stage, test, was conducted on the next day. The animals, on a water-deprivation schedule, were each allowed to make 100 licks on a water tube. On completion of the last response, the tone was activated and remained on until an additional 10 licks were performed. The basic measure, then, was simply the amount of time that was required to complete these 10 licks during the presence of the tone. This procedure proved to be extremely robust and produced two groups with virtually no overlap in data. "Prior habituation drastically attenuated the normal consequences of subsequent conditioning" (Carlton & Vogel, 1967, p. 348). Nevertheless, there is a problem here. A demonstration of latent inhibition requires that the effect of stimulus preexposure be on the subsequent association of the previously preexposed stimulus and another event – in this case between tone and shock. With this procedure, as in conditioned taste aversion, it is possible that stimulus preexposure may

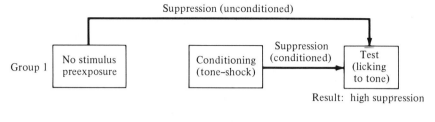

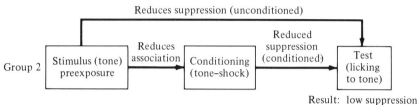

Figure 2. Two ways to reduce test suppression in Group 2 compared with Group 1: conditioned and unconditioned effects.

have directly affected test-stage drinking independent of affecting the association between tone and shock in the conditioning phase. The situation is illustrated in Figure 2. For Group 1, the nonpreexposed group, the amount of suppression of licking during the presence of the tone in test is determined by two factors: (1) the strength of the association or *conditioning* between tone and shock in the conditioning stage; the stronger the association, the greater the "fear" that is elicited by the tone in test and thus the greater the suppression of licking; (2) the *unconditioned* suppression of licking; a novel stimulus, particularly if it is intense, will tend to disrupt any ongoing activity. Therefore, in the nonpreexposed group, where the tone is relatively novel in the test, one also would expect an increase in suppression of licking. Thus, the total amount of lick suppression to the tone in the test would be a joint function of both conditioned and unconditioned effects. For Group 2, which received tone preexposures, the situation is more complex. Here the amount of suppression of licking, during the presence of the tone in the test, is determined by three factors: (1) as with the nonpreexposed group, the strength of the association between tone and shock in the conditioning stage; (2) however, this is presumably reduced as compared with the nonpreexposed group because the prior exposure of the tone interferes with the acquisition of this association; (3) test suppression to tone may be reduced *directly* by prior stimulus exposure, because a familiar stimulus will lose its unconditioned capacity to disrupt ongoing behavior.

Figure 2 illustrates the manner in which the reduced suppression in the stimulus-preexposed group may come from two different sources, conditioned and unconditioned, as described. The Carlton and Vogel study, like many others, does not give us the opportunity to decide whether stimulus preexposure affects the subsequent conditioned or associative strength of the stimulus when it is paired with shock or whether the preexposure directly affects the unconditioned strength of the stimulus.

In a second experiment, Carlton and Vogel (1967) used the same procedures but added some important groups to the original two-group design. Of most interest, there were two control groups that were not preexposed to the tone. During the conditioning stage, one of these groups received only shock (with no tone CS), and the other received neither the shock nor the tone. Thus, at test, the tone for both of these groups was completely novel, and one could assess the contribution of unconditioned suppression to the differences between the stimulus-preexposed and nonpreexposed groups. The data indicated that unconditioned suppression was not a factor in producing the differences between the latter two groups. The same point using different experimental designs has been made by Domjan and Siegel (1971) and Lubow and Siebert (1969), with both studies demonstrating that there are, indeed, two separable effects of preexposure of the to-be-conditioned stimulus. Baker and Mercier (1982) and Mercier and Baker (1985) provide one way of reducing the unconditioned component of the conditioned response – to give the nominally nonpreexposed group a few stimulus preexposures. In this manner, the unconditioned suppression of the stimulus is habituated and not allowed to enter into the test suppression score. Although the logic of this procedure is clear, it is also possible that the few stimulus preexposures have effects in addition to reducing unconditioned suppression, such as producing latent inhibition itself.

It should be clear that the problem of contamination of the CR with unconditioned responding is not limited to three-stage procedures. In addition to the preceding, any testing paradigm in which the unconditioned response to the nominal stimulus (sometimes called the orienting response or alpha response) resembles the CR may be a source of concern, as, for example, heart-rate conditioning, galvanic skin response (GSR) conditioning, and others. With the two caveats described in the preceding sections kept in mind, let us examine the different testing procedures that have been employed to demonstrate latent inhibition.

Avoidance conditioning

A common procedure for demonstrating latent inhibition is that of avoidance learning. In this procedure the subject must perform (or refrain from performing) a particular response in order *not* to receive shock. Conversely, failure to execute the response results in the animal receiving shock. Although there are several variations of the avoidance

Table 1. *Latent inhibition studies using two-way avoidance in rats*

Experiment	Number of Preex-posures	Duration of Preex-posures (sec)	PE-test delay (hr)	Strain	Sex
Ackil & Mellgren (1968)	0, 5, 60	5	0	Hooded	M
Ackil et al. (1969)	0, 30	5	0	Hooded	M
				Long Evans	M
Alleva et al. (1983)[a]	0, 36	5	48	Wistar	?[b]
Archer (1982a, Exp. 3)	0, 5	10	24	Sprague-Dawley	M
Asin et al. (1980, Exp. 1)	0, 30	10	0	Sprague-Dawley	M
Feldman (1977)	0, 100	5	0	Charles River	M
Hellman et al. (1983)	0, 30	5	0	Sprague-Dawley	M
	0, 30	5	0	Sprague-Dawley	M
Solomon et al. (1981)					
Exp. 2	0, 30	5	0	Sprague-Dawley	M
Exp. 3	0, 30	5	0	Sprague-Dawley	M
Exp. 1	0, 30	10	0	Sprague-Dawley	M+F[d]
Solomon et al. (1978)					
Exp. 2	0, 30	10,	0	Sprague-Dawley	M+F[d]
Solomon et al. (1980)	0, 30	5	0	Sprague-Dawley	M
Solomon & Staton (1982)	0, 30	5	0	Sprague-Dawley	?[b]
Weiner (1983)					
Exp. 2.1-2.5	0, 50	5	24	Wistar	M
Exp. 2.6	0, 50	5	0, 24, 48	Wistar	M
Exp. 2.7	0, 50	5	72	Wistar	M
Exp. 1 Appendix	0, 50	5	0, 24	Wistar	M
Exp. 2 Appendix	0, 50	5	0, 24	Charles River	M
Weiner et al. (1984)	0, 60	5	24	Wistar	M+F
Weiss et al. (1974)	0, 30	10	0	Long-Evans	M

[a] Study reported briefly in introduction (Alleva et at., 1983, p. 83).
[b] Sex of animals not indicated.
[c] N-2-chloreothyl-N-2-ethyl-bromobenzylamine.
[d] Numbers for each sex not indicated. No analysis for sex differences.
[e] Parachlorophenylalanine.

Other variables	Results
–	LI for 60 PE
Hippocampal and neocortical lesions	LI for nonoperated and neocortical-lesioned groups
–	LI
DSP4[c]	LI for stimulus-preexposed group with and without DSP4
Medial raphe lesions	LI for nonoperated group
–	LI
Amphetamine (1 mg/kg), tail pressure	LI for stimulus-preexposed groups with saline, amphetamine, or tail pressure; amphetamine plus tail pressure – no LI
Amphetamine (1 or 4 mg/kg)	LI for saline control and amphetamine (1 mg/kg); no LI for 4 mg/kg
Amphetamine (4 mg/kg), chloropromazine	LI for amphetamine plus chloropromazine; no LI for amphetamine alone
Haloperidol, barbiturate	LI for saline control and barbiturate group; no LI for haloperidol
PCPA[e]	LI for controls; no LI in PCPA group
Medial raphe lesions	LI, retardation but no summation
Dorsal raphe lesions	LI for controls and dorsal raphe lesions; no LI for medial raphe lesions
Amphetamine in nucleus accumbens or caudate putamen	LI for controls and amphetamine in caudate putamen; no LI for amphetamine in nucleus accumbens
–	LI
–	LI for 24- and 48-hr delay; no LI for 0 delay
–	LI
–	LI for 24-hr delay; no LI for 0 delay
–	LI for 0 delay; no LI for 24-hr delay
Handled, nonhandled	LI for handled males and females, no LI for nonhandled males
Septal lesions	LI for nonoperated groups; no LI for septal lesion group

paradigm (e.g., two-way shuttle, one-way, passive), to achieve latent inhibition requires that stimulus preexposure be followed by a test in which the preexposed stimulus needs to be associated with another event, such as a response or reinforcer. Consequently, studies of latent inhibition with avoidance learning generally use signaled avoidance tests. Typically, this is accomplished with an active two-way avoidance procedure. However, as noted, other procedures also have been employed. This section will examine the effects of stimulus preexposure on response acquisition in the two-way, one-way, and passive avoidance procedures.

Two-way avoidance

The first studies to demonstrate latent inhibition in two-way avoidance were those of Ackil and Mellgren (1968) and Ackil, Mellgren, Halgren, and Frommer (1969). Their procedure provided a standard that other successful latent inhibition studies have followed.

Table 1 lists 14 two-way avoidance studies with rats that have used similar procedures and have reported similar latent inhibition (LI) results. As can be seen, the typical effective number of preexposures ranges from 30 to 60. One study (Ackil & Mellgren, 1968) reported no latent inhibition with 5 tone preexposures, whereas another (Archer, 1982a, Exp. 3) did succeed in finding an effect with 5 tone preexposures. As elsewhere, there appears to be an interaction between number of preexposures and duration of each preexposure; Ackil and Mellgren (1968) used a 5-sec tone presentation, whereas Archer (1982a; Exp. 3) used a 10-sec presentation.

In addition to the studies using tone as the preexposed and test CS, Alleva, De Acetis, Amorico, and Bignami (1983) and Dore (1981) obtained latent inhibition in a two-way avoidance test after preexposure to a light stimulus.

The delay between preexposure and test varied across experiments from 0 to 72 hr, with apparently little effect on latent inhibition. However, one study (Weiner, 1983) systematically varied the delay. Data from several experiments in this series indicate a complex interaction between amount of delay and genetic strain of the rat. Thus, with Wistar rats, latent inhibition was obtained with 24- and 48-hr delay, but not with zero delay (Weiner, 1983). With Charles River rats, the opposite was found: latent inhibition for the zero-delay group but not for the 24-hr-delay group. Other variables that modulate the latent inhibition effect include handling and sex, which interact (Weiner, Schnabel, Lubow, & Feldon, 1985), such that handled males and females, as well as nonhandled females, exhibit normal latent inhibition, whereas nonhandled males do not. With the exception of some of the complex and not easily understood interactions, as reported earlier, the two-way shuttle procedure appears to be a robust one for demonstrating latent inhibition. Indeed, as can be seen from Table 1, it has found considerable use in assessing the effects of a number of drug and lesion manipulations.

The results of one study, however, do not appear to fit the homogeneous pattern described earlier. Olton and Isaacson (1968) used the same general procedure as Ackil et al. (1969), but did not obtain latent inhibition. Their failure to find latent inhibition is readily accounted for by noting that the zero-preexposure group was not given any apparatus familiarization experience, whereas, of course, the stimulus-preexposed group inevitably had such experience. Unlike the situation for one-way avoidance, such apparatus preexposure for the control group in two-way avoidance would appear to be essential for demonstrating latent inhibition. For a fuller discussion of this point, see the next section, on one-way avoidance.

In addition to the studies described earlier, in which preexposure consisted of discrete presentation of the to-be-warning signal of the conditioned avoidance test, several studies have looked at the effects of preexposure of the diffuse apparatus cues on the acquisition of signaled avoidance responding. This work comes primarily from McAllister and his students (Dieter, 1977; McAllister, McAllister, Dieter, & James, 1979). McAllister developed a modified version of the two-factor theory of avoidance learning and proposed that effective reinforcement in avoidance conditioning is based on the *difference* between the reduction in fear level that occurs when the warning signal and/or the shock are terminated and the level of fear that is elicited by the apparatus stimuli. These context cues, of course, acquire fear-producing properties as a result of their presence during shock. McAllister has argued, then, that preexposure of the apparatus should reduce the subsequent associability of those cues with shock and thereby increase the effective reinforcement value of signal and/or shock offset. Indeed, in two two-way avoidance studies that have examined this hypothesis, confirmatory evidence has been found (Dieter, 1977; McAllister et al., 1979): namely, apparatus preexposure improves avoidance learning – presumably by weakening the association between context cues and reinforcement.

Whereas all of the studies cited earlier examined stimulus preexposure effects on rats with the two-way avoidance procedure, there also have been several studies with mice (Alleva et al., 1983; Bignami, Giardini, & Alleva, 1985; Oliverio, 1968; Schnur & Lubow, 1976) and two studies with goldfish (Braud, 1971; Shishimi, 1985). All of the former, with mice, reported the basic latent inhibition effect with avoidance conditioning. The recent data from godlfish suggest that, in spite of Braud's results (1971), there is no convincing evidence for latent inhibition in this species (Shishimi, 1985).

One-way avoidance

The picture for latent inhibition in two-way avoidance is quite clear. Stimulus preexposure, as opposed to the absence of such preexposure, produces a significant deficit in the acquisition of the avoidance response. Furthermore, this deficit cuts across a variety of simple manipulations such as type of stimulus and delay between preexposure and test. For one-way avoidance, the results are more ambiguous, and, indeed, preexposure sometimes produces facilitated, rather than decremented, avoidance responding. To understand these results, let us first examine the methodology of one-way avoidance. In this procredure, the animal is placed in a compartment, and then, after the lapse of a short period of time, a discrete signal is presented, followed by shock. Together with the signal onset, either a door is opened, allowing the animal access to an adjacent shock-free compartment, or a platform above the grid floor is made available for the same purpose. Alternatively, the procedure may forgo the use of a discrete stimulus, and shock onset is proceded by the opening of the door or lowering of the platform. In either case, the animal may avoid the shock by making an appropriate response, or, when failing to avoid it, may make an escape response. On each subsequent trial, the animal is replaced into the first compartment, and the procedure is repeated. Thus, in one-way avoidance the animal always begins each trial in the same compartment (dangerous) and ends each trial in a different compartment (safe). This contrasts with the two-way or shuttle situation, in which the dangerous and safe locations are constantly interchanged.

Early descriptions of one-way avoidance emphasized the role of the CS as a warning signal. Termination of that signal was said to reinforce the avoidance-escape response (e.g., Miller, 1948; Mowrer, 1947). Later accounts (e.g., Bolles, 1972) have emphasized the role of apparatus cues, namely, those of the shock (start) compartment. Thus, in terms of latent inhibition designs, when the animal is in the start box, either one may preexpose a signal, such as a tone, and test with a signaled one-way avoidance procedure, or one may preexpose only the cues of the start box itself, and test with a nonsignaled procedure. However, it should be noted that the appropriate control groups for these two procedures are quite different. Depending on one's conception of what is learned in one-way avoidance, one might expect some of the typical latent inhibition manipulations to have very weak effects.

If, for example, the latent inhibition group is preexposed to a tone and the apparatus, and the control group is preexposed to the apparatus (which indeed is the typical procedure), then, if it is true that the apparatus cues are important for learning one-way avoidance, one might expect to find little or no difference between the groups, because they were both preexposed to the primary stimulus for subsequent learning.

In order to tease out the relative contributions to latent inhibition of apparatus cues and signals in one-way avoidance, and thereby also their importance for one-way avoidance learning itself, an additional control group should be employed, one that is neither preexposed to the apparatus nor preexposed to the signal.

The results of experiments by Solomon, Sullivan, Nichols, and Kiernan (1979) and Dore (1981, Exp. 2), which did not employ such a control group, are very much in line with the expectation of weak latent inhibition effects or no effects. On the other hand, Wilson, Phinney, and Brennan (1974) compared a group of adult rats that was given preexposure to tone plus start box with a group that was not preexposed to the tone nor to the apparatus. As one would expect from the previous discussion, latent inhibition was obtained: The preexposed rats learned the one-way avoidance more slowly than the control group.

Studies by Grant and Young (1971) and Grant and Grant (1973), with mice, would appear to indicate that preexposure to apparatus cues prior to one-way avoidance trials facilitates such learning. However, the treatment of the preexposure group was such that prior to avoidance training they were allowed to explore *both* sides of the apparatus. In addition, whereas they were preexposed to the apparatus, they were not preexposed to the tone that was later used in the signaled avoidance test. On the other hand, the control group was not preexposed to the apparatus nor to the tone. Therefore, the resulting facilitation of learning for the preexposed as compared with the nonpreexposed group is not surprising. The facilitated learning would be predicted from Lubow, Rifkin, and Alek's analysis (1976) of the relationship between environmental familiarity and stimulus familiarity at the time of test, in which it was shown that latent inhibition depends on preexposure to both the same stimulus and the same environment as used in test. If *either* stimulus or environment is novel *relative* to the other component at the time of test, then facilitation is found, as compared with the traditional latent inhibition manipulation of preexposure to both elements, or as compared with novelty of both elements. The preexposed group of Grant and Grant (1973) was familiarized to the apparatus, but not to the signal, whereas the nonpreexposed group received preexposure to neither the stimulus nor the test environment. This type of explanation may also be used to explain the apparatus preexposure facilitation effect in two-way avoidance learning (Dieter, 1977; McAllister et al., 1979).

Passive avoidance

In the passive avoidance conditioning procedure, an animal receives a shock in a particular location. On a subsequent test trial, the animal must refrain from entering the place of shock, or else risk being shocked again. The dependent measure is the time between availability of the shock and the animal's entrance to the previously shocked location. Longer elapsed times indicate greater success in the acquisition of the passive avoidance response. This procedure is particularly attractive when assessing a variable that might, in addition to affecting learning, have activity effects. By manipulating the variable in question in both an active and a passive avoidance acquisition task, one can, at least theoretically, differentiate between the contributions of learning and activity to the overall performance.

In the one study that examined latent inhibition of a passive avoidance response, Mellgren, Hunsicker, and Dyck (1975) reported two experiments in which the preexposed group exhibited poorer learning than did the nonpreexposed group.

It is of some theoretical importance that latent inhibition can be obtained with passive as well as active avoidance. These data would seem to rule out any theory of stimulus preexposure effects that relies on simple activity reduction. Although it is unfortunate that there is only a single passive avoidance study in the latent inhibition literature, the conclusion about the absence of general activity deficits that otherwise could be used to account for latent inhibition is further buttressed by the *many* studies, with a variety of learning paradigms, that have shown that latent inhibition is stimulus-specific.

Conditioned taste aversion

Barnett (1963), summarizing many years of observation, provided a number of field illustrations of food aversion in the scavenging rat. He concluded that there are two sources for such aversion, one being novelty of the food. Rats, like most children, prefer familiar foods to new foods. However, he also described how wild rats will sample novel foods in small quantities and then wait for a period of time; if sickness occurs, they will thereafter avoid that particular item.

While Barnett's descriptions were based on meticulous field observations, it was not until the mid-1950s that the conditioned taste, or food, aversion phenomenon was introduced into the laboratory by Garcia and his colleagues (Garcia, Kimmeldorf, & Koelling, 1955). Conditioned food aversion has since become a very popular research area, and it has been extensively reviewed (e.g., Barker, Best, & Domjan, 1977; Klosterhalfen & Klosterhalfen, 1985; Milgram, Krames, & Alloway, 1977; Royet, 1983) and catalogued (Riley & Baril, 1976; Riley & Clarke, 1977; Riley & Tuck, 1985).

By 1973, Lubow, in his review of the latent inhibition literature, already had identified nine studies that had demonstrated the latent inhibition effect with conditioned taste aversion (Domjan, 1971; Farley,

McLaurin, Scarborough, & Rawlings, 1964; Garcia & Koeling, 1967; Maier, Zahorik, & Albin, 1971; McLaurin, Farley, & Scarborough, 1963; Nachman, 1970; Revusky & Bedarf, 1967; Revusky, Lavin, & Pshirrer, cited in Revusky & Garcia, 1970; Wittlin & Brookshire, 1968). In all of these studies, preexposure of a particular taste quality interfered with the acquisition of a food aversion when that same taste was paired with a toxin that produced a gastrointestinal disturbance.

The changes in preferences that characterize taste aversion following flavor–toxicosis pairing are interpreted by many writers in terms of classical conditioning. The CS is the ingested flavor, the US is the source of the toxicosis (e.g., irradiation, lithium chloride injection), and the UR is the discomfort that accompanies the toxic state. Stimulus-specific changes of food preferences as a result of such pairings are treated as instances of conditioning with exceptionally long interstimulus intervals (cf. Klosterhalfen & Klosterhalfen, 1985; Logue, 1979). The fact that the post-US aversion is greater for a novel food that is paired with a toxin than for a preexposed food paired with a toxin is exactly what one finds in the traditional classical conditioning experiments that illustrate latent inhibition.

Many other studies, in addition to those cited by Lubow (1973a), and some of which have been reviewed by Best (1982), support the basic finding that taste aversion after flavor–toxicosis pairing is reduced by prior familiarization to that flavor.[1] Of the more than 50 such studies, these are some of the more recent examples: Archer, Järbe, Mohammed, and Priedite (1985); Conti and Musty (1986); Djuric, Markovic, Lazarevic, and Jankovic (1987); Franchina and Gilley (1986); Hall and Channell (1986); Holder, Leon, Yirmiya, and Garcia (1987); Kraemer and Ossenkopp (1986); Kruse and LoLordo (1986); Misanin, Blatt, and Hinderliter (1985). Conditioned taste aversion, like avoidance conditioning and conditioned suppression, has been one of the major vehicles for the study of latent inhibition. However, unlike those procedures, conditioned taste aversion allows for a wide variety of different stimuli to be employed as CSs and USs.

Preexposed stimuli

The most frequently preexposed flavor stimulus to which latent inhibition has been demonstrated is saccharin: Batson and Best (1982), Bond and Westbrook (1982), Franchina and Horowitz (1982), Kraemer and Ossenkopp (1986), to mention only some of the more recent studies. However, various other flavored stimuli also have been used to produce latent inhibition, such as alcohol (Revusky & Taukulis, 1975), coffee (Cannon, Best, Batson, Brown, Rubenstein, & Carrell, 1985), maple (Dawley, 1979), sucrose (Franchina, Domato, Patsiokas, & Griesemer,

1980), lab chow (Bernstein & Goehler, 1983; Mikulka & Klein, 1977), methadone (Lynch, Porter, & Rosecrans, 1984), casein (Best & Gemberling, 1977; Krane & Robertson, 1982), vinegar (Best, Gemberling, & Johnson, 1979; Kalat, 1974), black walnut (McFarland, Kostas, & Drew, 1978), vanilla (Kruse & LoLordo, 1986), salt (Kiefer, Phillips, & Braun, 1977; Miller & Holzman, 1981), grape juice (Revusky & Bedarf, 1967), milk (Revusky & Bedarf, 1967), chocolate milk (Kraemer & Roberts, 1984; Wittlin & Brookshire, 1968), whiskey (Franchina, Dyer, Gilley, Ness, & Dodd, 1985), and even water (Riley, Jacobs, & Mastropaolo, 1983).

Although most food aversion studies have employed flavor as the conditioned stimulus, other food-related stimuli have, on occasion, also been used; see Riley and Baril (1976), Riley and Clarke (1977), or Riley and Tuck (1985) for bibliographies. Similarly, latent inhibition has been demonstrated to these non-flavor, food-related stimuli. The most obvious examples of such a stimulus class are odors. Thus, Rudy and Cheatle (1977, 1978) have shown latent inhibition effects for lemon odor, and Westbrook, Bond, and Feyer (1981) for cineole, which has a eucalyptus-like fragrance.

Other classes of food-related objects and stimuli to which preexposure produces latent inhibition include food container quality and food color. Martin, Bellingham, and Storlien (1977), using chicks as subjects, found that it was difficult to establish a conditioned aversion to a familiarly colored food, as opposed to the same food with a novel color. Similarly, Mitchell, Kirschbaum, and Perry (1975) showed that rats will develop latent inhibition to a familiar food-holding container. Although rats will initially avoid the more novel of two food containers, with continued experience they will learn to eat from both equally. However, a single pairing of a toxicosis-producing injection with food consumption from the novel or familiar container will produce strong avoidance of the former, but not the latter.[2] Using a non-food-related odor, Millard (1982) found that adult male mice could easily be conditioned to avoid the odor of a male mouse only if that mouse was from a different litter. Odor from a mouse from the same litter as the subject mouse (i.e., a familiar odor) did not provide an adequate stimulus for conditioned aversion. In a second experiment, Millard (1982) used a split-litter rearing procedure and showed that the poor conditioning in the familiar odor group was due to postnatal experience.

In summary, it is clear that the latent inhibition effect in aversion studies is quite general. Latent inhibition is obtained with a wide range of flavors, from affectively negative, such as vinegar and salt, to affectively positive, such as saccharin, sucrose, and milk, as well as with other stimuli, both food-related and nonrelated. Furthermore, the effect is independent of the type of US used in the acquisition session.

Conditioned suppression

One of the most popular techniques used by experimental psychologists studying learning in nonhuman animals is that of conditioned suppression. Although, no doubt, earlier examples of the procedure can be found, credit for its invention is usually ascribed to Estes and Skinner (1941). Typically, the procedure requires three stages. In the first stage, either the animal is trained so as to produce a reasonably high response rate [e.g., on a variable-interval (VI) schedule] or the experimenter takes advantage of a response with a naturally high rate of occurrence (as water-tube licking by a thirsty rat). In both cases, the important requirement is the establishment of a steady, reliable rate of responding. In the second phase, a classical conditioning procedure is employed in which a CS, such as tone, is paired with a noxious US, such as shock. Depending on the intensity of the US, the pairings may be given either one time or several times in order to obtain the desired effect in the third phase. In this latter phase, the animal is reintroduced to the conditions of phase 1, and while responding, the tone CS is presented. To the extent that the tone has been associated with the noxious shock, the ongoing behavior, such as that of licking, will be reduced. As such, conditioned suppression is an index of classical conditioning even though the dependent measure is an instrumental response.

Using the conditioned suppression paradigm to study latent inhibition requires an additional stage to the three already mentioned, that of stimulus (the to-be-conditioned stimulus) preexposure prior to the classical conditioning stage.[3] The expected latent inhibition effect, namely, that stimulus preexposure reduces the amount of conditioning, is evaluated against a control group that receives all of the stages but is not given nonreinforced preexposure to the to-be-conditioned stimulus. Successful demonstration of latent inhibition requires greater suppression of responding in the test stage for the control group as compared with the stimulus-preexposed group.

A large number of studies have used this procedure to demonstrate latent inhibition. The earliest of these studies included Anderson, O'Farrell, Formica, and Caponigri (1969), Anderson, Wolf, and Sullivan (1969), Carlton and Vogel (1967), Leaf, Kayser, Andrews, Adkins, and Leaf (1968), Lubow and Siebert (1969), Siegel and Domjan (1971), and Rescorla (1971); many more have followed.

In order to obtain a clear picture of the generality of the latent inhibition effect within the conditioned suppression procedure, it is worth noting the different response and reinforcement systems that have been used to index suppression. The most commonly employed procedure has been that of bar pressing for food, with more than 25 studies

Table 2. *Latent inhibition in EB and NMR conditioning*

	Conditioning type	Preexposure (PE) stimulus (Hz)
Clarke & Hupka (1974)		
Exp 1[a]	NMR	Tone, 1000
Exp 2[a]	NMR	Tone, 1000
Frey et al. (1976, Exp 2)	EB	Tone, 1000
Hernandez et al. (1981)	EB	Tone, 1216
Moore et al. (1976)		
Exp 2	NMR	Tone, 1200
Exp 4	NMR	Light
Plotkin & Oakley (1975)	NMR	Tone, 1000
Prokasy et al. (1978)	NMR	Tone, 1000
Reiss & Wagner (1972, Exp. 1)	EB	Tone, 3510
		Vibrator, 60
Salafia & Allan (1980b)	NMR	Tone, 1000
Salafia & Allan (1982)	NMR	Tone, 1000
Scavio et al. (1983)	NMR	Tone, 1000
Siegel (1969a)	EB	Tone, 2000
Siegel (1969b)	EB	Tone, 500, 2000, 4000
Siegel (1970)	EB	Tone, 2000
Siegel (1971)	EB	Tone, 2000
Siegel & Domjan (1971)	EB	Tone, 2000
Solomon, Brennan & Moore (1974)	NMR	Tone, 1200
Solomon & Moore (1975)	NMR	Tone, 1200
Suboski et al. (1964)	NMR	Tone, 800

[a] Although not reported as Experiments 1 and 2, for our purposes it is convenient to treat the study with numbers of preexposure as Exp. 1 and with duration of preexposure stimulus as Exp 2.
[b] However, on the last day of preexposure all subjects received 60 pairings of light and shock. The entire procedure used a within-subject design. Thus, all subjects were given acquisition trials with tone, vibration, and light, with tone and clicker being counterbalanced for number of preexposures.

Number of PE over number of days	Stimulus duration (msec)	Stimulus intensity (dB)	Delay
100/2, 100/5, 250/5	250	70	0
500/5	250, 1000	70	0
150/3	500	80	24
232/3	500	60,75,90	0
450/5	450	75	0
450/5	450	–	0
125/5	200	98	24
15/1	250	30	24
12/1[b]	1100	86	24
1380/5[b]			
150	800	90	24
300	300	90	24
480/6	500	86	24
100/1, 1300/7	500	75	0
1300/7	500	75	0
550/6	500	82	0, 24
550/6	500	75	0
550/6	500	82	0
450/5	450	75, 95	0
450/5	450	76	0
280/4	600	72	24

demonstrating latent inhibition[4] (e.g., Baker & Mercier, 1982; Hall & Pearce, 1982; Hall & Minor, 1984; Mercier & Baker, 1985; Pearce, Nicholas, & Dickinson, 1982; Sharp, James, & Wagner, 1980). A second popular technique has employed direct measures of water-tube licking (e.g., Balaz, Capra, Kasprow, & Miller, 1982; Coulombe & White, 1982; Hall & Minor, 1984; Kasprow, Catterson, Schachtman, & Miller, 1984; Lubow, Wagner, & Weiner, 1982; Weiner, Lubow & Feldon, 1984). In addition, several studies have employed bar pressing for tap water (Kasprow, Schachtman, & Miller, 1985) or sweetened water (Ayres, Moore, & Vigorito, 1984; Dexter & Merrill, 1969; Witcher & Ayres, 1980). All of the latent inhibition studies using conditioned suppression have used rats as subjects.

Eyelid blink and nictitating membrane response conditioning

Eyelid blink (EB) and nictitating membrane response (NMR) conditioning procedures have found wide use in experimental psychology laboratories, the latter primarily with rabbits. The preparations are basically similar. A short-duration CS, usually a tone, is followed by a mildly noxious US, a brief air puff to the outer canthus of the eyeball in the case of human subjects, or, for rabbits, a brief shock presented close to the eye socket. With eyelid conditioning, the unconditioned and conditioned responses are blinks of the eyelid. With nictitating membrane response conditioning, conditioned and unconditioned responses are extensions of the nictitating membrane, located, at rest, at the extreme nasal corner of the eye.

Eyelid and nictitating membrane response conditioning procedures offer the most convincing sets of data for the latent inhibition effect, because they are two-stage procedures and allow for assessment of alpha responding. Ironically, the earliest experiments in these areas failed to provide any indication of a latent inhibition effect. However, most of those studies used eyelid conditioning with human subjects (Allen, 1967; Grant, Hake, Riopelle, & Kostlan, 1951; Grant, Hake, & Schneider, 1948; Perlmuter, 1966; Schnur, 1967). As we now know, at least with adult humans, a masking task is required if latent inhibition is to be demonstrated (e.g., Hulstijn, 1978; Lubow, Caspy, & Schnur, 1982; Schnur & Ksir, 1969). A full discussion of this issue can be found in the chapter on latent inhibition in humans. Among the lower animals there were also some early reports of failure to obtain latent inhibition, all with rabbits and the nictitating membrane response (Suboski, DiLollo, & Gormezano, 1964; Plotkin & Oakley, 1975; Prokasy, Spurr, & Goodell, 1978). Against this, there have been 15 studies that have provided positive evidence for latent inhibition (Clarke & Hupka, 1974;

Frey, Maisiak, & Dugue, 1976; Hernandez, Buchanan, & Powell, 1981; Moore, Goodell, & Solomon, 1976, Exp. 2 and 4; Reiss & Wagner, 1972; Salafia & Allan, 1980a, b; Scavio, Ross, & McLeod, 1983; Siegel, 1969a, b, 1970, 1971; Siegel & Domjan, 1971; Solomon, Brennan, & Moore, 1974; Solomon & Moore, 1975).[5] These studies are summarized in Table 2.

An examination of Table 2 indicates that with but two exceptions – Reiss and Wagner (1972), who used a vibratory stimulus as well as a tonal stimulus, and Moore et al. (1976, Exp. 4), who used a light stimulus – all of the studies reported in this section on NMR and EB conditioning with rabbits employed auditory stimuli. Furthermore, those stimuli invariably were pure tones and, with the exception of a study by Siegel (1969a) and one by Suboski et al. (1964), were always 1000 Hz or greater. The latter study employed a preexposed tone of 800 Hz and, as already indicated, failed to find a latent inhibition effect.

In conclusion, it seems safe to assert that within the EB and NMR conditioning procedures, latent inhibition exhibits itself as strongly stimulus-specific, at least when comparing a preexposed stimulus and a nonpreexposed stimulus of different qualities (Reiss & Wagner, 1972). Furthermore, latent inhibition shows an orderly graded effect within a given stimulus dimension, with the amount of latent inhibition being proportional to the degree of similarity between the preexposed and test stimuli (Siegel, 1969a). Finally, there is a suggestion that some stimuli may be less effective than others in inducing latent inhibition. In this case, low-frequency auditory stimuli appear to be less effective than high-frequency auditory stimuli. One might speculate as to the significance of such a phenomenon, but first it would be prudent to conduct a parametric study to determine if, in fact, the speculation is warranted.

Other classical defensive conditioning preparations

In addition to classical conditioning of the nictitating membrane and eyeblink responses there have been several isolated studies that have employed a variety of different defensive classical conditioning procedures. These studies are described in Table 3. The classical defensive studies included conditioning of the rabbit's pinna reflex (Lubow, Markman, & Allen, 1968), the rat's tail flexion (Chacto & Lubow, 1967), leg flexion in sheep and goats (Lubow & Moore, 1959; Lubow, 1965), and leg flexion in cats (Wickens, Tuber, & Wickens, 1983), as well as activity in goldfish (Shishimi, 1985, Exp. 1) and shuttle responses in goldfish (Shishimi, 1985, Exp. 3).

Two noteworthy points are immediately apparent from an inspection

Table 3. *Latent inhibition and classical conditioning preparations*

Type of conditioning	Experiment	Species	Preexposed stimulus
Defensive EB	See Table 2	Rabbit	See Table 2
NMR	See Table 2	Rabbit	See Table 2
Pinna reflex	Lubow et al. (1968, Exp. 1)	Rabbit	Tone
Pinna reflex	Lubow et al. (1968, Exp. 2)	Rabbit	Tone (varied)[a]
Tail flexion	Chacto & Lubow (1967)	Rat	Tone
Leg flexion	Lubow & Moore (1959, Exp. 2)	Goat & sheep	Rotor, flashing light
Leg flexion	Lubow & Moore (1959, Exp. 1)	Goat & sheep	Rotor, flashing light
Leg flexion	Lubow (1965)	Goat & sheep	Light patch
Leg flexion	Wickens et al. (1983, Exp. 1)	Cat	Noise or light
Activity	Shishimi (1985, Exp. 3)	Goldfish	Colored light[e]
Shuttle response	Shishimi (1985, Exp. 3)	Goldfish	Colored light
Shuttle response	Shishimi (1985, Exp. 4)	Goldfish	Red light[e]

[a] Different-intensity tones were presented to different groups.
[b] See text for explanation of preexposure procedures.
[c] Plus 60-msec overlap with shock US.
[d] In addition to an acquisition test, there was a test of retention administered 10 weeks after acquisition.
[e] Preexposed stimulus was the unreinforced stimulus in test (S-).

Number of preexposures/ days	Stimulus duration in preexposure (sec)	Delay	Results
See Table 2	See Table 2	See Table 2	Generally good LI
See Table 2	See Table 2	See Table 2	Generally good LI
0, 40, 80/1	0.6	0, 24 hr	LI
0, 40/1	0.6	0	LI
0, 20, 40/1	1	0	LI
0, 10/1	10	0	LI
0, 10/1	10	0	LI
0, 20, 40/1	4.1	0	LI
0, 540/63[b]	1[c]	24 hr[d]	LI
0, 1000/5	10	24 hr	No LI
0, 20/1, 40/2 80/4, 160/8	10	24 hr	No LI
0, 160/8	10	24 hr	No LI

of Table 3. First, defensive classical conditioning with mammalian species yields consistent latent inhibition effects, regardless of species of mammal (rabbit, rat, goat, sheep, and cat), type of defensive conditioning (eyeblink, nictitating membrane response, pinna reflex, tail flexion, leg flexion), or type of preexposed CS (tones of various intensities, noise, flashing light, light patch).[6] The second point to emerge concerns the apparent failure to find latent inhibition in a submammalian species, namely, the goldfish. A similar failure has been reported for appetitive conditioning in the honeybee (Bitterman, Menzel, Fietz, & Shäfer, 1983) and in a mollusc (Farley, 1987a, b). The studies with invertebrates are described more completely in the chapter on organismic variables affecting latent inhibition.

Miscellaneous classical conditioning preparations

There are several procedures that are basically classical conditioning but do not belong to the previous categories. In this section, three such paradigms will be discussed. One represents archetypal classical conditioning, namely, salivary conditioning. The other two are procedures that, at a minimum, may be labeled classical conditioning, because the US inevitably follows the CS, irrespective of the subjects' responses.

Conditioned defensive burying

A rat is placed in an apparatus whose floor is covered with loose, movable materials, such as sawdust. The rat then receives a single short, intense shock delivered from a fixed, low-lying prod that extends into the cage from one of the walls. Subsequently, the animal exhibits a set of behaviors involving spraying and moving the floor materials in the direction of the prod. This procedure has been studied extensively by Pinel and his associates (e.g., Pinel & Treit, 1979; Pinel, Treit, & Wilkie, 1980) and has been labeled "conditioned defensive burying" (Terlecki, Pinel, & Treit, 1979). These investigators consider this preparation to be an example of classical conditioning, in which the shock prod is the CS, and the shock is the US. Conditioning, then, consists of learning the association between the prod and the shock. The index of the strength of this association is the "burying" behavior elicited by the prod in the absence of the shock.

Oberdieck and Tarte (1981) have argued that if, indeed, this paradigm can be classified as classical conditioning, it should display a latent inhibition effect. Preexposure of the prod, prior to prod–shock pairing, should produce less burying behavior than if the prod were not preexposed. To test this prediction, they preexposed rats to the apparatus and bedding materials for 30 min per day over 4 days. One group was preexposed with the shock prod present (but inoperative), and the other with the prod absent. On the following day, the subjects were returned to the apparatus. When a rat contacted the prod, it received a single shock. Each rat was allowed to remain in the box for an additional 15 min. At the end of that period the animals were removed. The prod-preexposed group produced lower piles of bedding material and spent less time engaged in burying behavior than did the nonpreexposed group, thus indicating a latent inhibition effect. Similar results were obtained by Moderasi (1982), who found that the size of the effect was a function of the similarity of position of the preexposed and conditioned target stimulus. This latter finding is quite similar to the one obtained by Willner (1980) using a similar design, but with conditioned taste aversion.

Salivary conditioning

The salivary conditioning preparation allows for a relatively clear separation of unconditioned alpha responses and conditioned responses to the CS. With salivary conditioning, it would not be expected that a novel CS would evoke an *unconditioned* salivary response that would interfere with the valid measurement of *conditioned* salivation.

Within the salivary conditioning procedure, there has been only one study in which stimulus preexposure was manipulated (Herendeen & Shapiro, 1976). Unfortunately for our purposes, the data were incompletely reported and analyzed. The mean conditioned salivary response was presented only for the first six acquisition trials. In addition, there were no statistical comparisons between pairs of groups. An inspection of their Figure 1 suggests that there was poorer performance for the stimulus-preexposed group as compared with the nonpreexposed group. Nevertheless, their analyses, which included other groups, suggested that this difference was not reliable.

Some writers (e.g., Herendeen & Shapiro, 1976; Lubow, 1973a) have cited a paper by Konorski and Szwejkowska (1952) as one that shows latent inhibition of salivary response conditioning. However, a more careful reading of those studies indicates that the phenomenon being displayed probably was one of conditioned inhibition, not latent inhibition, because preexposure of the to-be-conditioned stimulus was conducted within a context of excitatory conditioning.

Thus, as with heart-rate conditioning in animals (discussed in a later section), the evidence for a latent inhibition effect is not very strong. This conclusion is made more compelling when one remembers the ease with which latent inhibition has been demonstrated with other paradigms, as, for instance, conditioned taste aversion, conditioned avoidance, and conditioned suppression.

Autoshaping

When a hungry pigeon is repeatedly presented with a lighted pecking key that is immediately followed by access to the food hopper, the pigeon will soon emit pecking responses to the illuminated disc. This procedure, called autoshaping, is generally accepted as an example of Pavlovian conditioning, because the appearance of the US (grain) is independent of any response that the subject makes to the CS (the illuminated key), yet the CS comes to elicit a response (CR) that was not elicited prior to the CS–US pairing. Autoshaping initially was reported by Brown and Jenkins (1968) and has since been the subject of extensive research; for recent reviews, see Balsam and Tomie (1981) and Locurto, Terrace, and Gibbon (1981).

Using the autoshaping preparation with pigeons, a number of investigators have examined the effects of uncorrelated CS and US presentations, CS alone, or US alone. Some of the earlier studies, however, did not employ a nonpreexposed control group, thus making it impossible to evaluate the amount and direction of preexposure effects (e.g., Gamzu & Williams, 1971, 1973; Wasserman, Franklin, & Hearst, 1974).

However, there have been five studies that may allow an assessment of latent inhibition (Mackintosh, 1973; Reilly, 1987; Tomie, Murphy, Fath, & Jackson, 1980; Tranberg & Rilling, 1978; Wasserman & Molina, 1975). The results have been quite mixed. Reilly (1987) and Tranberg and Rilling (1978) reported strong latent inhibition effects after nonreinforced preexposure to the illuminated key, whereas Wasserman and Molina (1975, Exp. 2) and Tomie et al. (1980) reported no such effect. Mackintosh (1973), at best, found a very weak effect. He noted that in the autoshaping test, the CS-preexposed group responded more slowly than did the nonpreexposed group, but only on the first 2 of 8 test days, and even then with marginal statistical significance. Table 4 describes the five studies in terms of numbers of subjects in the two comparison groups, total number of preexposures, duration of each preexposure, number of days of preexposure, intertrial interval between stimulus preexposures, whether magazine training occurred before or after preexposures, the dependent measure obtained during the autoshaping test procedure and, finally, whether or not latent inhibition was obtained.

Tomie et al. (1980) used a 4×2 factorial design with four different preexposure conditions occurring either in the same context as in the autoshaping test or in a different context. The preexpossure groups were as follows: (1) CS only; (2) US only; (3) a truly random control group in which both the CS and US were preexposed; (4) a nonpreexposed group. Tomie et al. (1980) not only failed to find latent inhibition but

Table 4. *Autoshaping and latent inhibition*

Experiment	Groups number of subjects 0 PE	Groups number of subjects PE	Total number of PEs	PE stimulus duration (sec)	Days of PE	PE ITI (sec)[a]	Hopper training before or after PE	Dependent measures	Results
Mackintosh (1973)	8	8	160	5	4	VT 60	Before	Responses per second	?
Reilly (1987, Exp. 1)	8[b]	8[b]	300	4	10	VT 30	Before and after	Trials to first CS	LI
								Trials with CS peck (first 5 sessions)	LI
								Trials with CS peck (last 5 sessions)	No LI
								Response rate (first 5 sessions)	LI? ($p < .08$)
								Response rate (last 5 sessions)	No LI
Tomie et al. (1980, Exp. 2)	12	12	1800	7.5	30	VT 45	After	Trials to criterion	Not relevant
								Trials with CS peck	
Tranberg & Rilling (1978)	12	12	500[c]	8	10[c]	VT 90	Before	Trials to first CS peck	LI
								Trials with CS peck	LI
								Response rate	LI
Wasserman & Molina (1975, Exp. 2)	4[d]	4[d]	126	8	7	VT 96	Before and after	Trials to first CS peck	No LI
								Trials to second CS peck	No LI
								Trials with CS peck	No LI

[a] ITI, intertrial interval; VT, variable time.
[b] Includes only the nonoperated subjects; but same results with hyperstriatal lesions.
[c] Minimum values.
[d] Same subjects in both groups, within-subject design.

also reported that the CS-only group showed better learning than did the nonpreexposed group. However, the meaning of these results for latent inhibition is not clear. First, one would expect better learning for a CS-preexposed group in which the context was changed from preexposure to test than for a group without such a change in context. To demonstrate latent inhibition, one needs a group that is preexposed and tested in the same environment. Although this latter group was part of the factorial design, the comparison nonpreexposed group was not present. In addition, the novel stimulus group of Tomie et al. (1980) was not preexposed to the apparatus for the same period of time as was the stimulus-preexposed group. Indeed, there was a difference of 30 days in the amount of apparatus preexposure for the two groups. As a result, the study of Tomie et al. (1980) is not relevant to an assay of latent inhibition.

That leaves four studies (Mackintosh, 1973; Reilly, 1987; Tranberg & Rilling, 1978; Wasserman & Molina, 1975), only two of which (Reilly, 1987; Tranberg and Rilling, 1978) showed a clear latent inhibition effect. Tranberg and Rilling (1978) noted that the two earlier studies may have used a lower-than-required number of preexposures. As can be seen from Table 4, they raised the number of preexposures from 160 or less to a minimum of 500. This may well have been the critical factor in their ability to produce latent inhibition. Indeed, the fact that the one other autoshaping study (Reilly, 1987) that reported latent inhibition also used a relatively large number of stimulus preexposures (300) adds some support to this contention. However, it is difficult to understand why this should be the case. Tranberg and Rilling proposed that because of "the biological preorganization of a pigeon's keypeck in the presence of food (Woodruff & Williams, 1976), a substantially greater number of preexposures may be required to detect the phenomenon [latent inhibition] in pigeons" (Tranberg & Rilling, 1978, p. 273). This suggestion does not seem viable. Although the meaning of "biological preorganization" is not entirely clear, it would appear that conditioned taste aversion could also be placed in this category. However, as will be recalled, one can demonstrate latent inhibition in that paradigm with only a *single* preexposure.

The first question, then, is this: Can the differences between Reilly (1987) and Tranberg and Rilling (1978), on the one hand, and Mackintosh (1973) and Wasserman and Molina (1975), on the other, be accounted for by differences in the numbers of preexposures? If the answer is yes, the reasons for latent inhibition requiring a larger number of preexposures in autoshaping, as compared with other paradigms, may be related more to appetitive versus aversive tests than to "biological preorganization."

Let us examine some differences between autoshaping and other conditioning preparations that may account for the apparent finding that autoshaping requires a comparatively extensive number of preexposures to produce latent inhibition. The simplest, and perhaps the least interesting, possibility is that pigeons preexposed to the stimulus in the autoshaping experiments do not receive nearly the numbers of stimulus presentations as are programmed by the experimenter. Mackintosh (1973) offered one such explanation for the relatively poor performance of an uncorrelated keylight–food-preexposed group compared with a keylight-alone-preexposed group – results, by the way, that are quite in the opposite direction from those reported by Tomie et al. (1980) and Wasserman and Molina (1975). Mackintosh conjectured as follows: "It is conceivable that the pigeons in the CS only group, exposed during Phase 1 to repetitive presentations of the CS, went quietly to sleep in the corner of the box, and were therefore not effectively exposed to as many presentations of the CS as were the birds in the CS/Food group, who were kept awake by the occasional presentations of food" (1973, pp. 85–86).

The same type of argument, irrespective of the food, can be applied to these other studies. Whether asleep or just facing in the opposite direction, a weakly illuminated, localized stimulus, presented to a free-moving animal, may not be registered by that organism on every trial. A large number of trials may be necessary to achieve the relatively low number of stimulus registrations required to produce latent inhibition.

Assuming, however, that the foregoing hypothesis is incorrect, what are some alternative explanations for the autoshaping data suggesting that to obtain latent inhibition, many more preexposure trials are required than in other learning paradigms? Perhaps there is a positive relationship between the number of trials to acquire the target response and the number of preexposure trials necessary to develop latent inhibition. It will be recalled, for instance, that in the taste aversion conditioning paradigm, where a single taste–poisoning episode produces conditioned aversion, a single preexposure is sufficient to produce latent inhibition. Within the autoshaping procedure, then, if it requires several hundred preexposures to produce latent inhibition, one should expect to find that normal acquisition similarly requires an extensive number of trials. This, in fact, is what is found in these studies. Hundreds of trials typically are required before asymptotic levels of responding are reached. This fact means that the subjects, in *all* groups, receive considerable passive exposure to the CS. It may well be that this passive stimulus exposure in the previously nonpreexposed group also serves to create a latent inhibition effect and thus obscures any differences between the groups.

There is yet another problem in the latent inhibition autoshaping literature. As can be seen from Table 4, the general procedure in these studies is to magazine-train the birds before preexposure to the CS. Because the magazine training is accomplished in the same context as the subsequent preexposure and test, it is possible that this procedure creates an excitatory context that allows the preexposed CS to develop conditioned inhibition rather than latent inhibition. Because all of the tests were of retardation,[7] which is common to both latent and conditioned inhibition, there is no way to differentiate which of these effects, conditioned inhibition or latent inhibition, actually occurred. To do this, on would also have to employ a summation test[8] (cf. Reiss & Wagner, 1972; Rescorla, 1971). This problem is equally present in other procedures where pretraining precedes stimulus exposure and both are conducted in the same apparatus.

Finally, there is an additional, highly plausible, explanation for the different results of these autoshaping studies – one related to the earlier discussion of the unconditioned responses elicited by the CS. In the Tranberg and Rilling (1978) experiment, in which a strong latent inhibition effect was obtained, the subjects were preexposed until there were 2 successive days of a zero operant response rate. It appears that the lighted key elicited some unconditioned responding and that this was extinguished before the subject entered the autoshaping test. The nonpreexposed group, therefore, began the test with the full complement of unconditioned keypecking responses and was thus exposed to the response-reinforcement contingencies before the preexposed group. These unconditioned responses also would be tallied as conditioned responses. That this may be the mechanism for producing the better performance in the nonpreexposed group as compared with the preexposed group is indicated by the fact that in both the Tranberg and Rilling (1978) and Reilly (1987) studies, the nonpreexposed groups took considerably fewer trials before making the *first* keypeck response than did the preexposed groups. Comparable data from Wasserman and Molina (1975), who did not report latent inhibition, indicate no differences in the mean numbers of trials to the first peck for the two groups. Although Mackintosh did not report these data, he did note that "no subject pecked the key more than a few times during the first phase of the experiment" (1973, p. 85).

In summary, of the several autoshaping studies reviewed, only Reilly (1987) and Tranberg and Rilling (1978) reported a reliable latent inhibition effect.[9] However, analyses of the procedures and data of these studies suggest that differences between groups can be accounted for by differences in extinction of an unconditioned response, as a result of which, in the test, the nonpreexposed group is more likely to encounter the response-reinforcer contingencies earlier than is the preexposed group – independent of any effects of preexposure on central associative processes. In addition, these unconditioned responses will inadvertently be summed with the conditioned responses to give an inflated estimate of conditioning strength. (For a similar analysis of such a possibility in classical conditioning of the pinna response to a preexposed auditory stimulus, see Lubow et al., 1968.) Alternatively, the poorer learning of the stimulus-preexposed group may reflect conditioned inhibition rather than latent inhibition.

If the differences in acquisition of the keypeck response between the stimulus-preexposed and the nonpreexposed groups can be accounted for by the response hypothesis that was suggested earlier, then latent inhibition has not been demonstrated in this preparation. It would follow that either autoshaping or the pigeon is at fault. Because latent inhibition has been demonstrated in so many different learning procedures, there would seem to be no reason to suspect that autoshaping is the source of the problem.

Conditioning of orienting responses

The orienting response (OR) is a many-splendored thing that has a variety of forms and functions, the particulars of which depend on the focus of the individual investigator. Traditionally, the OR is traced back to Pavlov's observations (1927) that an animal will turn toward any novel stimulus that is presented during a conditioning session. Pavlov labeled this "somatic activity," which serves to tune the organism's peripheral receptors to the source of the new stimulation, the "investigatory reflex." Subsequent Russian research has chosen to emphasize the autonomic and central nervous system aspects of the OR. With the translation of Sokolov's seminal works (1960, 1963) into English, and the extensive commentary they received (e.g., Berlyne, 1960; Lynn, 1966), the OR has become a lively topic of research in the West.

For Sokolov and his collaborators, the OR is defined as a complex set of physiological responses elicited by moderate- or low-intensity novel stimulation, whose function is to increase perceptual sensitivity. Responses that represent aspects of the OR include the following: EEG desynchronization (alpha blocking), changes in skin potential and conductivity, pupillary dilation, eye movement, peripheral vasoconstriction (as in the finger), cephalic vasodilation (as in the forehead), and heart-rate deceleration. More recently, analyses of EEG slow waves have suggested that event-related potentials such as P300 may be components of the OR (e.g., Blowers, Spinks, & Shek, 1986; Pritchard, 1981). The OR, according to Sokolov, is elicited by stimuli in any sensory modality and occurs independent of the direction of change, either by stimulus onset or offset or by an increase or decrease in stimulus intensity. Most important for the present discussion, the amplitude of the OR gets smaller with repeated stimulus presentations. This habituation of the OR is also subject to dishabituation, which may be brought about by some modification of the stimulus or by the passage of time; for recent reviews, see Ohman (1983) and Siddle, Kuiack, and Kroese (1984).

The OR is related to latent inhibition in two quite distinct ways. First, the OR may contribute to an understanding of the processes that serve as the basis of latent inhibition. Orienting response explanations of learning decrements following stimulus preexposure will be discussed in a subsequent section on theories of latent inhibition. In addition, however, there are empirical considerations. Is the OR conditionable, and, if so, does stimulus preexposure retard conditioning of the OR?

The data that exclusively concern GSR conditioning in humans are discussed in the section on adult human latent inhibition. In brief, the evidence for latent inhibition of GSR suggests such an effect, at least for short-interval responses. However, the data on conditioning of heart-rate deceleration, another index of OR activity, have been more problematic.

Heart-rate conditioning

Classical conditioning of heart-rate change involves a straightforward procedure in which a CS, often a tone, is paired with a US that, by itself, elicits a change in heartbeat rate. Typically, in this preparation, the US is a noxious shock. Although such conditioning has been extensively studied (cf. Obrist, Sutterer, & Howard, 1972), there have been only a few experiments that have attempted to demonstrate latent inhibition of conditioned heart rate. Subjects have included rabbits (Bostock & Gallagher, 1982; Gallagher, 1985; Gallagher, Meagher, & Bostock, 1987; Hernandez et al., 1981), rats (Fitzgerald & Hoffman, 1976; Ray & Brener, 1973; Wittman & DeVietti, 1981), pigeons (Cohen & MacDonald, 1971), and humans (Wilson, 1969). Whereas some of these studies have reported a latent inhibition effect (Bostock & Gallagher, 1982; Gallagher, 1985; Hernandez et al., 1981; Wilson, 1969; Wittman & DeVietti, 1981), others have found no differences between the preexposed and nonpreexposed groups (Cohen & MacDonald, 1971), or even a facilitory effect (Fitzgerald & Hoffman, 1976; Ray & Brener, 1973).

Unfortunately, within those studies that purport to demonstrate latent inhibition, there is much to criticize in terms of procedure and experimental design (Table 5). Bostock and Gallagher (1982), Gallagher (1985), and Gallagher et al. (1987), for instance, do not appear to have given the nonpreexposed group exposure to the apparatus equal to that given the stimulus-preexposed group. Hernandez et al. (1981), on the other hand, used a CS preexposure duration of 0.5 sec, considerably shorter than that used in any other latent inhibition study. Because they

Table 5. *Heart-rate conditioning and latent inhibition*

Experiment	Subjects	CS	Number of PEs	Stimulus duration (sec)	Time between last PE & 1st acquisition	Results[a]
Bostock & Gallagher (1982)	Rabbit	Tone	60[b]	5.0	0	PE < NPE
Cohen & MacDonald (1971)	Pigeon	Light	10, 40	6.0	1	PE = NPE
Fitzgerald & Hoffman (1976)	Rat	Tone	10, 50	6.3	0	PE > NPE
Gallagher (1985)	Rabbit	Tone	20	5.0	24	PE < NPE
Gallagher et al. (1987)						
Exp. 1	Rabbit	Tone	45[b]	5.0	0	PE < NPE
Exp. 2	Rabbit	Tone	30[b]	5.0	0	PE < NPE
Hernandez et al. (1981)	Rabbit	Tone	232	0.5	0	PE < NPE[c]
Ray & Brener (1973)	Rat	Tone	120	5.0	24	PE > NPE
Wilson (1969)	Human	Tone	9[d]	1, 5, 10	0	PE < NPE
Wittman & DeVietti (1981)	Rat	Tone	6	15.0	48	PE < NPE

[a] Stated in terms of conditioning performance, e.g., PE < NPE – the preexposed (PE) group showed less heart-rate deceleration during CS acquisition trials than did the nonpreexposed group (NPE) – purportedly LI.
[b] 15 PEs per day; the last 15 PEs were presented just prior to the CS–US pairings; the NPE group was also given these 15 tone preexposures.
[c] The direction of results depends on stimulus intensity and other conditions.
[d] There were also nine presentations of the US.

employed the same short CS duration during conditioning, they were forced to use a US-omission procedure in order to get a measure of conditioned heart rate. Such trials were embedded in each acquisition session, and they counted the number of beats in an 8-sec period *following* the CS (as in all of these studies, this measure is compared with a pre-CS rate). The use of the omission procedure to study latent inhibition is problematic because it may result in responses to changes in stimulus events independent of preexposure conditions.

In addition to the methodological problems just mentioned, there is the problem that we have encountered in various other preparations: the confounding of the unconditioned response to the CS (which in this case is typically a cardiac deceleration, and as such is an index of the OR) and the conditioned response to the CS (which is, again, a cardiac deceleration). One can find graphic evidence to support the presence of this confounding in the data of Hernandez et al. (1981) and Wittman and DeVietti (1981). Wilson (1964, 1969) made a similar point, even more strongly, in writing about heart-rate conditioning in adult humans: "It would appear that combining shocks with tones serves to reinstate and magnify the deceleration response initially elicited by the tone itself... deceleration spontaneously evoked by tone... is being strengthened and made more resistant to adaptation" (Wilson, 1964, pp. 294–295). "This conclusion clearly implicates the original response to the CS as a significant component in what ultimately develops as the conditioned response" (Wilson, 1969, p. 2).

In summary, in spite of the fact that six out of nine animal studies have reported poorer conditioned heart-rate deceleration for the CS-preexposed group than for the nonpreexposed group (Table 5), because of the methodological considerations, it would be inappropriate to identify this difference as representing an instance of latent inhibition. The issue might be clarified by using a design in which the stimulus preexposure intensity is varied and crossed with test intensity. However, particular attention must be paid to the fact that the unconditioned response to the CS changes its form as a function of stimulus intensity. With low and moderate intensities the response is one of deceleration and typifies an OR; at high intensities the response is acceleration and typifies a defensive reaction (Graham & Clifton, 1966).[10]

Conditioning of observing responses

In addition to the various changes in autonomic activity that characterize the OR, there are also the motor responses originally described by Pavlov as investigatory reflexes. Several studies have examined the effects of stimulus preexposure on these skeletal or voluntary motor aspects of the OR. Siegel (1971) reported one of the earliest of these studies. Using an eyeblink conditioning preparation with rabbits, he not only found the usual latent inhibition effect but also made direct observations of orienting responding during preexposure, conditioning, and extinction. The OR was defined as an increased opening of the eyelid in the presence of the tone. Note that the opening response is *antagonistic* to the conditioned response, which is closure of the eyelid. During the stimulus preexposure period, the stimulus-preexposed group emitted more ORs to the tones than did the nonpreexposed group in comparable no-stimulation periods. During the acquisition period, the preexposed group showed fewer ORs to the tones than did the nonpreexposed group.

Siegel's results (1971) show that, at least for one particular index, stimulus preexposure results in habituation of the OR. Although there were group differences in percentage of trials with ORs during the acquisition session, with the stimulus-preexposed group exhibiting fewer ORs than the nonpreexposed group, it is difficult to conclude that it was the conditioning of the OR that was retarded, because neither group provided evidence for such conditioning. In fact, both groups showed a decline in number of ORs as the number of CS–US pairing trials increased. However, it is also quite clear that conditioning of the eyeblink response did occur and that preexposure of the to-be-conditioned stimulus produced retardation of such conditioning.

Of considerable interest is the finding that a very small difference in OR frequency between the preexposed and nonpreexposed groups at the end of the preexposure period nevertheless was followed, in the conditioning session, by a very large difference in OR elicitation and in CRs that were antagonistic to the OR. Both of these facts, individually and together, eliminate any interpretation of latent inhibition effects in terms of simple OR habituation. Nor does the relationship between the OR and conditioning during the acquisition phase allow one to implicate the OR as an explanation of latent inhibition. To accomplish this, one would need to show that the stimulus-preexposed group had a *rearoused* OR in the acquisition session and that such OR rearousal either preceded or accompanied the formation of conditioned responding. An inspection of Siegel's graphic data does not suggest that the course of OR elicitation and the strength of conditioning occurred in a parallel manner. Clearly, then, whatever relationship there is between OR elicitation and conditioning, it is not a perfect one. Nevertheless, it is possible that the sought-after covariation is masked by the fact that Siegel's data are grouped in blocks of 25 trials. There is reason to expect that a rearousal of the OR, if it occurs at all, should take place *within* the first 25 trials.

Kaye and Pearce (1984a) and Holland (1980), using a conditioning preparation quite different from the one reported earlier, also directly investigated the strength of the voluntary OR (an overt skeletal response as opposed to a response of the autonomic nervous system) during Pavlovian conditioning (as well as during blocking; Kaye & Pearce, 1984b). In general, Holland reported that during appetitive conditioning with continuous reinforcement there was a *sustained* OR to a light CS. However, Kaye and Pearce (1984a) found that the OR *declined* with trials. They tried to resolve this discrepancy on the basis of number of preexposures before acquisition, number of training trials, and type of conditioning procedure, all of which differed between the two sets of studies.

In the Kaye and Pearce (1984a, Exp. 5) study, all subjects were given magazine training with food pellets. In order to obtain a pellet, the rat had to learn to push back a perspex flap that covered the magazine recess. Following this, in the next session, all subjects were given six nonreinforced 10-sec presentations of a light emanating from a small bulb located near the magazine. For our purpose, we shall consider only two of the four groups. One group was given four additional stimulus preexposure sessions, similar to the one just described (Group PE), and the other group was placed in the apparatus for the same periods of time, but without exposure to the light (Group NPE). Both groups were

42 *Latent inhibition and conditioned attention theory*

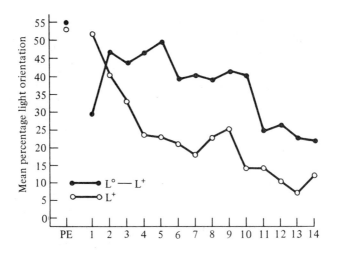

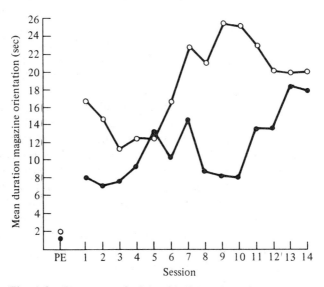

Figure 3. Percentage of trials with light orientation (upper panel) and mean duration of magazine orientation (lower panel) during the preexposure and conditioning sessions. (From Kaye & Pearce, 1984a.)

then given 14 conditioning sessions, 6 trials per session, in which the light preceded delivery of a single food pellet. An OR was defined "either as rearing in front of the light or making contact with the light with either snout or paws" (1984a, p. 92). The conditioned response was defined as the duration of time the animal spent with its head in the magazine in the presence of the CS. As with the Siegel study (1971), the OR and CR were antagonistic to each other (i.e., the presence of one precluded the presence of the other). The data for both ORs and CRs are shown in Figure 3.

First, it should be noted that there were no group differences in OR frequency during the first preexposure session (in which *both* groups received six light presentations). Although not shown in Figure 3, during the next preexposure period (the one that differentiated between the groups, with the stimulus-preexposed group receiving four additional sessions of light presentations, and the nonpreexposed group receiving no such additional sessions), the stimulus-preexposed group exhibited significant OR habituation. In addition, the fact that the conditioning session began with the preexposed group emitting significantly fewer ORs than the nonpreexposed group indicates that the OR for that group was habituated during the prior preexposure session. So far, then, the data resemble those of Siegel (1971). However, after the first conditioning session, there was a reversal in OR frequency, with the stimulus-preexposed group exhibiting *more* ORs than the nonpreexposed group, exactly the opposite from what was reported by Siegel. Nevertheless, as in the study by Siegel, conditioning, this time of the magazine-entering response, appeared to be more rapid for the nonpreexposed group than for the preexposed group. However, these differences were not statistically reliable. Because there was no significant difference in magazine conditioning between the two groups, one perhaps should not try to make too much of the fact that the preexposed group exhibited more ORs during the conditioning session than did the nonpreexposed group. Although, on the face of it, this would not be congruent with expectations from OR theories of latent inhibition, Kaye and Pearce (1984a) do predict such differences in OR strength. They hypothesize that ORs are elicited by conditions of unpredictability or change. Presumably, going from a series of nonreinforced stimulus presentations to reinforced presentations constitutes greater unpredictability than going from nothing to reinforced stimulus presentations. Nevertheless, that statement, concerning the conditions for producing the OR, by itself, does not allow one to join OR and latent inhibition in a meaningful manner.

In regard to the question whether or not stimulus preexposure will result in latent inhibition of the conditioning of the OR, it will be recalled that in the Siegel study (1971) there was no evidence of OR conditioning in either the preexposed group or the nonpreexposed group. In the Kaye and Pearce study (1984a), however, there was a reliable increase in OR responding from Session 1 to Session 2, thus allowing for the possibility that OR conditioning occurred. The fact that only the stimulus-preexposed group showed such an effect, and that there was greater OR elicitation in the preexposed group than in the nonpreexposed group, suggests a facilitation of OR conditioning rather than latent inhibition. However, this interpretation is made suspect because, as noted, the magazine CR and the OR are antagonistic responses.

Other studies that have used appetitive conditioning of the observing response to investigate latent inhibition include those by Hall and Channell (1985b), Hall and Schachtman (1987), Kaye and Pearce (1987a,b), and Kaye, Preston, Szabo, Druiff, and Mackintosh (1987). Overall, for the animal literature, the relationship between the OR and latent inhibition, particularly in regard to the question whether or not stimulus preexposure retards subsequent conditioning of the OR, is quite unclear. In the case of both heart-rate conditioning and the conditioning of observing responses, there have been several procedural problems, the most important of which has been the inability to independently identify or differentiate between the original (unconditioned) OR to a stimulus and the conditioned OR, as in the heart-rate conditioning studies, or to design tasks where the OR index and other nominally conditioned responses are not in conflict, as in the observing response studies. In addition to the foregoing considerations, there are data that are simply contradictory. Thus, whereas Siegel (1971) found more ORs during conditioning for the nonpreexposed stimulus group than for the preexposed stimulus group, Kaye and Pearce (1984a,b) found exactly the opposite. Finally, there are the data of Hall and Channell (1985b), who found an association between OR habituation and latent inhibition for the condition in which stimulus preexposure and context preexposure were administered conjointly, but a dissociation when they were presented independent of each other. In the latter case the OR remained habituated when an old stimulus was presented in the independently familiarized environment, but with the same test conditions, latent inhibition was abolished. A similar dissociation, occurring with a long retention interval, was reported by Hall and Shachtman (1987). On the other hand, Kaye and Pearce (1987a) found parallel effects of hippocampal lesions on the orienting response and latent inhibition.

Clearly, the issue of the relationship between habituation of the orienting response and the appearance of latent inhibition remains to be resolved. Additional studies will be required that can overcome some of the procedural inadequacies, as well as studies that employ different indices of the orienting response. In the meantime, it would be safe to conclude that there is an absence of a one-to-one correspondence between latent inhibition and habituation of the orienting response.

Discrimination learning

This section examines those latent inhibition experiments that have been conducted within a formal discrimination learning paradigm, such that at the time of test there have been two discrete stimuli, at least one of which has been preexposed for one of the groups. The discrimination procedure has to be differentiated from those paradigms that were discussed earlier, to which they are, indeed, orthogonal. Thus, one can study discrimination learning by using avoidance learning, conditioned suppression or conditioned taste aversion, and so forth. Nevertheless, discrimination learning will be treated as a separate problem because of the number of unique features contained within the paradigm, features that would complicate the exposition of the other paradigms.

Tests for latent inhibition using discrimination procedures can be divided into those that use classical conditioning and those that use operant or instrumental conditioning procedures. In addition, within those categories, one can differentiate between studies that preexpose the to-be-conditioned stimulus on a discrete, trial-to-trial basis (as have virtually all of the studies that have been reviewed up to this point) and those studies that give a single, prolonged period of preexposure (e.g., Channell & Hall, 1983; Gibson & Walk, 1956; Lubow, Rifkin, & Alek, 1976). The terms *discrete preexposure* and *continuous preexposure* will be employed to identify these procedures. There are, in addition, those discrimination tests that can be identified as simple, in which only two stimuli are presented, one reinforced (S^+) and one not reinforced (S^-), as well as those discrimination tests in which one of the stimuli, usually the preexposed one, is placed in a nonreinforced compound (SP^-) while the single novel stimulus is reinforced (S^+). Studies using this latter procedure are usually aimed at determining whether or not stimulus preexposure produces conditioned inhibition. These two procedures will be distinguished from each other by the terms *simple discrimination* and *compound discrimination*. Another factor describing discrimination studies is whether the reinforcement used in the test is appetitive or aversive.

Any of the dichotomous factors listed earlier can take a value independent of any other factor's value. Although all possible combinations can be used, in practice there are significant clusters. For example, almost all simple instrumental discrimination studies use appetitive reinforcement. The distribution of the discrimination studies across the various combinations of categories is shown in Table 6. As can be readily seen, there are four major areas of concentration. These are described next.

Discrete stimulus preexposure followed by a classical conditioning test with a compound stimulus (N^+ versus NP^-) and a noxious US

In this cell, the preexposed stimulus (P) is compounded with the novel stimulus (N) to serve as a predictor of nonreinforcement, and N^+, by itself, is paired with a noxious shock US. As noted, this condition provides the typical procedure for producing conditioned inhibition. The

Table 6. *Latent inhibition and discrimination learning*

	Stimulus	Discrete preexposure		Continuous preexposure	
		Simple test P+ vs. N− or P− vs. N+	Compound test N+ vs. NP− or NP+ vs. N−	Simple test P+ vs. N− or P− vs. N+	Compound test N+ vs. NP− or NP+ vs. N−
Classical test	Appetitive				
	Noxious		Reiss & Wagner (1972) Rescorla (1971) Solomon, Brennan, & Moore (1974) Solomon, Lohr, & Moore (1974) Solomon & Moore (1975)		
Instrumental Test	Appetitive	Baker & Mackintosh (1977, Exp. 1) Harrison (1979) Mason & Lin (1980) Mellgren & Ost (1971) Robbins et al. (1982)	Baker & Mackintosh (1977, Exp. 2) Burton & Toga (1982) Halgren (1974)	Bell & Livesey (1977) Channell & Hall (1981) Green et al. (1977) Hall & Channell (1983) Gibson & Walk (1956)[a] Hall (1979) Lubow, Rifkin, & Alek (1976)	
	Noxious				

[a] One example of a class of studies.

presence of conditioned inhibition is indexed by lower conditioned responding to NP⁻ as compared with a group that has not received preexposure to stimulus P. If, however, conditioned responding to NP⁻ is greater for the preexposed group than for the nonpreexposed group (i.e., if there is *poorer* discrimination between N⁺ and NP⁻), then conditioned inhibition accruing from the nonreinforced stimulus preexposure is not implicated, and the results can be taken as another manifestation of latent inhibition. The studies in this cell already have been noted in the sections on EB and NMR conditioning (Reiss & Wagner, 1972; Solomon, Brennan, and Moore, 1974; Solomon, Lohr, & Moore, 1974; Solomon & Moore, 1975) and conditioned suppression (Rescorla, 1971). The conclusions are quite clear: Discrete preexposure of a to-be-conditioned stimulus does *not* endow that stimulus with the properties of a conditioned inhibitor. Rather than facilitating discrimination, as would be required if P were a conditioned inhibitor, such preexposure interferes with the normal acquisition of the discrimination, suggesting that the discrimination decrement is due to a latent inhibition effect.

Discrete stimulus preexposure followed by an appetitive instrumental conditioning test with a compound conditioned stimulus

The conclusion of the last paragraph was confirmed by studies in which an instrumental appetitive test was used to assess differences between groups with and without stimulus preexposure in the acquisition of the N⁺ versus NP⁻ discrimination (e.g., Baker & Mackintosh, 1977, Exp. 2; Halgren, 1974).

The other study that shares this particular cell with Baker and Mackintosh (1977) and Halgren (1974) is the only one that used a compound stimulus test in which the preexposed stimulus, a tone, was presented together with a light stimulus and was reinforced (NP⁺), whereas the light, by itself, was not reinforced (N⁻) (Burton & Toga, 1982). Although sham-operated preexposed rats showed considerably poorer discrimination scores than did nonpreexposed animals, the effect arose because of overresponding to N⁻. As in the Halgren study (1974), responding to the NP⁺ tone was at a maximum from the beginning of testing and was, no doubt, a result of earlier discrimination training. The net result is that the data, though interesting (particularly because of the reversal for a lateral-septal-lesioned group), are quite uninterpretable in terms of latent inhibition.

Table 7. *Latent inhibition with discrete preexposure and a simple instrumental appetitive discrimination test*

Experiment	Species	Preexposure Stimulus	Number	Stimulus duration (sec)	Delay (hr)
Mason & Lin (1980)	Rat	T, L	0, 80/10[a]	60	24
Mellgren & Ost (1971)					
Exp. 1	Rat	T, L	0, 90/6[b]	60	24
Exp. 2	Rat	T, L	0, 90/6[b]	60	24
Exp. 3	Rat	T	0, 75, 150 300[c]	60	24
Harrison (1979)	Rat	noise-1 & noise-2[e]	108/3	5	48[f]
Baker & Mackintosh (1977, Exp. 1)	Rat	T	160/4	30	24
Robbins et al. (1982)	Rat	T, L	?	?	?

[a] 80 preexposures to each stimulus; preexposures contiguous.
[b] 90 preexposures to each stimulus; preexposures contiguous.
[c] Over 5, 10, and 15 days. Total time in apparatus was equated.
[d] Major analyses were performed on ratios. Some comments on independent responding to the S+ and S− were also made.
[e] Same noise, but from different locations within the apparatus.
[f] Minimum.
[g] See text.

Latent inhibition testing procedures

Stimuli and reinforcement conditions	Test Successive or simultaneous	Type	Dependent variable	Result and comment
T$^+$L$^-$ L$^+$T$^-$	Succ.	Lever press	Disc. ratio	LI
T$^+$L$^-$ L$^+$T$^-$	Succ.	Lever press	Disc. ratio^d	LI
T$^+$L$^-$ L$^+$T$^-$	Succ.	Lever press	Disc. ratio^d	LI; no effect of deprivation level
T$^+$L$^-$ L$^+$T$^-$	Succ.	Lever press	Disc. ratio^d	LI; not much effect of increasing numbers of preexposures
See text	"Simult."^{g}	Lever press	Percentage adjacent responses	LI; between-experiment comparison
T$^+$	Succ.	Lick	CS–preCS	LI
T$^+$L$^-$ L$^+$T$^-$	"Simult."^{g}	Nose-poke	Percent correct	LI

Discrete stimulus preexposure followed by an appetitive instrumental test with simple CSs

Studies that fall into this category are listed in the first column of Table 7. Table 7 describes the basic experimental designs and various parameter values: type of preexposed stimulus; number of preexposures; duration of preexposure; whether both the to-be-reinforced stimulus and the not-to-be-reinforced stimulus are preexposed, or just one of the two; and, relatedly, whether the preexposed stimulus is used as the reinforced stimulus in the test or as the nonreinforced stimulus in the test.

One of the first studies of latent inhibition and discrimination learning (Mellgren & Ost, 1971, Exp. 1, 2, and 3) used a procedure in which *two* qualitatively different stimuli were preexposed. This was followed by a successive discrimination test in which one of the stimuli appeared as an S^+, and the other as an S^-. Specifically, in the test, bar-press responding in the presence of one of the stimuli produced an appetitive reinforcer, whereas responding in the presence of the other stimulus produced no such reinforcement (a go/no-go discrimination procedure). In all three experiments, the discrimination scores were poorer for the stimulus-preexposed group as compared with the nonpreexposed group. Although no specific details were given, the authors suggested that the poorer discrimination resulted, in general, from higher responding to the nonreinforced test stimuli – a result similar to those reported in the previous section. Retardation of appetitive discrimination learning as a result of stimulus preexposure was also reported by Baker and Mackintosh (1977, Exp. 1). Preexposure to a tone retarded the subsequent acquisition of a licking response when the tone was paired with water.[11]

In a study that was not designed to look for latent inhibition effects, Harrison (1979) devised an unusual procedure for demonstrating rapid discrimination learning. In this experiment (Exp. 1), rats were trained to discriminate the position of a noise source. One sound source was located next to one lever, and a second source was located next to the other lever. During training, the animal was reinforced only when it pressed the lever that was adjacent to the activated sound source. A lever press to the nonadjacent lever produced a time-out. Such a procedure yields extremely rapid learning in rats (Harrison, 1979) and monkeys (Downey & Harrison, 1972), sometimes from the very first trial. In a separate experiment, Harrison (1979, Exp. 3) examined the effects of stimulus preexposure of the two spatially differentiated sounds on the previously demonstrated rapid acquisition of discrimination. Even though the interpretation of the data depended on a cross-experimental comparison, and there were only small numbers of subjects, the dramatic differences, presented in Figure 4, are worth noting. As can be

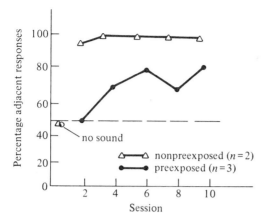

Figure 4. Percentage adjacent responses (correct) as a function of session for a stimulus-preexposed group and a nonpreexposed group. (Adapted from Harrison, 1979, Exp. 3 and Exp. 1, respectively.)

seen, the nonpreexposed group learned the discrimination almost immediately, whereas the group preexposed to the two sound sources learned the discrimination with extreme difficulty. It would seem, then, that stimulus preexposure has a profound effect when used in conjunction with this type of discrimination procedure. However, one must be cautious in interpreting these data in terms of reduced stimulus associability for the stimulus-preexposed group. The stimulus-preexposed group may simply have extinguished unconditioned exploratory responses to the sound source, which because of its proximity to the lever also resulted in extinction of lever-pressing responses. On the other hand, the nonpreexposed group would begin the test with the full complement of these responses.

A similar discrimination study was conducted by Robbins, Everitt, Fray, Gaskin, Carli, and de la Riva (1982). However, instead of preexposing sounds at different locations, a tone and a light were preexposed. In the subsequent discrimination test, a nose poke to one feeder was reinforced when the response was made in the presence of the light, and a poke to the other feeder was reinforced when the response was made in the presence of the tone. Although there was a strong latent inhibition effect that was still present after 12 sessions of discrimination training, the differences between groups developed slowly, appearing only at the fifth session. This is in contrast to Harrison's study (1979), in which the group differences were apparent in the first training session. The differences between experiments may well have been due to the extinction of unconditioned responding in the Harrison study (1979), but not in the study of Robbins et al. (1982).

It should be noted that in both the Harrison study, (1979) and the study of Robbins et al. (1982) the discrimination test was nominally of the successive variety (i.e., only one of the stimuli was presented on any given trial). However, the procedure also could be considered as a simultaneous procedure because of the presence of redundant features (i.e., the *absence* of any specific stimulus was perfectly correlated to the *position* at which *no* reinforcement was available). This, of course, is quite different from the go/no-go procedures, in which there is one response and one source of reinforcement, and one stimulus indicates the availability (while another indicates the unavailability) of reinforcement (e.g., Halgren, 1974; Mason & Lin, 1980; Mellgren & Ost, 1971). It would appear that the procedure using discrete stimulus preexposure, followed by an appetitive instrumental conditioning test with *simultaneously* presented stimuli, should be studied more extensively. In particular, it might be the procedure of choice in drug studies, because, as noted by Robbins et al. (1982), it "is not likely to be confounded by acquisition impairments," nor, we might add, by changes in activity levels.

Continuous stimulus preexposure followed by a simple instrumental appetitive test

All of the discrimination studies that we have considered up to this point used relatively short preexposure periods in which the to-be-targeted stimuli were presented on a discrete trial basis. Typically the preexposure period might be 1 hr in duration, with the stimulus presented for 60 trials, each trial having a 30-sec duration. The paradigm under current consideration is quite different in that the preexposed stimulus is presented for lengthy periods of time, on the order of weeks or months, and is continuously available for inspection by the animal. When the preexposed stimulus is used as the discriminative stimulus *and* when such procedure typically results in *facilitation* of subsequent learning, the phenomenon is called "perceptual learning." Indeed, according to a theoretical model proposed by Hebb (1949), one would expect that prior exposure to a set of stimuli would increase the discriminability of those stimuli and thus enhance subsequent discrimination learning. A number of investigators have tested this hypothesis directly (e.g., Bennett & Ellis, 1968; Forgus, 1958), although some arrived at similar predictions from a different direction (e.g., Gibson & Walk, 1956). What is interesting from our point of view is that the early studies all chose to use prolonged periods of stimulus preexposure.

The use of long periods of stimulus preexposure (which was probably deemed efficacious on the basis of the idea that if stimulus preexposure causes facilitation, then prolonged stimulus preexposure should cause considerable facilitation) dictated, for simple purposes of economy, that such preexposure be conducted in the home cages. Testing, however, which typically involved a Grice discrimination box (Grice, 1948), necessarily had to be conducted in an apparatus different and apart from the home cage.

These types of studies, for the most part, reported facilitation of discrimination learning following stimulus preexposure. There were, however, several exceptions to this rule, in which no differences were found between the preexposed and nonpreexposed groups. Absence of facilitation has been attributed to such modulating factors as whether the stimuli were painted on a flat background or were manipulable cutouts (e.g., Bennett & Ellis, 1968; Gibson, Walk & Tighe, 1959; Walk, Gibson, Pick, & Tighe, 1959) and whether or not the preexposed stimuli were present during feeding (e.g., Kerpelman, 1965). An extensive experimental literature has developed around these procedures and manipulations. For early reviews, see Epstein (1967) and Gibson (1969); for a recent review, see Hall (1980). These reviews confirm the general conclusion, as stated earlier, that prolonged preexposure of to-be-discriminated stimuli, when the preexposure is carried out in the home cage and the test is conducted in a different apparatus, results in facilitated learning.

Thus, there are two independent research literatures, latent inhibition and perceptual learning, both of which involve passive preexposure of the to-be-discriminated stimuli, and each of which gives fairly consistent effects, but in opposite directions: decremented learning for latent inhibition, and facilitated learning for perceptual learning. This apparent contradiction is resolved by an examination of the procedures used in each of the paradigms.

An analysis of the latent inhibition and perceptual learning paradigms suggests important differences in the operations relating preexposure to testing. In the perceptual learning paradigm, experimental and control groups are reared in their home environments, and the experimental group is exposed to the to-be-tested stimuli in that environment. Testing itself occurs in a new environment. Thus, the experimental group is tested with a familiar (old) stimulus in a new environment, whereas the control group is tested with a new stimulus in a new environment. The paradigm is shown, in terms of the conditions *at the time of test*, in Table 8. For perceptual learning, the critical comparison is between the S_oE_n (old stimulus, new environment) and S_nE_n (new stimulus, new environment) conditions. Typically, the former group learns the discrimination task much faster than the latter group ($S_oE_n > S_nE_n$) (e.g., Gibson & Walk, 1956; Oswalt, 1972).

As opposed to this, in the latent inhibition paradigm, *both* the experimental and the control groups are preexposed to the apparatus in which later testing will occur. It is during this period of apparatus preexposure

Table 8. *Relationship between stimulus and environmental novelty at time of testing*

Paradigm	Experimental group	Control group
Perceptual learning	$S_o E_n$	$S_n E_n$
Latent inhibition	$S_o E_o$	$S_n E_o$

Note: S = stimulus, E = environment, n = new, o = old.

that the subjects in the experimental group are also preexposed to the critical stimulus. During testing, the stimulus-preexposed group is confronted with an old stimulus in an old environment ($S_o E_o$), whereas the nonpreexposed group faces a new stimulus in an old environment ($S_n E_o$). With such conditions, learning proves to be slower for the former than for the latter ($S_o E_o < S_n E_o$) (e.g., see any of the latent inhibition studies).

In both the perceptual learning and the latent inhibition paradigms it would appear that a condition of *contrasting* novelty or familiarity of the environment with that of the stimulus provides a better condition for learning than does the absence of such a contrast. The apparent conflict in results, then, is a result of comparisons on the basis of simple stimulus preexposure conditions rather than on the basis of the contrast as described earlier. In perceptual learning, the stimulus-preexposed group is the contrast group ($S_o E_n$) during testing, and in the latent inhibition paradigm it is the nonpreexposed group that is the contrast group ($S_n E_o$) during testing. Both learn faster than their appropriate control groups.

Numerous experiments, beginning with a study by Lubow, Rifkin, and Alek (1976) involving rats and young children, have supported the validity of this analysis (see chapter 3, "Role of Context"). That the contrast effect can predict the direction of results in stimulus preexposure studies and account for the major differences between the perceptual learning and latent inhibition studies does not, of course, mean that there are no other modulating variables. Thus, within perceptual learning, where long-term stimulus preexposure precedes a test in a different environment ($S_o E_n$ versus $S_n E_n$), facilitory effects are not always obtained. For instance, as already noted, it would appear that it is important for the preexposed stimuli to have some three-dimensional or manipulable quality, as opposed to being flat, two-dimensional representations (Gibson et al., 1959; Bennett, Rickert, & McAllister, 1970; Bennett & Ellis, 1968; Walk et al., 1959). However, in spite of such a modulating variable, the effect of which is to produce an absence of group differences, it is quite clear that the stimulus–environment novelty contrast hypothesis accounts for the major results.

In addition to the Lubow, Rifkin, and Alek (1976) experiments, support for the contrast hypothesis comes from two studies that used prolonged stimulus preexposure in the home cage and a home-cage test and found retarded learning for the preexposed stimulus: Bateson and Chantrey (1972) with monkeys; Franken and Bray (1973) with rats. More recently, the superiority of S_nE_o and S_oE_n conditions over S_nE_n and S_oE_o has been replicated by others using home-cage preexposure of the to-be-discriminated stimuli for the S_oE_n condition (Channell & Hall, 1981; Hall, 1979; Hall & Channell, 1983).

If one accepts stimulus–environment novelty contrast as the prime determiner of whether retardation or facilitation is achieved in the test, then another interesting question arises: In the S_oE_o condition, which results in latent inhibition, is it necessary that the stimulus preexposure and the environment preexposure occur together? In other words, does the stimulus have to be preexposed *in* the environment in which the test is to occur? The studies that have been reviewed thus far used just that procedure. What happens, however, if the to-be-targeted stimulus is preexposed in one environment (A) and the animal *also* is familiarized (independent of the preexposed stimulus) to another environment (B) that will subsequently be the test environment? Numerous recent experiments have found that the latent inhibition effect is dependent on stimulus preexposure occurring in the same environment in which the test will take place. These studies are examined in chapter 3.

Other preexposure–discrimination paradigms

There are several other literatures that are relevant to the area under discussion. However, because either the preexposure conditions or test conditions are very different from those that characterize latent inhibition and perceptual learning literatures, those studies will not be discussed. Let it suffice simply to identify the areas: (1) the effects of "mere" exposure on preference (for a recent review of the animal literature, see Hill, 1978; and for the human literature, Zajonc, 1980); (2) the effects of enriched or impoverished environments (e.g., Krech, Rosenzweig, & Bennett, 1966), including dark rearing (e.g., Corrigan & Carpenter, 1979); (3) the effects of very early stimulus preexposure on imprinting (e.g., Chantrey, 1974; Stewart, Capretta, Cooper, & Littlefield, 1977; Kovach, Fabricius, & Fält, 1966); (4) the effects of various reinforcement conditions, especially the effects of nondifferential reinforcement during stimulus preexposure on subsequent learning (e.g., Andelman & Sutherland, 1970; Hall & Channell, 1980; Winfield, 1978). In addition, it is worth noting that within this last category there is a subset of studies in which preexposure conditions resemble those used with masking tasks in the human literature. In these studies, rats learn a simple discrimination between two stimuli, with reinforcement consistently applied when there is a correct response to one stimulus and consistently withheld when the animal responds to the other stimulus. Simultaneously with this standard discrimination procedure, a second pair of stimuli is presented that is not correlated with reinforcement. When these previously irrelevant stimuli are subsequently presented in a discrimination test in which they are now relevant (i.e., consistently reinforced), there is a tendency to learn the discrimination more slowly than does a control group that has not received prior exposure to those stimuli (Babb, 1956; Jeeves & North, 1956; Waller, 1970).

Summarizing the discrimination studies, it is clear that there is a wide variety of studies using a formal discrimination learning task that have provided substantial evidence for strong latent inhibition effects. These studies have been particularly illuminating in differentiating between those conditions of stimulus and context preexposure that produce facilitated learning (perceptual learning) and those that produce retarded learning (latent inhibition).

Summary

Now that all of the paradigms in which latent inhibition has been studied have been reviewed, what are the empirical generalizations that can be induced? There are, indeed, several:

1. Latent inhibition is a phenomenon that is readily accessible in a wide variety of testing situations. Although the most popular paradigms use a noxious US in the CS–US acquisition stage (e.g., avoidance conditioning, conditioned taste aversion, conditioned suppression, and various classical conditioning procedures), there is ample evidence from conditioned observing responses and various discrimination learning procedures that appetitive reinforcement in the acquisition stage is also capable of allowing latent inhibition effects to appear. In addition to being independent of quality of reinforcement, there is a considerable body of data indicating that the classical conditioning/instrumental conditioning distinction does not interact with the effects of stimulus preexposure.

2. In spite of the convincing evidence, individual experiments, especially in the area of conditioned suppression, conditioned taste aversion, heart-rate conditioning, and autoshaping, must be conducted and analyzed with caution in order to differentiate between the various effects of an unconditioned alpha response to the CS and the effects of the conditioned response.

3. When the foregoing considerations are taken into account for those

studies that purport to show latent inhibition, together with those that indicate the absence of a latent inhibition effect, it appears that there is little evidence for a latent inhibition effect in heart-rate conditioning and salivary response conditioning. However, this conclusion, which would suggest that autonomic conditioning, in general, is not subject to latent inhibition, must be qualified by noting that the number of studies reported in this area is relatively small.

4. The same arguments as presented earlier also apply to autoshaping. Here, however, the reasons for the absence of compelling evidence for latent inhibition might better be attributed to the species employed in the experiment, rather than to the autoshaping procedure itself. Because all of the valid studies of latent inhibition of the autoshaped response have involved pigeons, and there is no independent evidence of latent inhibition in the pigeon, it would seem reasonable to suggest that this might be a species problem rather than a testing problem.

5. The foregoing suggestion of species differences is buttressed by the failures to find latent inhibition in molluscs, honeybees, and goldfish. If this generalization withstands additional experimentation, then latent inhibition will remain a behavioral characteristic of mammalian species, with simple stimulus preexposure effects in lower mammals and in young children, but more complex effects in human adults. This last point will be discussed in the section on species differences in chapter 4 and, more completely, in chapter 5.

3 Variables affecting latent inhibition

Similarity of preexposed stimulus and test stimulus

Studies of the influence of stimulus preexposure on subsequent learning usually assume that these effects are stimulus-specific. That is, preexposure to stimulus A should *not* retard the subsequent acquisition of an association between stimulus B and another event. This assumption is particularly critical if one maintains the idea that the subsequent decrement in the acquisition of the association between A and another event is a result of some previous associative *learning* during the preexposure phase. Associative learning, by definition, presumes some degree of stimulus specificity. Indeed, the apparent absence of such specificity in the learned helplessness effect, at least in rats (Maier & Seligman, 1976), in which preexposures of a motivationally significant stimulus is administered, by itself, might raise the question whether or not one is dealing with an associative learning phenomenon.[1] For latent inhibition, across different paradigms, the results are quite clear. The decremental effects of preexposure of the to-be-conditioned stimulus on subsequent acquisition of a new association are, without doubt, stimulus-specific.

Such stimulus specificity may be demonstrated in several ways. (1) A within-subject experimental design may be employed whereby the animal is preexposed to stimulus A and tested on both stimulus A and stimulus B. When appropriately counterbalanced, slower learning to the familiar stimulus compared with the novel stimulus serves as evidence for the stimulus specificity of latent inhibition. Such a design was employed by Lubow and Moore (1959), Reiss and Wagner (1972), and Wickens et al. (1983). (2) A between-subjects design in which one group is preexposed to stimulus A, and another to stimulus B, and both are tested with A may also be used. Carlton and Vogel (1967, Exp. 2) employed this design, but without counterbalancing. Both sets of experiments indicated that latent inhibition is stimulus-specific. (3) A third procedure for assessing stimulus specificity, but in a dimensional manner, is that of a generalization test. In this paradigm, in its simplest form, all groups are preexposed to stimulus A and then subdivided and tested with stimuli that vary across groups in their similarity to A. A generalization gradient or incomplete generalization of latent inhibition provides evidence for the stimulus specificity hypothesis. Demonstrations of this effect have been provided for tonal frequency (Siegel,

1969b) and stimulus intensity (Crowell & Anderson, 1972; Lantz, 1976). The larger group of studies, however, come from the conditioned taste aversion literature.

Although Klein, Mikulka, and Hamel (1976) failed to demonstrate any generalization of latent inhibition, thus obtaining complete stimulus specificity for sucrose concentration, Dawley (1979) did find a "gradient" across different flavors. Whereas the previously cited studies examined the generalization of latent inhibition by either preexposing or not preexposing a single flavor and conditioning different groups to either the preexposed flavor or related, but novel, flavors, another approach was provided by Braveman and Jarvis (1978), Miller and Holzman (1981), Tarpy and McIntosh (1977), and Zahorik (1976). In these studies, rats were preexposed to a number of flavors. The intent was to determine if latent inhibition would develop to a new flavor (presumably not related to the preexposure flavors) that was paired with illness.

Zahorik (1976) failed to find generalization of latent inhibition after preexposure to a variety of diets. Similarly, both Miller and Holzman (1981) and Braveman and Jarvis (1978), who preexposed three and four different solutions, failed to find generalization of latent inhibition under these conditions, but did find that there was a dissociation between neophobia reduction and latent inhibition – an important result considering the concern expressed earlier about possible confounding of these two sources of test data. However, generalization of latent inhibition was obtained by Tarpy and McIntosh (1977), who preexposed nine different solutions.

In general, these date indicate that for qualitatively different stimuli there is a considerable stimulus specificity for the latent inhibition phenomenon. However, as one might well expect, when there are small changes in a single dimension of a stimulus, such as intensity, there may be sizable generalization effects.[2]

Number of stimulus preexposures

The effect of number of stimulus preexposures on the development of latent inhibition was reviewed by Lubow (1973a, pp. 401–402) for a variety of learning paradigms and species: "The conclusions... are quite straightforward, yet somewhat startling in that the data from which they are drawn are almost completely consistent between experiments. Of 14 different groups from 6 experiments, only 1 (Cantor & Cantor, 1966) shows an inhibitory effect with less than 17 nonreinforced preexposures. On the other hand, within the very narrow range of 16–20 nonreinforced preexposures, suddenly 5 of 6 groups from 5 different

Table 9. Effects of number of flavor preexposures on conditioned taste aversions

Experiment	Number of preexposures	Duration of each preexposure (min)	Preexposed stimulus	Results
Domjan (1972)	0, 6[a]	0, 2.8, 6.7, 13.3	Saccharin	LI for 6.7- & 13.3-min PE; no difference between them
Elkins (1973)	0, 1, 3, 10, 20	Days–continuous	Saccharin	Graded LI in acquisition and extinction
Farley et al. (1964)	0, 1, 2, 4, 8	Days–continuous	Saccharin	LI from 1 to 8 PE; not graded
Fenwick et al. (1975)	0, 6, 13, 20	5	Sucrose	Graded LI in acquisition and extinction
Franchina et al. (1980)	0, 1, 2, 4, 8	30[d]	Sucrose	Graded LI
Franchina & Horowitz (1982)	0, 1, 3	30[d]	Sucrose	LI for 3, but not 1, PE
Gillette & Bellingham (1982, Exp. 2)	2, 10, 20	10	Salt-saccharin[f]	Graded LI
Kalat & Rozin (1973, Exp. 1A)	0, 1, 3	20	Sucrose	Same LI for 1 & 3 PE
Kalat & Rozin (1973, Exp. 1B)	0, 1, 7	20	Casein	Same LI for 1 & 7 PE
Klein et al. (1978, Exp. 1)	0, 1, 3	30	Sucrose	No LI
Mohammed & Archer (reported in Archer, 1982b)[b]	0, 1, 2, 3	–[e]	Saccharin	Graded LI
Robbins (1979)[b]	0, 1, 10	20	Sucrose	On first extinction trial, LI only for 10 PE; with continuous extinction, 1 & 10 PE extinguish *more slowly* than 0 PE
Smotherman et al. (1980, Exp. 2)	2, 5, 10[c]	20	Sweet milk	No difference between 2 & 5 PE; no aversion (LI) in 10 PE
Smotherman et al. (1980, Exp. 4)	5, 6, 7, 8, 9, 10[c]	20	Sweet milk	No difference between 5, 6, 7, & 8 PE; significantly more LI in 9 & 10 PE

[a] Unless noted, there were 24-hr intervals between flavor preexposures.
[b] Subjects were deer mice.
[c] Note that there was no 0 PE group.
[d] 12 hr between preexposures.
[e] Not reported.
[f] Compound solution.

experiments show inhibitory effects. This consistency is all the more surprising when one remembers the diversity of the 12 studies in both species and learning paradigm."

Had the preceding quotation been written more recently, the basic conclusions would remain the same, but with two major qualifications. First, it is now impressively documented that at least within the conditioned taste aversion paradigm, it is possible to demonstrate latent inhibition with just a single stimulus preexposure. Studies that have demonstrated poorer flavor avoidance after a single flavor preexposure, as compared with no such preexposure prior to flavor–toxicosis pairing, include those by Batson and Best (1982), Best and Gemberling (1977), Best et al. (1979), Bolles, Riley, and Laskowski (1973), Bond and DiGuisto (1975), Bond and Westbrook (1982), Domjan and Bowman (1974), Kalat (1977), Kalat and Rozin (1973), Nachman and Jones (1974), Siegel (1974), and Westbrook, Provost, and Homewood (1982), as well as others. Similar effects following a single odor familiarity trial have been obtained by Rudy and Cheatle (1977, 1978) and Westbrook et al. (1981).

It is important to note that these single-preexposure trials had stimulus availability durations of many minutes, and although the availability of the stimulus was constant, the actual contact was intermittent. Thus, although nominally there was a single stimulus preexposure of long duration, in fact the procedure can result in many discrete, short-duration stimuli.

At the opposite extreme from studies employing a single flavor preexposure before conditioning are those studies that use continuous preexposure over many days. One set of experiments that employed this type of extensive training was concerned with the question whether or not it was possible to condition a taste aversion to a solution as familiar as water. Given the fact that a single flavor preexposure significantly attenuates taste aversion, one might expect that the very extensive experience with solutions such as water would make it impossible to condition an aversion to it. Several studies have shown that although under these conditions latent inhibition is indeed very pronounced, it is never complete (i.e., some conditioned aversion to water is obtained) (Elkins, 1974; Nachman, 1970; Riley et al., 1983).

A more complete picture of the effect of number of preexposures on the magnitude of latent inhibition can be seen in Table 9, which describes those studies that have systematically varied the number (or amount) of flavor preexposures. The table lists the number of preexposures, the duration of each preexposure, and the magnitude of the latent inhibition effect.

It is clear from an inspection of Table 9 that although latent inhibition was a consistent finding some studies did not find a graded latent inhibition effect with an increase in number of preexposures (Farley et al, 1964; Kalat & Rozin, 1973), and others did find such an effect (Elkins, 1973; Fenwick, Mikulka, & Klein, 1975; Franchina et al., 1980). The reasons for these discrepancies are, no doubt, related to the use of different conditioning parameter values.

Within the conditioned suppression testing paradigm in which stimulus availability and stimulus "contact" are coincident, the range of stimulus preexposures across latent inhibition studies went from the zero preexposure of the standard control groups to relatively low numbers such as 1, 2, 5, or 10 (Baker, 1976; Crowell & Anderson, 1972; DeVietti & Barrett, 1986a; Domjan & Siegel, 1971; Lantz, 1976, Exp. 2). However, the typical study, which simply compares a preexposed group with a nonpreexposed group, will use anywhere between 20 and 80 preexposures. These values almost always prove to be effective for demonstrating latent inhibition. At the high end of the continuum, May, Tolman, and Schoenfeldt (1967) used 4,176 preexposures to a tone prior to initiating suppression training. Five studies have explicitly varied the number of stimulus preexposures (DeVietti et al, 1987; Domjan & Siegel, 1971; Gray, 1978; Lantz, 1973; Lubow, Wagner, & Weiner, 1982; Wright & Gustavson, 1986).

In summary, these conditioned suppression studies support the general notion that latent inhibition is a function of number of preexposures. In addition, they suggest that with relatively few preexposures there is no latent inhibition effect and that the effect quickly reaches an asymptote, perhaps after only 20 or 30 stimulus preexposures. The exceptions to this generalization occurred in studies by Crowell and Anderson (1972) and DeVietti and Barrett (1986a) in which a single noise or tone preexposure was sufficient to produce latent inhibition when tested immediately, 24 hr, or 169 hr later (Crowell & Anderson, 1972) or just 24 hr later (DeVietti & Barrett, 1986a). However, it is important to note that the single preexposure was 270 sec in duration in the Crowell and Anderson study and 60 sec in the DeVietti study[3] – considerably longer than any of those reported in the aforementioned experiments. Once again, as in the latent inhibition/conditioned taste aversion literature, there appears to be a powerful interaction between number of preexposures and duration of preexposure.

In addition to the date from conditioned taste aversion and conditioned suppression studies, there have been several standard classical conditioning experiments that have explored the relationship between the number of stimulus preexposures and the magnitude of the latent inhibition effect (Chacto & Lubow, 1967; Clarke & Hupka, 1974; Lubow, 1965; Lubow, Markman, & Allen, 1968; Siegel, 1969a). As in the other preparations in which latent inhibition was obtained, the magnitude of the effect was a function of number of preexposures. However, it appears that more preexposures are required to obtain the effect in the standard classical conditioning procedure than in conditioned taste aversion or conditioned suppression. This may be due to the fact that the former procedures typically use stimuli of very short duration compared with the latter procedures. Indeed, this may account for two failures to obtain latent inhibition in spite of quite extensive numbers of preexposures (Plotkin & Oakley, 1975; Suboski et al., 1964). On the other hand, as suggested for autoshaping (see chapter 2), this finding may be related to the fact that procedures such as nictitating membrane response conditioning and eyelid conditioning also require a large number of CS–US pairings to produce asymptotic conditioning performance.

Duration of the preexposed stimulus

As with many other variables, there are wide ranges of parameter values for duration of preexposed stimulus, and as already mentioned, much of the variance can be accounted for by the type of preparation employed. Thus, typically with conditioned taste aversion, whereas the number of stimulus preexposures is low, the duration is long. In other preparations (e.g., classical conditioning, conditioned avoidance, conditioned suppression, etc.), the converse is true. Nevertheless, within each paradigm a range of durations is employed, and almost inevitably the duration of the preexposed stimulus is the same as the duration of that stimulus when paired with the US in the acquisition stage.

In conditioned suppression, the preexposed stimulus has ranged in different experiments from 3 sec (Lubow, Schnur, & Rifkin, 1976, Exp. 1) or 10 sec (e.g., Carlton & Vogel, 1967; Lubow & Siebert, 1969) to 120 sec (e.g., Kremer, 1972; Logan & Schnur, 1976) and even 180 sec (Gray, 1978) and 270 sec (Crowell & Anderson, 1972). The choice of duration would appear to be less dependent on theoretical considerations than on the particular kind of suppression technique employed. The shorter durations are used with the high-rate lick responding; the longer durations are used with the relatively lower response rate of bar pressing.

A comparison of latent inhibition effects across experiments that used different durations of the to-be-conditioned stimulus during preexposure strongly suggests that there is an interaction between number of preexposures and stimulus duration, as mentioned in the previous section. Thus, whereas generally one does not obtain latent inhibition with less than 15 or 20 preexposures (Lubow, 1973a), the effect can be obtained under these conditions if one uses relatively long stimulus durations (e.g., Crowell & Anderson, 1972; DeVietti & Barrett, 1986a;

DeVietti, Emmerson, & Wittman, 1982; DeVietti, Wittman, & Comfort, 1980; DeVietti, Wittman, Emmerson, & Thatcher, 1981; Gray, 1978; Lantz, 1973). Conversely, whereas, generally, short-duration stimuli are ineffective, this may be overcome by increasing the number of stimulus preexposures (e.g., Lubow, Rifkin, & Alek, 1976; Reiss & Wagner, 1972; Siegel, 1971; Solomon & Moore, 1975). A more formal demonstration of the interaction between number and duration of stimulus preexposures has been provided by DeVietti et al. (1987).

The problem of separating number of stimulus preexposures from duration of preexposure in the conditioned taste aversion procedure has already been discussed. However, there has been one study that has accomplished this feat. Domjan (1972, Exp. 3) fitted rats with oral-cavity cannulae and then preexposed them to the target flavor. A fixed-volume saccharin solution was infused at three different rates so that flavor exposure was for periods of 13.3, 6.7, or 2.8 min. In addition, there was a zero-preexposure group. Following a saccharin–LiCl pairing, preference for saccharin versus water was measured. Saccharin preexposure attenuated aversion learning, but only for the longer preexposure durations.

Another important finding related to stimulus duration concerns its interaction with the time between preexposure and conditioning. A number of experiments have indicated that the duration of stimulus preexposure modulates the amount of latent inhibition when there is a long time interval between stimulus preexposure and CS–US acquisition, but not when there is a short interval (e.g., Best & Gemberling, 1977; Bond & Westbrook, 1982). This conclusion will be more fully documented in the section on retention of latent inhibition.

In conclusion, the data from both conditioned suppression and conditioned taste avoidance point in the direction of a very potent effect of stimulus duration on latent inhibition, such that as duration increases, so does the amount of latent inhibition. Furthermore, there appears to be a pronounced interaction between the number of stimulus preexposures and the duration of each preexposure, suggesting that the *total amount* of experience with the preexposed stimulus is a major modulator of the size of the latent inhibition effect, and that this, in turn, interacts with the temporal interval between preexposure and acquisition.

Intensity of the preexposed stimulus

As indicated earlier, there is a generalization of latent inhibition when there are small changes in stimulus intensity between preexposure and acquisition–test. The question that concerns us here, however, is whether or not there is an independent effect of preexposure intensity on the amount of latent inhibition. Because it is well established that CS intensity does affect conditioning strength, one has to be careful in the design of these experiments to differentiate between the effects of the intensity of the preexposed stimulus and the effects of the intensity of the conditioned stimulus.

To assess the effects of the intensity of the preexposed stimulus independent of CS conditioning intensity, a counterbalanced design is required with preexposure stimulus intensity crossed with conditioned stimulus intensity. In addition, one needs nonpreexposed groups that are given the acquisition–test with each of the preexposed stimulus intensities of the other groups.[4] Most studies of the effects of preexposed stimulus intensity on latent inhibition have failed to employ such a design (Hernandez et al., 1981; Lantz, 1976; Solomon, Brennan, & Moore, 1974). There have, however, been two studies in which the complete factorial design has been used (Crowell & Anderson, 1972, Exp. 1; Schnur & Lubow, 1976, Exp. 2). Both experiments used auditory stimulus intensity and conditioned suppression, and both found that latent inhibition was an increasing function of the intensity of the preexposed stimulus.

Indirect evidence supporting this contention was recently reported by Midgley, Wilkie, and Tees (1988). In that study, the same stimulus was preexposed to two groups. The subjects in one group had lesions of the superior colliculus, and those in a second group were sham-operated. In the former, orienting responses either were not elicited or were habituated rapidly, as compared with the latter group, in which there were substantial orienting responses that were habituated slowly. Thus, one can assume that the preexposed stimulus was more salient for the sham-operated group than for the lesioned group. Nevertheless, the sham-operated group, in a subsequent test, acquired conditioned suppression more slowly than did the lesioned group, thereby suggesting that there was a greater latent inhibition effect for a more salient stimulus than for a less salient stimulus.

Interstimulus interval

The interstimulus interval (ISI) describes the time between stimulus presentations during the preexposure period. These times, of course, can be constant or varied. However, the usual procedure when examining the effects of ISI is to keep the interval constant within subjects and to vary it across groups. In one such study, Lantz (1973) compared a nonpreexposed group with groups that had an interval of 2, 10, 30, or 150 sec between the offset of one stimulus and the onset of the following stimulus. Latent inhibition, in a conditioned suppression test, was obtained only for the 30- and 150-sec interstimulus interval groups, with the latter showing poorer conditioning (i.e., more latent inhibition) than the former.

The interval between stimulus preexposures also was varied by Crowell and Anderson (1972), but with considerably more extreme values than those employed by Lantz (1973): 45 sec versus 24 hr. Although latent inhibition was demonstrated, there was no effect of ISI.

From the first study it would appear that latent inhibition is sensitive to the effects of interstimulus interval, at least at the low end of the scale, with the amount of latent inhibition increasing with the increase in time interval between stimulus presentations in the preexposure period.[5] This conclusion is supported by Schnur and Lubow (1976) with data from an avoidance conditioning study.

An exception to this statement was reported by DeVietti and Barrett (1986b), who found no effect of time between stimuli. The discrepancy between their study and those of Lantz (1973) and Schnur and Lubow (1976) may be accounted for by the differences in time between preexposure and acquisition. In the latter studies there was zero delay, whereas in the DeVietti study there was a 48-hr delay between preexposure and acquisition. It remains to be seen whether or not this interaction between ISI and preexposure–acquisition interval will hold up in a within-experiment design. However, perhaps from the vantage point of general adaptiveness, such an interaction should not be unexpected. The preexposure interstimulus interval information should become of decreasing significance to the organism as the delay between the last preexposure and the to-be-acquired association increases.

Retention of latent inhibition

If the latent inhibition phenomenon is indeed learned and is not simply a manifestation of traditional habituation processes (i.e., reduced sensitivity to the conditioned stimulus as a result of prior exposure to that stimulus), then the decremental effects of stimulus preexposure should be retained over a reasonably long period of time. To measure the retention of latent inhibition, in the three-stage procedure, the manipulation of the temporal interval should involve the interval between the last stimulus preexposure and the CS–US pairing. To use the interval between conditioning and test would, of course, confound the retention of stimulus preexposure effects per se with the retention of the conditioning trial effects, with both contributing to the assessment of retention of latent inhibition. However, there are good theoretical reasons for examining the conditioning–test interval. This section will review both sets of data, as well as the interaction between the stimulus preexposure–acquisition interval and stimulus duration.

Time between stimulus preexposure and acquisition

The earliest study that addressed the question of retention of the familiarization effects in the conditioned taste aversion paradigm was that of McLaurin et al. (1963). Compared with a nonpreexposed control group, flavor-preexposed groups that had a delay between the end of preexposure and the beginning of acquisition of either 3 or 6 days showed marked latent inhibition effects. Similarly, Kalat and Rozin (1973) found that latent inhibition was completely retained after a 21-day delay. On the other hand, Kraemer and Roberts (1984, Exp. 4), McIntosh and Tarpy (1977), and Elkins and Hobbs (1979) demonstrated that there was considerable dissipation of the effect after long delays.

A study by Robbins (1979), with deer mice, although manipulating the appropriate interval for measuring the retention of latent inhibition, did not clarify the picture. Neither a 4-day-delay group nor a 20-day-delay group showed latent inhibition when preexposure consisted of a single presentation of sucrose. With an increase in the number of familiarization trials, latent inhibition was produced after a 4-day delay.

The six retention studies are summarized in Table 10. Although the overall pattern is far from clear, it would seem reasonable to conclude that, with one set of procedures, a single 20-min flavor preexposure has a profound debilitating effect on the subsequent flavor–poison association even with a delay of as much as 21 days. However, with other procedural variables, the effects of delay between preexposure and acquisition are not as pronounced (McIntosh & Tarpy, 1977; McLaurin et al., 1963; Robbins, 1979). The critical variables for modulating the retention of latent inhibition have yet to be identified.

Within the conditioned suppression paradigm, two studies varied the interval between stimulus preexposure and acquisition (Crowell & Anderson, 1972; James, 1971). There was no effect of interval, whether varied in the order of minutes (James, 1971) or days (Crowell & Anderson, 1972).

With few exceptions, the effect of delay between stimulus preexposure and acquisition has not been investigated with procedures other than conditioned suppression and conditioned taste aversion. However, one may note that in all of the other standard classical conditioning experiments demonstrating latent inhibition, for which purpose only two stages are required, the delay between the last stimulus preexposure presentation and the first acquisition trial took only one of two values: zero or 24 hr. Although it is true that all of the failures to find latent inhibition used a 24-hr delay (Lubow et al., 1968; Plotkin & Oakley, 1975; Prokasy et al., 1978; Suboski et al., 1964), other studies using that delay succeeded in demonstrating the latent inhibition effect (Frey, Maisiak, & Dugue, 1976, Exp. 2; Reiss & Wagner, 1972, Exp. 1; Scavio et al.,

Table 10. *Retention of latent inhibition*

Experiment	Retention interval	Amount of preexposure	Preexposed stimulus	US	Conditioning–testing interval (days)	Type of test	Results
Elkins & Hobbs (1979)	0	15 days	Saccharin	Cyclophosphamide	1	2-bottle	LI
	20						No LI
	100						No LI
Kalat & Rozin (1973)	1	20 min	Casein hydrolysate	LiCl	2	2-bottle	LI
	21	20 min					LI
	—	0					—[a]
Kraemer & Roberts (1984, Exp. 4)	1	20 min/day 3 days	Saccharin	Scopolamine	1	2-bottle	LI
	21	5 days[b]					LI, less
McIntosh & Tarpy (1977)	1	5 days[b]	Saccharin	LiCl	1	2-botle	x[c]
	24						
McLaurin et al. (1963)	0	6 days[b]	Saccharin	X-ray	0[d]	2-bottle	LI, most
	3	6 days[b]					LI, less
	6	6 days[b]					LI, less
	—	0					—[a]
Robbins (1979)[e]	4	1, 20 min	Sucrose	LiCl	2	1-bottle	No LI[f]
	4	1, 20 min					No LI[f]
	4	10, 20 min each					LI
	—	0					—[a]

[a] Comparison group against which presence or absence of LI is evaluated; where no such group is indicated, comparisons are between those groups listed.
[b] Continuous access to the preexposed fluid.
[c] No comparison group against which to measure LI; see text.
[d] Subjects given access to test fluids immediately after CS-US pairing, but amount consumed was measured once per day over 7 days. First measure presumably about 12 hr after pairing.
[e] Deer mice were subjects.
[f] As measured on first test-extinction trial; with subsequent extinction trials, all three preexposed groups showed more aversion (i.e., increased resistance to extinction) than the zero-PE control group.

1983; Wickens et al., 1983). A study by Siegel (1970) directly compared the effects of the two different delays. He found strong latent inhibition for both groups, with no effect of the delay variable, either directly or interacting with number of preexposures.

Interaction between duration of the preexposed stimulus and time between preexposure and conditioning

There have been several conditioned aversion experiments that have used a single-preexposure procedure and varied the duration of that preexposure and/or the time between the stimulus preexposure and the stimulus–poison pairing (Table 11). These studies, using either flavors or odors as the critical stimuli, present a fairly convincing picture of an interesting interaction between the two variables, one that suggests two quite different processing mechanisms for the stimulus preexposure effect. For purposes of exposition, let us call the first preexposure S_1, and the second exposure, the one paired with the US, S_2.

The single-preexposure procedure was used by Bond and Westbrook (1982), who manipulated the duration of the preexposure by allowing rats to drink either 1.6 ml or 9.7 ml of saccharin either 23.5 hr or 3.5 hr before the saccharin–poison pairing. As compared with a group that did not receive a saccharin preexposure, the most latent inhibition was shown by the 9.7-ml–23.5-hr group, and there was no latent inhibition in the 1.6-ml–23.5-hr group. Both 3.5-hr groups showed an intermediate amount of latent inhibition; there were no differences between them. Thus, this study demonstrates the interaction between the amount of stimulus preexposure and the delay between the preexposure and conditioning episodes, such that preexposure duration has an effect with long intervals, but not with short intervals. Inspection of the studies listed in Table 11 supports this generalization. At short intervals (i.e., 3.5 hr and less), irrespective of S_1 duration, there is a preexposure effect that produces latent inhibition (Best & Gemberling, 1977, Exp. 1; Bolles et al., 1973; Bond & Westbrook, 1982, Exp. 1; Bond & DiGiusto, 1975; Kalat & Rozin, 1973, Exp. 2; Siegel, 1974, Exp. 1A; Westbrook et al., 1982, Exp. 1B). However, at longer intervals, S_1 duration plays a critical role in determining the latent inhibition effect (Best & Gemberling, 1977, Exp. 2 & 3; Bond & Westbrook, 1982, Exp. 1; Johnson, 1979; Westbrook et al., 1981). It should be noted that this interaction is in the opposite direction from that reported for ISI and stimulus preexposure–acquisition interval.

Time between acquisition and test

As noted earlier, varying the time between the CS–US acquisition stage (which was preceded by the preexposure of the to-be-conditioned stimu-

Table 11. *Effect of time interval between a single stimulus preexposure (S_1) and the conditioning event (S_2–US) as modified by S_1 duration*

Experiment	Exp. number	S_1[a] duration (min)	S_1–S_2 interval	S_2 duration (min)	S_2–US interval (min)	S_1S_2	Results and comments
Best & Gemberling (1977)	1	—	—	5	30	Casein hydrolysate	The greater the interval, the more disruption of taste aversion, i.e., more LI
		5[b]	0.25 hr	5	30		
		5[b]	1.50 hr	5	30		
		5[b]	3.50 hr	5	30		
	2	—	—	5	30	Casein hydrolysate	As in Exp. 1, increase in disruption through 3.5-hr interval; with greater intervals, a graded decrease in LI, but still significant amount of LI after 1-week interval
		5[b]	0.25 hr	5	30		
		5[b]	3.50 hr	5	30		
		5[b]	7.50 hr	5	30		
		5[b]	23.50 hr	5	30		
		5[b]	1.0 wk	5	30		
	3	—	—	5	30	Casein hydrolysate	Considerable LI for 3.5-hr group; LI, but significantly less than in 3.5-hr group, for 11.5-hr group
		5	3.50 hr	5	30		
		5	11.5 hr	5	30		
Bolles et al. (1973)		—	—	10	4 hr	Saccharin	—
		10	3.5 hr	2.5	0.5 hr		LI
		—	—	2.5	0.5 hr		—
Bond & DiGiusto (1975)	2a	4.5 ml	3.5 hr	1.5 ml	0.5 hr	Saccharin	Less aversion (LI)
		—	—	4.5 ml	4 hr		Less aversion
		—	—	1.5 ml	0.5 hr		Most aversion
	2b	1.5 ml	3.5 hr	4.5 ml	0.5 hr	Saccharin	Most aversion (No LI)
		—	—	1.5 ml	4 hr		Least aversion
		—	—	4.5 ml	0.5 hr		Most aversion

Study	Exp					CS	Outcome
Bond & Westbrook (1982)	1	1.6 ml	23.5 hr	3 ml	30	Saccharin	As much aversion as control
		9.7 ml	23.5 hr	3 ml	30		Considerable LI
		1.6 ml	3.5 hr	3 ml	30		Reduced LI
		9.7 ml	3.5 hr	3 ml	30		Reduced LI
		–	–	3 ml	30		–^{e}
Domjan & Bowman (1974)	1	–	–	2.5	30	Casein hydrolysate	–^{e}
		10	3.5 hr	2.5	30		LI
	2	–	–	2.5	30	Weak saccharin	–^{e}
		10	3.5 hr	2.5	30		LI, but less than in Exp. 1
	3	–	–	2.5	20	Saccharin	–
		10	6 hr	2.5	20		No LI
Kalat & Rozin (1973)	1	–	–	2.5	30	Casein hydrolysate	–
		20	24 hr	2.5	30		Considerable LI
		20	3 wk	2.5	30		Considerable LI
	2	–	–	2.5	30	Sucrose or salt or casein	–^{e}
		10	3.5 hr	2.5	30		Considerable LI for all PE flavor groups
		–	–	10.0	4 hr		–
		–	–	10.0	24 hr		–
Rudy & Cheatle (1977)^{d,f}		10	20	10	40	Lemon odor	No aversion
		10	45	10	15		No aversion
		10	50	10	10		Aversion
		10	30	10	40		No aversion
		10	50	10	40		No aversion
		–	–	10	60		–^{e}
		–	–	5	5	Lemon odor	–^{e}
		30	1 hr	5	5		No aversion
		30	4 hr	5	5		No aversion
		30	24 hr	5	5		No aversion

Table 11. (cont.)

Experiment	Exp. number	S_1 duration (min)	S_1–S_2 interval	S_2 duration (min)	S_2–US interval (min)	S_1S_2	Results and comments
Rudy & Cheatle (1978)[c]	1[d]	—	—	5	5	Lemon odor	Significant aversion for condition where S_1 and S_2 duration was 5 min, and where S_2, not preceded by S_1, was 5 min; significant attenuation of aversion when S_1 was 30 min or when S_2, not preceded by S_1, was 30 min
		—	—	5	90		
		—	—	30	90		
		5	85 min	5	5		
		30	85 min	5	5		
	2[d]	—	—	10	80	Lemon odor	As S_1 duration increased, the magnitude of the aversion decreased; similarly for S_2 duration; no effect of S_1 plus S_2 as compared with S_2 alone
		—	—	15	75		
		—	—	20	60		
		10	85 min	5	5		
		15	85 min	5	5		
		20	85 min	5	5		
	4[d]	—	—	10	480	Lemon odor	No aversion
		—	—	10	90		Aversion
		—	—	10	60		Aversion
		10	460 min	10	10		No aversion
		10	70 min	10	10		Aversion
		10	40 min	10	10		Aversion
		10	20 min	10	70		No aversion
		10	20 min	10	40		No aversion
Siegel (1974)	1A	30	24 hr	30	30	Coffee	LI
		—	—	30	30	Coffee	–[e]
		30	24 hr	30	30	Vinegar	LI
		—	—	30	30	Vinegar	–[e]

Study	Exp					Stimulus	Result
Westbrook et al. (1982)	1A	—	—	2	30	Either saline or sucrose	—[e]
		8	27.5 hr	2	30		LI, even more than 3.5-hr group
		8	3.5 hr	2	30		LI, considerable
	1B	—	—	2	30	Either saline or sucrose	—[e]
		2	27.5 hr	2	30		No LI
		2	3.5 hr	2	30		LI
Westbrook et al. (1981)	1	—	—	2	2	Eucalyptus odor	The single short PE almost abolishes the aversion; i.e., considerable LI
		—	—	2	180		
		2	3 hr	2	2		
	2	2	4 hr	2	2	Eucalyptus odor	LI for 4-hr S_1–S_2 interval, but not for longer intervals
		2	28 hr	2	2		
		2	76 hr	2	2		
	3	—	—	2	2	Eucalyptus odor	No LI
		40 sec	24 hr	2	2		No LI
		2	24 hr	2	2		LI
		6	24 hr	2	2		LI
		18	24 hr	2	2		—[e]
		—	—	2	2		
	6	10	24 hr	2	2	Eucalyptus odor	LI Another set of groups were given same conditions, but repeatedly exposed to context between S_1 and S_2 presentations; for these groups, LI only in S_1–2-min, 4-hr S_1–S_2-interval group
		2	4 hr	2	2		LI
		10	4 hr	2	2		LI
		2	24 hr	2	2		No LI
		—	—	2	2		—[e]

[a] Unless indicated otherwise, all times are in minutes.
[b] Rats were allowed to consume 5 cc; see Best & Gemberling's footnote 1 (1973, p. 255).
[c] Experiment 3 used adult rats and varied only S_2 duration, (without preexposure), using saccharin for 3, 10, or 30 min before injecting LiCl; flavor aversion increased from 3 to 10 min and decreased from 10 to 30 min.
[d] Subjects were 8-day-old rat pups.
[e] Comparison group against which presence or absence of LI was evaluated; where no such group is indicated, comparisons were between those groups listed.
[f] Design combined from several studies reported in Rudy & Cheatle (1977, pp. 174–181).

lus) and the test stage does not provide a clear representation of memorial effects of stimulus preexposure. Although several studies have indicated that latent inhibition can survive extensive delays between acquisition and test (e.g., 10 weeks in an experiment by Wickens et al., 1983), only a relatively few have actually manipulated the acquisition–test interval.

In this regard, Kraemer and Roberts (1984, Exp. 3) reported that after a 21-day test interval, as compared with a 1-day interval, latent inhibition was present, but considerably reduced, a finding replicated by Kraemer and Ossenkopp (1986). Although this type of data may be interpreted in terms of retrieval interference rather than encoding failure (see chapter 7), the absence of a control group for neophobia sensitization (i.e., a CS, US unpaired group) makes such an interpretation premature (Kraemer, Hoffmann, & Spear, 1988). Relatedly, it should be noted that Bouton (1987) used conditioned suppression and varied the retention interval from 90 min to 7 days. He found that the amount of latent inhibition was independent of the length of the conditioning–test interval.

Role of context

The term "context" usually refers to all of those environmental stimuli that are relatively constant during the course of a particular procedure. Unless specifically noted, context can be considered as synonymous with apparatus, environment, or background cues. The major exception to this definition arises from a distinction that can be made between tonic and phasic context stimuli (Archer, Sjoden, & Nilsson, 1985), with the former referring to the constant background, as noted earlier, and the latter referring to those stimuli that occur only in the presence of the CS and/or US, or at some other *discrete* time. With this distinction, phenomena such as blocking, overshadowing, compound conditioning, and others become special cases of context manipulation. The distinction is introduced here primarily because Archer and his associates have adopted it in their "noisy"–flavor studies of contextual control of various aspects of taste aversion conditioning, including latent inhibition, acquisition, and extinction. In this paradigm, to be discussed in more detail shortly, the noise is perfectly correlated with drinking behavior, being produced by the banging of ball bearings in the metal drinking tube when the rat licks.

In most learning experiments the constancy of the phasic background cues is referenced against the transience of the conditioned stimulus and the reinforcement. Indeed, it is the adventitious association of the CS and/or US with the apparatus-produced stimuli that has taken on a heavy explanatory burden in learning theory. For example, the Rescorla and Wagner (1972) and Wagner (1978) theories of classical conditioning make assumptions about CS–US pairings that include the development of associations between the US and the context. Strength of conditioning in test is presumed to be a result of a summation of associative strength (excitatory or inhibitory) to both the nominal CS and the background stimuli. Alternatively, Medin (1975) and Spear (1973) utilize context as a cue for retrieval of associations by the nominal CS. There are other applications as well, and the foregoing brief description is meant to give only a flavor of current usages of context.

In virtually all latent inhibition studies, the context, unless specifically an experimental variable, is constant throughout the various phases of the procedure. Within a three-stage experimental design for latent inhibition, such as in conditioned taste aversion, there are three different occasions on which the context *may* be manipulated: in preexposure, in conditioning, and in test. However, most studies that have investigated the effects of context changes on latent inhibition have varied the contexts from preexposure to conditioning, while leaving the test and conditioning environments the same.

In an earlier review of the latent inhibition literature, Lubow (1973a) devoted one short paragraph to the effects of changing the place of acquisition–test from that of stimulus preexposure:

There is another aspect to stimulus specificity, the place of preexposure in relation to the place of testing. A series of studies by Anderson and his associates, using conditioned suppression of bar pressing, were designed to test the effects of place of preexposure, conditioning and testing on the strength of fear conditioning and latent inhibition (Anderson, Cole, & McVaugh, 1968; Anderson, O'Farrell, Formica, & Caponigri, 1969; Anderson, Wolf, & Sullivan, 1969). It appears that latent inhibition is increased when the place of preexposure is the same as the place of testing (standard procedures). However, latent inhibition can be achieved when the two places are distinctly different (May et al., 1967). (Lubow, 1973a, p. 403).

Since the time of that review, the role of context, or background stimuli, has taken on considerable theoretical importance, deriving, as already noted, primarily from certain postulates and assumptions of the Rescorla-Wagner model of conditioning (Rescorla & Wagner, 1972) and elaborations of that model (Wagner, 1976, 1978, 1979). It is no surprise, then, to find that the number of latent inhibition papers dealing with the effects of context has risen accordingly.

The various experiments that have addressed the role of context in modulating latent inhibition can be conveniently divided into several categories. The largest of these is concerned with the effects of changing the context from one stage to another. In addition, there is a group of studies that examines the effects of exposure of the context prior to the stimulus preexposure stage, and a group that looks at the effects of context presentations after the stimulus preexposure stage but before the CS–US acquisition stage. These three categories of context manipulation will be reviewed separately.

Context change, S_oE_o versus S_oE_n

As with other variables, the effects of context change on latent inhibition have been studied primarily with the conditioned suppression and conditioned taste aversion procedures. Major exceptions are two studies by Lubow, Rifkin, and Alek (1976), one with rats and one with children. These studies are important for three reasons: (1) They provide one of the first clear demonstrations of the effects of context change on latent inhibition. (2) They clarify the discrepancy between those studies of stimulus preexposure that produce facilitation of subsequent learning (perceptual learning) and those that produce a decrement in subsequent learning (latent inhibition).[6] (3) They introduce a terminology that is useful in describing the various experimental manipulations.

Beginning with terminology, it will be recalled (see Table 8) that four different conditions at the time of acquisition–test can be identified: After the organism receives preexposure to a given stimulus in a particular environment, then at the time of acquisition–test the preexposed stimulus and preexposed environment may be the same as those that were preexposed, in which case there is a condition of an old stimulus in an old environment (S_oE_o). Both may be changed from the preexposure condition, creating a condition of a new stimulus in a new environment (S_nE_n). Or either the stimulus or the environment may be changed, with the other remaining constant. This produces the condition of either an old stimulus in a new environment (S_oE_n) or a new stimulus in an old environment (S_nE_o).

The typical latent inhibition experiment compares the S_oE_o condition to the S_nE_o condition, the outcome of which is poorer learning for those animals in the S_oE_o group. When context is changed from preexposure to text–acquisition, the comparisons of interest become S_oE_n versus S_oE_o and S_oE_n versus S_nE_o.

As was described, Lubow, Rifkin, and Alek (1976) combined the basic elements from both the perceptual learning and the latent inhibition paradigms into a single experimental design. For the preexposure phase of the experiment, rats were assigned to one of two environments, circular (C) or rectangular (R), and to one of two olfactory stimuli, lemon (L) or wintergreen (W). For the testing phase, the subjects were further subdivided for assignment to either the familiar or novel environment and to either the familiar or novel stimulus. Thus, there were four major groups and 16 subgroups.

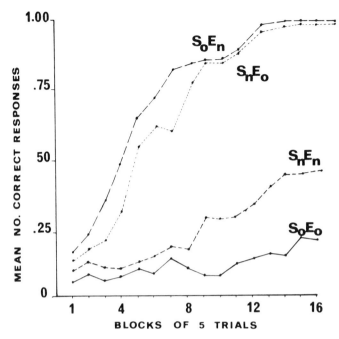

Figure 5. Mean number of correct responses per trial as a function of trial blocks for four major conditions at the time of test: S_oE_n, S_nE_o, S_nE_n, S_oE_o. (From Lubow, Rifkin, & Alek, 1976.)

Both the circular-cage environments and the square-cage environments contained two small tunnels. The odor source, which the animal could not reach, was attached to the outside end wall of each tunnel.

On Day 1 of the experiment, each subject was placed in the preexposure environment, which it inhabited for 14 days. During this period, the preexposed olfactory stimulus was presented twice daily for a total of 2 hr each day. The positions of the two stimuli were varied randomly.

Discrimination testing took place over a period of 4 days following preexposure. Before each test trial, the test odor source (either familiar or novel) was attached to one of the tunnels, a neutral stimulus was attached to the other, and the barrier was raised. A correct response, reinforced with a food pellet, was defined as entry into the cylinder to which the test odor container was attached.

The acquisition of the discrimination for the four major groups (collapsed across subgroups) is shown in Figure 5. As can be seen, the most rapid learning occurred in the S_nE_o and S_oE_n groups, with very poor learning in the S_oE_o and S_nE_n groups. The pattern of the data is very similar to that obtained in a study of young children that used the same experimental design (Lubow, Rifkin, & Alek, 1976). These experiments, as noted in chapter 2, clearly demonstrate that the retarded learning that characterizes latent inhibition ($S_oE_o < S_nE_o$) and the facilitated performance that characterizes the perceptual learning paradigm ($S_oE_n > S_nE_n$) are not contradictory.

The most effective conditions for learning are those in which there is a contrast between the novelty of the environment and the novelty of the stimulus. The fact that the same pattern of results was obtained for children and for rats under diverse circumstances indicates that the contrast variable has broad external validity. More specifically, in regard to latent inhibition, it is quite evident that a change in context from the stimulus preexposure stage to the acquisition–test stage disrupts latent inhibition. Other studies, published after the Lubow, Rifkin, and Alek (1976) demonstration of the context effect, have supported the same conclusion. Thus, the studies of Channell and Hall (1981), Hall (1979), and Hall and Channell (1983) all provide evidence for the superiority of the S_nE_o and S_oE_n conditions over S_nE_n and S_oE_o in discrimination learning tasks. The picture is quite similar with conditioned suppression testing procedures. A change of context markedly reduces or abolishes latent inhibition (Bouton, 1987; Bouton & Bolles, 1979; Dexter & Merrill, 1969; Hall & Minor, 1984; Lovibond, Preston, & Mackintosh, 1984).[7]

The situation with conditioned taste or odor aversion is somewhat more complicated. First, it should be noted that all of the studies of context change cited in the preceding paragraphs manipulated tonic context (i.e., *place* of preexposure and conditioning). Similar procedures, ones that manipulate changes in tonic environment, using conditioned taste aversion have consistently failed to find a disruption of latent inhibition (Best & Meachum, 1986; Hall & Channell, 1986, Exp. 2; Kurz & Levitsky, 1982).

However, when phasic context rather than tonic context is changed from the stimulus preexposure to the conditioning session, disruption of latent inhibition is evident. Archer, Mohammed, and Järbe (1983) reported such results for phasic context. Saccharin or water was preexposed in either a "noisy" bottle or a "silent" bottle, followed by a "noisy" saccharin–LiCl pairing. Test consumption of saccharin, also from a "noisy" bottle, indicated a pronounced latent inhibition effect for the group in which preexposure, conditioning, and test had the same phasic context conditions: noise. However, the group that underwent a change from a "silent" bottle in preexposure to a "noisy" bottle in conditioning showed virtually no latent inhibition. Archer and his associates have replicated this basic finding of phasic contextual control of latent inhibition on a number of occasions (Archer, Järbe, Mohammed, & Priedite, 1985; Archer, Mohammed, & Järbe, 1986; Mohammed, Callenholm, Järbe, Swedberg, Danysz, Robbins, & Archer, 1986; Mohammed, Jonsson, Soderberg, & Archer, 1986). Westbrook, Bond, and Feyer (1981) found the same effect. Specifically, rats were preexposed to an odor a single time before conditioning to that odor. The preexposure was either in the same context (saccharin solution) that would be present at the time of conditioning or in a different context (no saccharin). Thus, groups S_oE_o and S_oE_n were created, where environment was defined in terms of phasic characteristics, presence or absence of a saccharin context, rather than an apparatus. The Westbrook et al. (1981) data indicate that changing the local context from preexposure to conditioning significantly disrupts latent inhibition (also see Mitchell et al., 1975; Mitchell, Winter, & Moffitt, 1980).

In summary, there is ample evidence from the discrimination and conditioned suppression testing paradigms that latent inhibition is disrupted by a change in tonic context from the stimulus preexposure stage to the conditioning stage. With conditioned taste aversion the same effect is obtained with a change of phasic context, but evidently not with tonic context. It may be that phasic context is *relatively* more important than tonic context for pure classical conditioning, and tonic more than phasic for instrumental conditioning (and perhaps for those testing procedures where the classically conditioned association is expressed by an instrumental act, as, for example, conditioned avoidance and conditioned suppression). One might expect that phasic context would be more critical for classical conditioning, because classical conditioning, by definition, is static in space, with the significant events, CS and US, *coming to* the organism. With instrumental conditioning, on the other hand, the organism *moves through* the environment and seeks out the critical events. In such a situation, global apparatus cues may be more significant.[8]

Although the several conditioned suppression studies that have shown tonic context control of latent inhibition would seem to cast doubt on such a speculation, two points are warranted: First, it is suggested that there might be a *relative* difference in importance of phasic and tonic contexts in classical and instrumental conditioning. And because there have been no studies outside of conditioned taste aversion that have manipulated the phasic context cues, no conclusions on this point can yet be reached. Second, it may be important to differentiate between the acquired excitatory or inhibitory properties of the tonic contexts per se and the associations, if any, that are established between those contexts and the nominal CSs and/or USs. Clearly, the foregoing speculations refer to the second class of processes.

Independent familiarization of stimulus and context (S_o, E_o)

The typical experiment that examines the effects of context change on latent inhibition compares the conditions S_oE_o and S_nE_o, in which stimulus preexposure and environment preexposure (tonic context) occur conjointly (i.e., the stimulus is preexposed *in* the environment). The question can be raised whether or not this procedure is necessary to obtain latent inhibition. One could preexpose the stimulus and the test environment independent of each other (S_o, E_o, as opposed to S_oE_o). Such a procedure has been used with appetitive discrimination (Channell & Hall, 1981, Exp. 3) and with conditioned suppression (Hall & Minor, 1984; Lovibond et al., 1984). The results indicate that independent preexposures of stimulus and environment do not interfere with subsequent learning about those stimuli within that environment (i.e., this procedure does not promote latent inhibition).

Similarly, with conditioned taste aversion, Hall and Channell (1986, Exp. 3) demonstrated such context specificity of latent inhibition. Groups of rats were preexposed alternately to chamber A without saccharin and chamber B with saccharin. In the conditioning session, half the subjects received the saccharin–LiCl pairing in the same chamber in which saccharin was preexposed, and half in the other chamber. A comparison of saccharin consumption for these groups with appropriate control groups indicated better conditioning in the A–B condition than in the A–A condition (i.e., context specificity of latent inhibition). Ironically, as already noted, Hall and Channell (1986, Exp. 2) failed to demonstrate context specificity when using the more standard procedures for producing S_oE_o and S_oE_n conditions at the time of CS–US pairing, rather than S_o, E_o versus S_oE_o. It should be noted that the Hall and Channell (1986, Exp. 3) procedure was designed to optimize the discrimination between the A and B environments by allowing the subjects multiple exposures to each environment, with flavor exposure in only one of them. Such discrimination training may be *the* necessary condition for showing tonic context control of latent inhibition in the conditioned taste aversion paradigm.

Context preexposure that continues after stimulus preexposure

There is yet another variation that can be played on the context theme. Wagner's explanation (1978) of latent inhibition requires that the CS be preexposed and conditioned in the same context in order for latent inhibition to occur. As just shown, there is ample support for this requirement. However, Wagner's theory also demands that a context extinction procedure should eliminate latent inhibition. In other words, if the subject is preexposed to a stimulus in context A and then experiences context A on a number of occasions *without* the CS, subsequent conditioning in context A should proceed normally (i.e., there should be no latent inhibition). Although Wagner himself has referred to some data to support this view (Wagner, 1979, p. 67), more detailed reports can be found in other studies that have addressed this issue more fully (Baker & Mercier, 1982; Hall & Minor, 1984; Westbrook et al., 1981).

Baker and Mercier (1982), in a series of six conditioned suppression experiments, showed that latent inhibition could be reduced, at least moderately, by following stimulus preexposure with context exposure. However, the effect appeared to be quite fragile and was dependent on the type of reinforcement schedule used in the test. Thus, with a partial reinforcement test schedule (50%), context extinction failed to produce a reduction in latent inhibition (Exp. 1 and 5), whereas with a continuous reinforcement schedule in test, context extinction did reduce latent inhibition (Exp. 2, 3, and 4). Experiment 6 in the series formally compared the two test procedures and confirmed these findings. As an explanation of the ineffectiveness of partial reinforcement in test to bring out the context-extinction-induced reduction of latent inhibition, Baker and Mercier (1982) suggested that "with partial reinforcement [context] extinction is not effective because the nonreinforced trials during conditioning may reinstate the animals' memories of the previous nonreinforced experience with the CS" (p. 414).

Hall and Minor (1984), however, failed to find *any* supporting evidence for the contention that context extinction reduces or eliminates latent inhibition. In a series of six studies they used conditioned suppression of bar pressing for food, as in the Baker and Mercier (1982) experiments, as well as conditioned suppression of licking. Nevertheless, some evidence that prior extinction of the context–stimulus association disrupts latent inhibition was provided by Westbrook et al. (1981, Exp. 6). They demonstrated that if the phasic saccharin-solution context was exposed for a few minutes following the odor–saccharin pairing, then the effects of context familiarity (i.e., increased latent inhibition) were mitigated.

Context exposure prior to stimulus–context preexposure

Finally, rather than preexposing a stimulus in a context and then extinguishing that context, as was done in the preceding studies, one can ask what would happen if the context itself were preexposed *prior to* the CS preexposure. If CSs or USs form associations with the contexts in which they are presented, then preexposure of the contexts prior to the introduction of the stimulus should reduce those associations (i.e., provide latent inhibition of the to-be-conditioned context). The Rescorla-Wagner model of conditioning would, for example, derive the US preexposure effect (Randich & LoLordo, 1979; retardation of the acquisition of a conditioned response to previously presented US) by appealing to the formation of context–US associations during the preexposure period that subsequently *block* the development of the CS–US association. It follows from this that if a context–US association is acquired during the US preexposure period, then exposure of the context *prior* to the presentation of the US in that context (the US preexposure phase) should produce latent inhibition of that association as revealed in the subsequent CS–US test phase; that is, the US preexposure effect should be attenuated. Balaz et al. (1982) and Mowrer (1987) provided evidence for the latent inhibition of such context–US associations.

More to the point, there has been one study of the effect of context preexposure *prior to* preexposure of the to-be-conditioned stimulus in the same context. Unlike the situation with the US preexposure effect, Hall and Channell (1985a) reported that such context preexposure potentiated latent inhibition.

Second-stimulus effects

In the preceding section we noted that change of context, at least from preexposure to acquisition, reduces the amount of latent inhibition, as compared with keeping the context constant. There are other manipulations that *may* be related to the context change effect and that also attenuate latent inhibition. For example, pairing a target stimulus (S_1) with a second stimulus (S_2) during preexposure may result in reduced latent inhibition, as will be described later. These two procedures, context change and S_1–S_2 pairing, resemble each other not only because of similar outcomes (less latent inhibition) but also because the S_1–S_2 procedure typically involves a change from preexposure to acquisition. In this case the S_2 is removed in acquisition, and a new stimulus (S_3, usually a shock) is substituted for it. One might conceive of S_2, particularly because it is temporally contiguous to and contingent on S_1, as forming a mini-context and, from that point of view, expect a similar disruption of latent inhibition when S_2, the mini-context, is changed from preexposure to acquisition.

This same type of analysis would apply also to those studies that employ preexposure of a stimulus compound (S_1S_2) and test for the acquisition of associative strength to the elements, or those studies that use a stimulus preexposure procedure in which S_1 is predicted by a prior stimulus or response (S_2–S_1).

This section will examine the effects of all three types of manipulations: S_1–S_2, S_2–S_1, and S_1S_2.[9] One might expect that these mini-context manipulations would have a more profound latent-inhibition-attenuating effect than would the normal or maxi-context manipulations, because, unlike the latter, the former are uniquely correlated with the target stimulus, both temporally and contingently.

Table 12. *Experimental design and data from Lubow, Schnur, and Rifkin (1976)*

Group	Preexposure conditions	Acquisition	Test conditions	Suppression ratio Exp.1	Exp.2
1	–	S_1–US	S_1	.05	.02
2	S_1–S_2	S_1–US	S_1	.35	.05
3	S_1, S_2	S_1–US	S_1	.46	.30
4	S_1	S_1–US	S_1	.47	.25

S_1–S_2

One of the first studies to use the S_1–S_2 manipulation was that of Lubow, Schnur, and Rifkin (1976). In two separate conditioned suppression experiments, groups of rats were given one of four preexposure conditions: (1) preexposure only to the apparatus; (2) preexposure to the to-be-conditioned stimulus, with each preexposure immediately followed by a second stimulus (S_1–S_2); (3) preexposure to the to-be-conditioned stimulus and to the second stimulus, but unpaired (S_1, S_2); (4) preexposure only to the to-be-conditioned stimulus (S_1).

The basic design and the results from the two studies are shown in Table 12. An inspection of Table 12 indicates that the profound latent inhibition effect, demonstrated by comparing Groups 1 and 4, was attenuated by the addition of a second stimulus in preexposure, but only when that stimulus immediately followed S_1 (Group 2, S_1–S_2), not when the two were presented independent of each other (Group 3, S_1, S_2).

Unfortunately, there were several shortcomings in that study. First of all, again, a single test trial probably confounded unconditioned and conditioned suppressions. This problem is only partially alleviated by noting that the first test stimulus presentation was, in Experiment 1, the second postpreexposure experience with S_1 for the zero-preexposure group (*after* one conditioning trial), and in Experiment 2 it was the third postpreexposure experience with S_1 for the zero-preexposed group (after two conditioning trials). More important, however, most of the critical comparisons involve groups that had the *same* amount of stimulus preexposure (Groups S_1, S_1–S_2, S_1, S_2). As such, a difference in unconditioned suppression was not likely to have been a factor.

Szakmary (1977a, Exp. 1) directly tested the hypothesis that it might be the removal of S_2 in test that attenuates latent inhibition in the S_1–S_2 group. He compared two S_1–S_2 groups. One was a typical S_1–S_2 group, in that the S_2 was omitted in the acquisition–test session. In the other group, the S_2 was not omitted, but rather was coincident with the shock. Both groups showed the same degree of latent inhibition attenuation when compared with appropriate control groups.

Further confirmation of the Lubow, Schnur, and Rifkin (1976) findings that the S_1–S_2 preexposure preparation reduces latent inhibition comes from additional experiments by Szakmary (1977a,b, 1978).[10] Using a conditioned suppression of bar-press-for-food procedure, he compared the effects of eight 30-sec noise–light pairings and eight specifically unpaired presentations on the acquisition of conditioned suppression over four 30-sec noise–shock pairings (Szakmary, 1977b, Exp. 1). There was significantly more suppression in the S_1–S_2 paired group than in the S_1, S_2 unpaired group. In addition, he found no difference between the two groups for first trial test suppression (i.e., the first trial on which noise and shock were paired), thus providing evidence that the subsequent differences in suppression ratios were *not* functions of differences in unconditioned suppression to the noise. In contrast with the Lubow, Schnur, and Rifkin (1976) study, Szakmary used a *long*-duration S_1 in preexposure and test; furthermore, stimulus duration was the same in both the preexposure and acquisition–test sessions. A second experiment (Szakmary, 1977b, Exp. 2) showed that the paired S_1–S_2 preexposure group did not differ significantly from a zero-preexposure group.

In addition to the foregoing, Szakmary (1977a, Exp. 2) reported some preliminary evidence that whereas increasing the number of preexposures (2, 4, 16) to unpaired light and tone increased latent inhibition to the light, as, of course, would be expected, similarly increasing the number of preexposures to paired light–tone resulted first in a decrease in latent inhibition and then an increase in latent inhibition (i.e., an inverted U-shaped function). Other data from his laboratory, however, indicated that the differences between paired and unpaired stimulus preexposures may not be reliable. Szakmary (1978) found that eight light–tone pairings yielded a nonsignificantly smaller latent inhibition effect than that found in a comparable nonpaired preexposed group. In fact, in two of these studies (Exp. 1 and 2, but not 3), he found significantly more conditioned suppression for the unpaired as compared with the paired preexposure groups. However, the meaning of these data is not clear, because the conditioning session used the S_1–S_2 sequence as the conditioned stimulus. Thus, for the paired group, where the preexposed stimulus was identical with the conditioned stimulus, one would, indeed, expect to find more latent inhibition than for the unpaired group, where the preexposed and conditioning stimuli were quite different. In summary, the Lubow, Schnur, and Rifkin (1976) and Szakmary (1977a,b, 1978) studies provide evidence that S_1–S_2 pairing during preexposure reduces the amount of latent inhibition, as compared with a group preexposed to S_1 alone. However, it should be noted that a study by Mercier and Baker (1985), which also used conditioned suppression, reported no such effect. The reasons for this discrepancy are not apparent.

Table 13. *Experimental design and results of Frey et al. (1976)*

Group	Number and type of PEs	Results
1	0	0^b
2	150 tone	−
3	150 tone–flash	0
4	150 tone–ICS^{+a}	0
5	150 tone–ICS$^-$	+
6	150 tone–flash & ICS$^+$	+
7	150 tone–flash & ICS$^-$	+

[a] ICS, intercranial stimulation; ICS$^+$, appetitive, ICS$^-$, aversive
[b] Group 1 served as a reference group. If there was no difference between a given group and Group 1, this is indicated by 0; if facilitation, then +; if poorer performance relative to Group 1, then −.

Although it would appear that the EB and NMR preparations would be well suited to studying the preexposure effects of a second stimulus (S_2) that immediately follows the target stimulus (S_1) on subsequent conditioning, the studies in this area have been disappointingly few. One study that did describe the effects of a variety of S_1–S_2 procedures, albeit for reasons quite foreign to our own, was reported by Frey et al. (1976). They employed seven independent groups, each of which received, over a period of 3 days, a different tone preexposure condition. These conditions are described in Table 13.

As can be seen, S_2 took a variety of forms. The tone was immediately followed by a light flash or by appetitive or aversive intercranial stimulation (ICS, as determined by an independent bar-pressing test) or by a combination of flash and intercranial stimulation in which the two overlapped in time. Beginning 24 hr later the rabbit subjects began a 5-day series of eyelid conditioning trials in which the tone CS was paired with shock. The results are summarized in Table 13. Taking Group 1, the nonpreexposed control group, as a reference point, the data fall into three categories; facilitation (+) of conditioned eyeblink responding, inhibition (−), or no change (0), as compared with the nonpreexposed group. Thus, although a strong latent inhibition effect was obtained (Group 2 performing more poorly than Group 1), the addition of an S_2 completely wiped out the effect by raising the conditioning performance either to the level of the controls (Groups 3 and 4) or to a level superior to that of the controls (Groups 5, 6, and 7). In general, it appears that the stronger the S_2 (flash *plus* ICS), the greater the attenuation of latent inhibition.

Salafia and Allan (1980b) also found that high-intensity electrical stimulation of the hippocampus immediately following each of 150 tone preexposures reliably reduced the amount of latent inhibition. However, in a second experiment (Salafia & Allen, 1980a) they reported the opposite effect with a lower level of hippocampal stimulation. Unfortunately, from these two studies, one cannot make a generalized statement concerning the effects of hippocampal S_2 intensities, because the second experiment used 300 preexposures, as compared with 150 in the first.

The S_1–S_2 effect also appears in the conditioned taste aversion literature. When a preexposed flavor was immediately followed by a second flavor, latent inhibition was markedly reduced, as compared with a simple flavor preexposure group (Best et al., 1979). In addition, Best et al. (1979) reported that prior exposure of S_2 reduced its ability to attenuate latent inhibition, as did an increase in the temporal interval between S_1 and S_2.

In contrast to the studies of Best et al. (1979), which emphasized the latent-inhibition–attenuating role of a flavor S_2 immediately following S_1, a series of experiments by Rudy, Rosenberg, and Sandell (1977) focused on the effects of an S_2 (placement in a black box) when it preceded the conditioning event. Such a procedure also resulted in improved aversion conditioning for the flavor-preexposed group, which the authors attributed to a disruption of latent inhibition. As in Best et al. (1979), prior familiarization with this S_2 caused it to lose its effectiveness. However, this series of six experiments also demonstrated that S_2 was effective in disrupting latent inhibition only when it preceded the CS–US pairing, not when it was paired with S_1. A series of studies by Westbrook et al. (1982) also employed a variety of S_2 manipulations and failed to find the S_1–S_2 effect reported by Best et al. (1979).

How can the markedly different results of Best et al. (1979) and Westbrook et al. (1982) be reconciled? Westbrook et al. (1982) suggested that the difference might be due to the fact that, in their studies, the preexposed stimulus was presented for a 2-min period, whereas in the Best et al. (1979) studies it was presented for a 5-min period; however, other procedural differences were (a) the duration of the distractor stimulus (2 min versus 5 min), (b) CS flavor (sucrose versus vinegar), and (c) the distractor flavor (salt, quinine, or hydrochloric acid versus vanilla), respectively.

It is possible that a combination of these factors was responsible for the different results. The procedure of Best et al. (1979), in which 5-min preexposure of vinegar was immediately followed by a 5-min exposure to vanilla, may have produced a phenomenal compounding of the two flavors, one that was not achieved by the Westbrook et al. (1982) procedures. In the Westbrook et al. (1982) studies, the 2-min preexposure to sucrose typically preceded the 2-min distractor stimulus by 30 min. In the one condition in which the presentations were contiguous, three different distractors were employed, a situation that would not encourage compounding. Needless to say, the procedure used by Rudy et al. (1977)

to produce an S_2 (placement in a black box) also would not provide an opportunity for S_1S_2 compounding. If, indeed, the Best et al. (1979) procedures created a compound, whereas the Westbrook et al. (1982) and Rudy et al. (1977) procedures prevented such compounding, then one should turn directly to studies of compounding for possible further clarification of the issue. The relevant studies will be examined in a subsequent section.

An additional consideration may also be important here. The relative lack of strength of the S_1–S_2 effect in the conditioned taste aversion paradigm may have been due to the number of S_1–S_2 pairings, which typically was very low, even 1. Although this should have dramatically affected compounding, there may have been other effects that were controlled by the number of S_1 and S_1–S_2 presentations. Indeed, some theories, such as conditioned attention theory (Lubow, Weiner, & Schnur, 1981), emphasize the role of number of preexposures.

Although Mackintosh (1983) has questioned the replicability of the S_1–S_2 effect, there appears to be ample evidence supporting its existence, including a recent study by Matzel, Schachtman, and Miller (1988). Indeed, Kaye, Swietalski, and Mackintosh (1988) also reported the S_1–S_2 effect, but attributed its occurrence to generalization decrement processes, a conclusion not supported by Szakmary's (1977a) data. Additional evidence for the S_1–S_2 effects was provided by Dickinson (1976), who demonstrated that a tone that was paired with food (S_2) was subsequently associated more quickly with shock than was the same tone that was preexposed by itself. Finally, confirmation of the basic S_1–S_2 effect has been provided by the many studies that have demonstrated the Hall-Pearce effect.

Hall-Pearce effect

A special case of S_1–S_2 preexposure occurs when S_2 is a shock and when that same shock, albeit at a higher intensity, is used in the acquisition stage. Common wisdom suggests that such early S_1–low-shock pairings should enhance subsequent S_1–higher-shock associations. However, the results of a number of experiments indicate that a latent-inhibition-like effect accrues even under such unfavorable circumstances (Ayres et al., 1984;[11] Brown-Su, Matzel, Gordon, & Miller, 1986; Hall & Pearce, 1979, 1982; Kasprow et al., 1985; Kaye, Preston, Szabo, Druiff, & Mackintosh, 1987; Schachtman, Channell, & Hall, 1987; Sigmundi & Bolles, 1982; Swartzentruber & Bouton, 1986).[12]

This type of support for the role of S_2, in the S_1–S_2 preexposure preparation, in attenuating latent inhibition first came from a series of studies by Hall and Pearce (1979). The aim of those experiments was to demonstrate that latent inhibition occurs even if the S_2 in preexposure is a shock and that same shock, at higher intensities, is used in acquisition. The principal results, in fact, make *two* points. Latent inhibition does occur when S_2 is a weak shock, but the amount of latent inhibition is *reduced* compared with an S_1-alone preexposed group. Hall and Pearce (1979) have also shown that latent inhibition can be obtained with the S_1–S_2 procedure when the preexposure is conducted off baseline, and even when the intensity of the S_2 shock is raised to ensure that there is some conditioned suppression in the preexposure stage.

In a follow-up study, Hall and Pearce (1982) again found that some latent inhibition accrued when the to-be-conditioned stimulus was paired with a mild US, a finding that has been replicated by Ayres et al. (1984),[13] Kasprow et al. (1985), Schachtman et al. (1987), and Swartzentruber and Bouton (1986). Interestingly, Hall and Pearce (1982) also found that the associability of the S_1 that was paired with weak shock could be restored by omission of the expected shock during the preexposure period. Perhaps related to this is the suggestion by Kasprow et al. (1985) that the Hall-Pearce negative transfer effect is specific to the situation where a qualitatively similar S_2 is used in the preexposure and test phases. This suggestion, however, is not congruent with the data from the Lubow, Schnur, and Rifkin (1976) and Szakmary (1977a,b) studies, in which the S_2 of the preexposure stage was a neutral stimulus, whereas the acquisition stage S_2 was a shock.

Finally, it should be noted that the basic Hall-Pearce phenomenon, the loss of some associability even with S_1–S_2 pairing, where S_2 is the US in the acquisition test, not only appears with excitatory conditioning, as described earlier, but also appears as a loss of associability by a conditioned inhibitor (Pearce et al., 1982). This, of course, is also true of the standard latent inhibition effect. The preceding finding, together with the recent report that the Hall-Pearce effect is context-specific even when the familiarity of the test context is controlled (Swartzentruber & Bouton, 1986), as it is in latent inhibition, strongly confirms that the Hall-Pearce effect represents a special case of latent inhibition.

One additional study deserves mention in this section, although it does not fit neatly into the S_1–S_2 design category. Scavio et al. (1983), using the NMR rabbit preparation, found that as few as three *standard* CS–US pairings preceding as many as 480 CS-alone presentations were sufficient to completely block the acquisition of latent inhibition. Although that study used the same US shock intensity in preexposure and in test, the fact that three presentations should have had such a dramatic effect makes it less surprising that Ayres et al. (1984) failed to find any latent inhibition with their S_1–S_2 procedure (CS–weak-shock pairing preceding CS–stronger-shock pairing) when using the rabbit NMR preparation. In those experiments, the number of preexposures of CS–weak-shock pairings varied from 25 to 450 across experiments.

Some comments on primacy

That a mere three CS–US pairings that *precede* the CS-alone stage will totally abolish latent inhibition, whereas the same CS–US pairings presented *after* the CS-alone stage have no such effect (as witnessed by the fact that latent inhibition exhibits itself in the acquisition–test stage over scores, if not hundreds, of trials), would seem to establish an important role for *primacy*. The primacy principle simply asserts that first training is disproportionately strong compared with subsequent training. The importance of primacy has been noted by Konorski in regard to transfer of excitatory responses to inhibitory conditioned responses, and vice versa, as well as a more general principle (Konorski & Szwejkowska, 1952). More recently, the same principle of primacy has been promulgated by Wickens et al. (1983), based on classical conditioning studies with cats. A primacy effect with rats also has been reported in studies of the truly random control procedure. The fact that the truly random control procedure frequently produces a CS with some excitatory properties (e.g., Kremer, 1971, 1974; Kremer & Kamin, 1971) has been attributed, in part, to the chance ordering of CS–US pairings and US-alone trials. If the CS–US pairings *precede* US-alone trials, then the CS will attain excitatory strength (Ayres, Benedict, & Witcher, 1975).

It would be interesting to know whether or not the power of primacy varies across species. The fact that the Hall-Pearce negative transfer effect is obtained in rats (Ayres et al., 1984; Hall & Pearce, 1979), but not in rabbits (Ayres et al., 1984), suggests either that there is a species difference, perhaps related to primacy, or that the differences in S_1–S_2 effects are related to the type of conditioning preparation, conditioned suppression, and NMR. In regard to species differences, it is of interest to note that for another S_1–S_2 paradigm, sensory preconditioning, it has also been suggested that there may be a phylogenetic modulation (Thompson, 1972).

S_2–S_1 effects

What if we turn the S_1–S_2 situation around and ask what happens to a target stimulus when it is preceded by (predicted by) another stimulus (S_2–S_1)? Relatively few studies have addressed this question, and the answers are anything but clear. Szakmary (1978) found that a predicted preexposed target stimulus accrued more latent inhibition than a nonpredicted one. However, the experimental design confounds prediction in preexposure with change from preexposure to acquisition–test. When one group gets S_2–S_1 preexposure and another S_2, S_1, and both are given S_2–S_1 shock training trials, there should be more latent inhibition in the paired group than in the unpaired group simply because of greater similarity of the preexposed and conditioning stimuli. Conversely, for the same reasons, if both of these groups are subsequently conditioned with the S_1-shock, the unpaired group should show more latent inhibition.[14]

Although no experimenter using the EB or NMR paradigm has sought to investigate the effects of predicting the preexposed target stimulus on subsequent conditioning, a modified version of this approach appears, inadvertently, in the backward conditioning literature. There have been two studies that have compared backward conditioning, in which the US precedes the CS, with CS-alone presentations, as they affect subsequent forward conditioning with the *same* CS and US (Plotkin & Oakley, 1975; Siegel & Domjan, 1971). In both experiments the backward procedure proved to be *more* detrimental than the CS-alone procedure for the acquisition of a forward CS–US association. Thus, one might conclude that, at least in the NMR and EB preparations, predicting the target stimulus in the preexposure phase does not attenuate latent inhibition; quite the opposite, it appears to increase the subsequent detrimental effect. Of course, the generality of this statement must be qualified by noting that the backward conditioning procedure uses the same S_2 (US) in both preexposure and test. However, Batson and Best (1982), with conditioned taste aversion, obtained results in exactly the opposite direction: S_2 preceding S_1 did disrupt latent inhibition. They began by addressing the problem of understanding the *proximal US preexposure effect*. This phenomenon describes the fact that a US preexposure that precedes a CS–US pairing interferes with the establishment of an associative bond between the CS and US. For example, whereas the pairing of a flavor with LiCl injection will produce profound conditioned aversion to the flavor, if such a pairing is preceded within a few hours or less by a similar LiCl injection, there will be a significant reduction in the subsequent conditioned aversion (Best & Domjan, 1979; Domjan & Best, 1977, 1980). Batson and Best (1982) reasoned that if the proximal US manipulation is effective because it interferes with the processing of the CS in the CS–US pairing, then a US preceding a *preexposed* flavor should likewise interfere with the processing of that stimulus and thereby disrupt the normal effects of such preexposures, namely, attenuate latent inhibition. Indeed, they showed that latent inhibition was attenuated by the prior US exposure and that the attenuation was greatest for the shortest interval between LiCl injection and flavor preexposure.

A problem raised by the Batson and Best (1982) study concerns the nature of the S_2 that precedes the preexposed stimulus presentation. In their study, The S_2, LiCl, was the same as the US employed in the conditioning episode. Is this identity a necessary condition for disruption of latent inhibition? For the basic distractor US preexposure effect itself,

Table 14. *Median suppression ratios for actively preexposed groups, passively preexposed groups, and nonpreexposed groups*

Preexposure condition	Exp. 3	Exp. 4
Active	.25	.05
Passive	.48	.40
None	.05	.02

Source: Lubow, Schnur, & Rifkin (1976, Exp. 3 and 4).

there is considerable evidence that the proximal (S_2) and the nominal US do not have to be the same in order to disrupt conditioned taste aversion; see Domjan (1980, pp. 325–326) for a listing of eight such studies. On this basis, one would expect that the Batson and Best (1982) results also were not dependent on using similar USs in preexposure and conditioning. One must ask, then, why with NMR and EB procedures we find that the S_2–S_1 manipulation, as compared with S_1 alone, retards subsequent conditioning, whereas with conditioned taste aversion the opposite occurs: There is disruption of latent inhibition (i.e., improved learning). There are several possibilities, all based on the intrinsic nature of the paradigm. Of particular importance are differences in number and duration of stimulus preexposures and the S_2–S_1 interstimulus interval.

One set of studies in which S_2 neither was included as part of the CS in the acquisition stage nor was related to the US in the acquisition stage was reported by Lubow, Schnur, and Rifkin (1976). In those experiments the preexposed stimuli were produced by bar-press responses. The responses themselves can be thought of as producing proprioceptive stimuli that *precede* the onset of the S_1 target stimulus. In the Lubow, Schnur, and Rifkin (1976) study, the preexposure period was differentiated for three groups on the basis of whether the preexposed stimulus, a light or tone, was produced by a lever press, was passively received by a yoked group, or was absent. There were no significant differences between groups for the number of bar presses, and therefore, of course, none for the number of stimulus presentations. Two days after the last preexposure, animals were given one tone–shock pairing. On the following day, a single lick-suppression test trial was presented. The experimental design and the critical data, conditioned suppression of drinking during the test presentation, are presented in Table 14.

Two points are quite clear: First, there was the usual latent inhibition effect. In both experiments, the passive preexposed group showed less conditioned suppression than did the nonpreexposed group. Second, in both experiments, the group that actively produced the preexposed stimulus had more conditioned suppression (i.e., less latent inhibition) than did the passively preexposed group.

Possible criticisms of these studies in regard to either the use of a one-trial test or the use of an extended test CS duration, as compared with the preexposed stimulus duration, are obviated because of the fact that the major comparisons are between the actively preexposed group and the yoked, passively preexposed group. Thus, for both groups, the amount of unconditioned suppression due to stimulus novelty should be the same. Similarly, the changes in stimulus duration for the two preexposed groups from the preexposure stage to the conditioning stage and to the test stage should be the same.

S_1S_2: preexposing a compound stimulus

In addition to S_2 preceding or following the S_1 target stimulus during preexposure, it may also overlap with S_1, in which case we have preexposure of a compound stimulus. Several studies have investigated the effects of such compound preexposures on the acquisition of a response to the compound itself or to the elements, or to both. Those studies used either conditioned suppression (Baker & Mercier, 1984; Lubow, Wagner, & Weiner, 1982; Mackintosh, 1973; Mercier & Baker, 1985; Rudy, Krauter, & Gaffuri, 1976) or conditioned taste aversion procedures (Forbes & Holland, 1980, 1985; Gillette & Bellingham, 1982; Holland & Forbes, 1980).[15]

The first experiment to explore the effects of compound preexposure was reported by Mackintosh (1973, pp. 82–83). In that study, separate groups were preexposed to a light, to a noise–light (NL) compound, or to nothing. Light served as the CS in the CS–US acquisition stage. In addition to obtaining the usual latent inhibition effect, it was found that the preexposed NL compound group produced significantly more suppression (i.e., less latent inhibition) than did the element-preexposed group.

Rudy et al. (1976) repeated the foregoing design and, in one experiment, failed to obtain the results reported by Mackintosh but did replicate the effect in their second experiment. However, studies by Baker and Mercier (1984) and Mercier and Baker (1985) have provided little or no evidence for attenuation of latent inhibition by compound preexposure compared with element preexposure. This failure shows up most clearly in their Experiment 1, which was designed to replicate the design of Mackintosh (1973). As can be seen in Table 15, the basic comparison was between a group preexposed to 32 30-sec click stimuli and a group preexposed to a similar click stimulus, but compounded with light. Both groups showed significant and similar amounts of latent inhibition when compared with a control group that received only 4 click preexpo-

Table 15. *Latent-inhibition–conditioned suppression studies with compound stimulus preexposure*

Experiment	Group	Preexposure	Acquisition	Test	Suppression ratio
Lubow, Wagner & Weiner (1982, Exp. 2)	1a	TL	TL	L	.38
	2a	T	TL	L	.19
	3a	L	TL	L	.38
	4a	–	TL	L	.19
	1b	TL	TL	T	.33
	2b	T	TL	T	.43
	3b	L	TL	T	.31
	4b	–	TL	T	.36
Mackintosh (1973)	1	L	L	$-^c$	$.42^a$
	2	LN	L	$-^c$	$.27^a$
	3	–	L	$-^c$	$.16^a$
Mercier & Baker (1985, Exp. 1)	1	C	C	$-^c$	$.31^a$
	2	CL	C	$-^c$	$.31^a$
	3	$-^b$	C	$-^c$	$.16^a$
Rudy et al. (1976, Exp. 1)	1	N	N	$-^c$	$.40^a$
	2	LN	N	$-^c$	$.43^a$
	3	–	N	$-^c$	$.15^a$
Rudy et al. (1976, Exp. 2)	1	N	N	$-^c$	$.42^a$
	2	LN	N	$-^c$	$.28^a$

^{a} Data estimated from journal graphs.
^{b} Actually four preexposures; see text.
^{c} Single acquisition-test stage.

sures. Similar findings were reported in three additional experiments in that series. Mercier and Baker (1985) provided evidence that in spite of the absence of an S_1S_2 effect on latent inhibition, the animals did form an association between the two elements of the compound during the preexposure period.

The discrepancy between the Mackintosh (1973) and Mercier and Baker (1985) data is not easily resolved. Procedural differences, such as type of stimuli interacting with number of preexposures, may account for the presence or absence of an attenuated S_1S_2 latent inhibition effect.

An experiment by Lubow, Wagner, and Weiner (1982, Exp. 2), in which different groups were preexposed to tone, to light, to tone–light compound, or to nothing, did not clarify the situation, because all groups were given acquisition training with the compound. However, the study is of interest if one entertains the hypothesis that the superior conditioning of NL-preexposed groups compared with L-preexposed groups obtained by Mackintosh (1973) and Rudy et al. (1976) may have been due to the *change* of stimulus conditions from preexposure to acquisition. The compound-preexposed group underwent a greater change than did the element-preexposed group, because acquisition was conducted only with the element. In the Lubow, Wagner, and Weiner (1982) experiment, the conditions were reversed. The compound-preexposed group also received acquisition with the compound. Indeed, under these circumstances, the compound-preexposed group showed either the same or less latent inhibition than the element-preexposed group.

Related data were reported by Schnur (1971), who preexposed either a tone or a light and then gave different groups acquisition trials to tone, to light, or to tone–light compound and tested on either tone or light. Tone-preexposed groups and light-preexposed groups showed the same degree of latent inhibition independent of whether acquisition was with the element or the compound. A similar study with somewhat different results was reported by Carr (1974). An inspection of Table 15 may serve to clarify the point.[16]

In this table, the basic designs of the relevant experiments, divided into preexposure, acquisition, and test stages, are outlined. In addition, the last column indicates the suppression ratios that were obtained in test; note that when the test stage is absent, as in the Mackintosh (1973) and Rudy et al. (1976) experiments, the test and acquisition stages were one and the same. In Table 15 it can be seen clearly that compounding during preexposure reduced the amount of latent inhibition, as compared with simple element preexposure; compare Group 2 to Group 1 in the Mackintosh (1973) study, and see Experiment 2 of Rudy et al. (1976). That this occurred in spite of a change of stimulus conditions from preexposure to acquisition is indicated by the fact that changes from single-stimulus preexposure to compound stimuli in acquisition did *not* have the same effect as the reversed procedure. A comparison of groups 1a to 3a and 1b to 2b in the Lubow, Wagner, and Weiner (1982, Exp. 2) study indicates that the change condition either had no effect or had an effect in the opposite direction (i.e., decreasing the amount of suppression, compared with the no-change group). Together, these data suggest that if there is an attenuation of latent inhibition as a result of compounding during preexposure, it is not merely an artifact of stimulus change from preexposure to acquisition. Similarly, Forbes and Holland (1980) found that preexposure to the elements retarded the acquisition of a compound–element discrimination, whereas preexposure to the compound had no such effect. However, the effects of the element preexposure were "largely the result of differences in the loss of suppression in the nonreinforced elements, a pattern also observed by Halgren (1974)," but quite opposite to that reported by Holland and Forbes (1980) using similar stimuli and preexposure procedures.

Gillette and Bellingham (1982) also examined the effects of compound preexposure, but not in comparison with element preexposure. After preexposure to a compound flavor solution, each of the groups was subdivided into three on the basis of whether one of the elements of the compound or the compound itself would be paired with the toxin. Preexposure to the compound solution produced the most attenuation of aversion for the compound-conditioning group, and this was a positive function of the number of preexposures. In fact, with only two preexposures to the compound, aversion conditioning to the elements was similar to that of the compound; but with an increase in the number of compound preexposures, latent inhibition developed only to the compound, not to the elements. These results suggest that with repeated preexposure, the compound develops into a unique configural stimulus and becomes differentiated, phenomenally, from the two elements.

From these studies one can conclude that compound stimuli are relatively protected from the debilitating effects of exposure prior to conditioning, as compared with elemental stimuli. This retardation of the acquisition of latent inhibition to a preexposed stimulus compound, as compared with a stimulus element, then, might reconcile the differences between the Best et al. (1979) and the Westbrook et al. (1982) studies, as discussed in the section on S_1-S_2 effects. The S_2 stimulus procedures of the latter studies maintained the individual identity of the target stimulus (S_1), and the S_2 and thus produced normal latent inhibition. The S_2 procedures of the former, on the other hand, served to promote compounding and thus protected the target stimulus against latent inhibition.

Summarizing the effects of a second stimulus on latent inhibition, it seems reasonable to conclude that the S_1-S_2 sequential presentation and the S_1S_2 compound presentation during preexposure both serve to attenuate latent inhibition to S_1. As for the S_2-S_1 manipulation, there is insufficient evidence to reach a firm conclusion, at least when S_2 is a flavor. On the other hand, when S_2 is a toxin, perhaps necessarily the same as that used in the conditioning stage, latent inhibition to S_1 is disrupted.

4 Organismic variables affecting latent inhibition

In general, the effects of organismic variables have not been subjects of investigation in regard to latent inhibition. The few exceptions include the variables of age, sex, and handling, with the former receiving the greatest attention by far. These areas are represented by a small number of experimental investigations, and some conclusions may be reached from the controlled comparisons. In addition, however, there is a body of information that is derivable from the fact that experiments on latent inhibition have been performed in a variety of species. It is therefore possible to examine, between experiments, effects that, with considerable caution, may be attributable to species differences. Prudence is dictated by the fact that across studies there is almost complete confounding of species and testing procedures.

Age

Only two studies have looked at the effect of age on latent inhibition within the conditioned suppression paradigm (Cone, 1974; Wilson & Riccio, 1973). This, of course, is not surprising, because the basic technique, whether tube licking or bar pressing, requires a relatively mature motor system. Other procedures, such as odor preference, are more amenable to dealing with the inherent problems of testing immature organisms.

In a sketchy report that is difficult to evaluate, Cone (1974) reported an age-related latent inhibition effect in a one-trial conditioned suppression test. With rats that ranged in age from 30 to 365 days, all groups showed latent inhibition to a light CS, except the 90–120 day-old group. Rats from 30 to 160 days of age, similarly preexposed and tested, but with a tone CS, all showed latent inhibition. As suggested earlier, one-trial test procedures are particularly subject to confounding of unconditioned and conditioned suppression, thus making it difficult to assess the amount of latent inhibition.

A study by Wilson and Riccio (1973) did not clarify the situation. Lick-trained rats aged 23, 30, and 90 days were either preexposed to 60 tones or not preexposed. Immediately after this stage, three tone–shock pairings were administered, followed by six extinction test trials. The first test trial indicated no difference between preexposed and nonpreexposed groups. However, differences did develop on subsequent trials and indicated the typical latent inhibition effect. The size of the effect appeared to be independent of age. Even though the authors suggested that there was reduced latent inhibition in the 23-day-old group compared with the older group, this was supported neither by visual inspection of their graphed data nor by a report of a significant interaction between age and preexposure. In addition, the absence of a difference between the nonpreexposed, conditioned group and a control group that was not preexposed and given *unpaired* CS–US presentations questions any interpretation of the data in regard to latent inhibition.

Latent inhibition as a function of age had, until recently, received surprisingly little attention in the conditioned flavor/odor aversion literature. Given the importance of the age variable for inducing latent inhibition in humans and the suitability of flavor or odor preexposure for use with very young, as well as adult, animals, plus the general popularity of the paradigm, one would expect to find a number of systematic developmental studies of latent inhibition within this area. However, there have been only eight relevant experiments (Franchina et al., 1980; Franchina & Horowitz, 1982; Klein, Mikulka, Domato, & Hallstead, 1977; Kraemer, Hoffmann, & Spear, 1988; Misanin et al., 1985; Misanin, Guanowsky, & Riccio, 1983; Peterson, Valliere, Misanin, & Hinderliter, 1985; Rudy & Cheatle, 1977.[1] These studies are described in Table 16.

Franchina et al. (1980) and Franchina and Horowitz (1982) found that 19-day-old rats exhibited latent inhibition, even after one 30-min preexposure (Franchina et al., 1980). Indeed, Franchina and Horowitz (1982) observed more latent inhibition in 19-day-old rats than in 90-day-old rats. However, Klein et al. (1977) failed to find a latent inhibition effect in 23-day-old subjects, in spite of the fact that they used four 30-min preexposures. The major difference between the two studies appears to be the time interval between flavor preexposure and flavor–toxicosis pairing. In the Franchina experiments, that interval was 12 hr, and in the Klein et al. (1977) study it was a minimum of 24 hr. That this variable is critical is attested by the findings of Rudy and Cheatle (1977). On the one hand, they obtained latent inhibition in rat pups as young as 2 days old. On the other hand, they showed that whereas such an effect was displayed when the preexposure–conditioning interval was 1 hr, it was absent when that interval was increased to 4 hr. However, with older pups, 7 days of age, latent inhibition was present even with a 24-hr preexposure–conditioning interval. A comparison of the latter finding with that of the Klein et al. (1977) study, where there was failure to obtain latent inhibition, again indicates a discrepancy. This, however, can be resolved by noting that the latter used sucrose flavor, and the former used lemon odor. Sucrose may share common stimulus properties with the mother's milk. This would give

Table 16. *Effects of age on acquisition of latent inhibition*

Experiment	Age at preexposure (days)[a]	Number of preexposures	Duration of preexposure (hr)	Preexposed stimulus	PE–S$_2$–US interval (hr)	Results
Franchina et al. (1980)	19	0, 1, 2, 4, 8	0.5	Sucrose	12	LI for 1, 2, 4, & 8 PE; no differences between PEs
Franchina & Horowitz (1982)	19 90	0, 1, 3	0.5	Sucrose	12	LI for 3 PE, for both 19 & 90-day-old; more LI for young
Klein et al. (1977)	23 65	0, 4	0.5	Sucrose	24 hr or 28 days	LI in adults at both intervals; no LI in pups in either interval
Kraemer, Hoffman, & Spear (1988)	6 12	1	3 min	Saccharin	1	No LI LI after 20-hr conditioning-test interval; not after 10-day interval
Misanin et al. (1983)	24–25 99–105	0, 3	0.5	Saccharin	24	LI for young and old; more LI for young
Misanin et al. (1985) Exp. 2	21–22 93–104 723–733	0, 1	0.5, 3	Saccharin	24	Young: LI for both PE durations Adult: LI only for long PE duration, but smaller for young Old: No LI
Exp. 3	20–25 92–98 711–743	0, 1	3	Saccharin	24	Young: LI Adult: LI[b] Old: No LI

Exp. 4	19–23 150–170 680–720	0, 1	48	Saccharin	24	Young: LI Adult: LI Old: LI
Peterson et al. (1985)	20–25 145–155 800–840	0, 2	1	Saccharin plus maple odor[c]	24	Young: LI Adult: LI Old: LI weak
Rudy & Cheatle (1977)	2 8 7 2 8	0, 1	0.5	Lemon odor	4 24 1 1	No LI LI LI LI

[a] Age of the animal is the age, in number of days, at the time of the last preexposure.
[b] LI was the same for young and adult, unlike in Exp. 2, where there was a stronger effect for young. The difference between the two experiments was attributed to the fact that Exp. 3 used a relatively weaker LiCl treatment than Exp. 2.
[c] In the Peterson et al. study (1985), one of several groups was preexposed to saccharin taste and maple odor, and another was not preexposed. Both were conditioned to the taste–odor compound and tested for saccharine–water preference.

the nonpreexposed group some familiarity training and thus make it similar to the sucrose-preexposed group. In addition, latent inhibition was quite weak even in the adult groups, appearing only on the second of four tests. Furthermore, a 30-min exposure to the flavored solution (Klein et al., 1977) may be functionally less than a 30-min preexposure to the odor stimulus (Rudy & Cheatle, 1977).

As one might expect, there also is an interaction among age, stimulus preexposure, and time between conditioning and test. This interaction was evident in a recent study by Kraemer, Hoffman, and Spear (1988). Six-day-old rat pups did not show a latent inhibition effect after a very short stimulus preexposure period (3 min), whereas 12-day-old rat pups did exhibit latent inhibition, but only with a conditioning–test interval of 20 hr, not with an interval of 10 days.

Experiments by Misanin and his colleagues (Misanin et al., 1983, 1985; Peterson et al., 1985) all confirm that latent inhibition of conditioned taste aversion is obtainable in weanling rats and, indeed, is stronger than for adult rats. Furthermore, they provide evidence that latent inhibition in very old rats is either weak or absent (see Table 16 for details).

In summary, one can conclude that rats at a very young age, several days, are capable of processing stimuli and showing the familiarization effect that is characterized by subsequently poorer associability of the preexposed stimulus. This latent inhibition effect can be maximized in infant rats by using non-milk-related stimuli, preferably odors, extending the duration of preexposure, and minimizing the interval between stimulus preexposure and stimulus–toxicosis pairing. There is some evidence that under equal conditions, neonatal rats (2 days) show less latent inhibition than do older pups (8 days) (Rudy & Cheatle, 1977). However, at a somewhat older age, weanling rats (20–25 days) exhibit more latent inhibition than do young adults (90–110 days) (Misanin et al., 1983, 1985). The trend is reversed again with very old rats (700+ days), who exhibit much less, if any, latent inhibition compared with younger rats (Misanin et al., 1985; Peterson et al., 1985). It would appear, then, that the function relating age to latent inhibition, at least with the conditioned taste aversion procedure, is an inverted U, with the most latent inhibition occurring at about the age of 20 days.

To complete this section, it should be noted that age appears to be an important variable in determining latent inhibition in humans. Whereas young subjects, below about 6 years of age, show normal latent inhibition effects (Kaniel & Lubow, 1986; Lubow, Rifkin, & Alek, 1976), older subjects require that the preexposed stimuli be presented such that attention is not directed to them. The need to use a masking task in order to elicit latent inhibition in older children and adult humans has been demonstrated in several studies (e.g., Ginton, Urca, & Lubow, 1975; Hulstijn, 1978; Lubow, Caspy, & Schnur, 1982; Schnur & Ksir, 1969). The role of masking and its interaction with age in producing latent inhibition in humans will be reviewed more fully in chapter 9.

Sex and early handling interaction

Except in our own laboratory, the effect of the sex of the subject on latent inhibition has not been a topic of investigation; and in the vast majority of experiments with rats, the subjects have been male. Similarly, the effects of early handling on latent inhibition have been almost completely ignored. Two studies have examined the interaction between these two variables, one with conditioned suppression (Weiner, Feldon, & Ziv-Harris, 1987) and one with two-way active avoidance (Weiner et al., 1985). In both studies the handling condition consisted of temporarily removing the rat pups from their mothers and individually placing them in a small box containing wood shavings for 3 min. This procedure was repeated for 20 days, from 2 days until weaning at age 22 days. During this time the nonhandled pups remained undisturbed with their mothers. In the two studies, it was found that whereas both handled and nonhandled females showed the normal latent inhibition effect, as did handled males, nonhandled males failed to develop latent inhibition (Weiner et al., 1985; Weiner, Feldon; & Ziv-Harris, 1987).

Species

Whereas the preponderance of latent inhibition studies have been conducted with rats as subjects, various other species have also been examined for the presence of the effect: mollusc (Farley, 1987 a,b); honeybee (Abramson & Bitterman, 1986; Bitterman et al., 1983); goldfish (Braud, 1971; Shishimi, 1985); pigeon (Reilly, 1987; Tranberg & Rilling, 1978); mouse (Alleva et al., 1983; Bignami et al., 1985; Oliverio, 1968; Schnur & Lubow, 1976); rabbit (Lubow et al., 1968; Moore et al., 1976); sheep and goat (Lubow, 1965; Lubow & Moore, 1959); cat (McDaniel & White, 1966; Wickens et al., 1983); dog (Herendeen & Shapiro, 1976); monkey (Mineka & Cook, 1986); human (Ginton et al., 1975; Hulstijn, 1978; Schnur & Ksir, 1969).

Conditioning in a mollusc

It has been shown that the nudibranch mollusc *Hermissenda crassicornis* is capable of simple contiguity-based associative learning (e.g., Crow & Alkon, 1978; Farley & Alkon, 1982), as well as contingency learning (Farley, 1987a, b). Typically, in these experiments the CS is an increase in light intensity, and the US is rapid rotation. Such a procedure produces a conditioned suppression of approach to the light. Using this type of preparation, Farley (1987a) demonstrated that 50 preexposures to the light, followed after 2 min by the CS–US pairing session, did not produce a latent inhibition effect when the animals were tested 24, 48, or 72 hr later. In other words, compared with a nonpreexposed group, there was no difference in the level of conditioned suppression. The same absence of an effect was found when the preexposure–conditioning interval was increased from 2 min to 24 hr (J. Farley, personal communication). From these data it is reasonable to conclude that this invertebrate organism, which is capable of simple associative learning, does not exhibit latent inhibition.

Conditioning in the honeybee: appetitive and aversive

An experiment by Bitterman et al. (1983) employed a classical appetitive procedure and used honeybees as subjects. In these studies, classical conditioning was achieved by blowing a scented (e.g., geraniol) air stream across the honeybee's "face" for 6 sec. After the third second, the US, stimulation of the antenna and proboscis with sugar water, was applied. The unconditioned response was an extension of the proboscis to the US, and the conditioned response was an extension of the proboscis during the first 3 sec of the CS presentation. With such a procedure, Bitterman et al. (1983) demonstrated that asymptotic conditioning levels were achieved within two to three trials. However, whereas they found a marked retardation of acquisition after eight unpaired presentations of the CS and US, and a small but statistically insignificant effect of eight US-alone preexposures, eight 6-sec presentations of the CS, by itself, had no effect whatsoever on subsequent acquisition of the conditioned response.

More recently, however, Abramson and Bitterman (1986) have claimed, in contrast to the findings in their earlier study, that latent inhibition can be demonstrated in the honeybee by using aversive conditioning rather than the previously used appetitive procedures. Because the findings of the 1986 study run counter to the prevailing evidence with other lower organisms and with the honeybee itself, it is important to examine their procedures and conclusions.

Conditioning of honeybees was accomplished during feeding by pairing a CS, either a vibration of the substrate or an air stream, with a brief shock pulse. The shock could be avoided by withdrawal of the proboscis from the food solution, "although the typical response to the CS (after conditioning), as to the US (when it was not avoided), was flying up from the target for a few seconds" (Abramson & Bitterman, 1986, p. 184). In all of their experiments, the flying-up response was the dependent variable that reflected conditioning. Experiment 1 was a simple demonstration of the effects of CS preexposure. One group received ten 5-sec preexposures to the CS while feeding; the other group received no such preexposure. In the acquisition session, the CS was again presented while feeding, and after 5 sec was followed by a shock. The CS-preexposed group acquired the conditioned response (flying up from the feeding station during the CS presentation) significantly more slowly than did the nonpreexposed group.

Experiment 2 in the series used a discrimination test, with the counterbalanced CS being either the vibration or air stream. Three groups were preexposed to the to-be-CS^+, to the to-be-CS^-, or to nothing. The results indicated that preexposure to the to-be-CS^+ retarded conditioning to the CS^+, as in Experiment 1, but had no effect on responding to CS^-. This latter finding once again supports the conclusion that the effects of CS preexposure are stimulus-specific. Similarly, preexposure to the to-be-CS^- had no effect on CS^+ conditioning, but did affect CS^- conditioning. However, the effect was in the opposite direction from what would be expected. The group preexposed to the to-be-CS^- showed *more* suppressed responding to the CS^- in the test phase than did the nonpreexposed group. This finding suggests that the Abramson and Bitterman (1986) procedures produced conditioned inhibition or some other form of conditioned interference, rather than latent inhibition. Indeed, their third experiment indicates that a preexposed stimulus will retard responding in a traditional summation test of inhibitory and excitatory stimulus properties.

In conclusion, then, the data from Bitterman's laboratory indicate that honeybees do not show latent inhibition in an appetitive test (Bitterman et al., 1983), nor, in spite of their claim to the contrary, have these authors presented convincing evidence for latent inhibition in an aversive conditioning test (Abramson & Bitterman, 1986).

Conditioning in the goldfish

In a study with goldfish as subjects, Braud (1971) compared a group with 60 preexposures to a 25-sec-duration light and a group with no such preexposure. Avoidance conditioning was initiated 72 hr after the last stimulus preexposure and consisted of 20 light–shock trials. The nonpreexposed group successfully avoided the shock on 37% of the trials, whereas the stimulus-preexposed group achieved only 14% avoidance.[2]

The Braud (1971) study has recently been criticized by Shishimi (1985), who failed to find a latent inhibition effect in goldfish. In both classical activity conditioning and aversive conditioning, Shishimi (1985) found that preexposure to a colored light affected responsivity to *both* the preexposed color and a novel color. This nonspecific stimulus effect, which cannot be identified as latent inhibition, was attributed to arousal reduction. In regard to Braud's apparently successful demonstration of latent inhibition in goldfish, Shishimi commented that "it is not convincing for several reasons. One is that the course of acquisition was not shown. A second is that the test performance of the nonpreexposed animals was so good, given the small amount of training and the size of the animals in relation to the space through which they were required to shuttle. A third is that the test performance of animals for which the conditioned stimulus had previously been paired either with food or with long trains of inescapable shock was even better than that of the nonpreexposed animals" (Shishimi, 1985, p. 317). Although the series of experiments by Shishimi (1985) strongly suggested that goldfish do not exhibit latent inhibition, and this fits well with the data from other "lower" animals (mollusc and honeybee), it is necessary to examine a number of reasons why the typical latent inhibition effect may not have been found with the goldfish, reasons that are not directly related to phylogeny.

To begin with, the experiments with goldfish (Shishimi, 1985) used a 24-hr delay. Perhaps, in lower species, the latent inhibition effect cannot bridge such a temporal interval, as it most certainly can for higher species. More important, however, are the particular procedures that Shishimi employed. For example, in the first experiment of that series (Shishimi, 1985, Exp. 1), goldfish were preexposed 1,000 times over a 5-day period to either the same 10-sec colored light that would be used in acquisition or to a different-colored light. The test, however, consisted of discrimination training where the stimulus paired with shock (S^+) was a tone, and the nonreinforced stimulus (S^-) was the tone plus the colored light. For one group the colored light was the same as the one preexposed, and for the other group it was a different-colored light. The dependent variable was the number of conditioned activity responses per trial. The data indicated no differences between the groups. These results were interpreted as indicating a failure to show either conditioned inhibition or latent inhibition. In the first case, conditioned inhibition would have been manifested if the "same color" preexposed group had shown facilitated discrimination compared with the "different color" group. On the other hand, latent inhibition would have been manifested if the "same color" preexposure group had exhibited reduced associability compared with the "different color" group, and therefore an impaired discrimination score.[3]

A different study in that series (Shishimi, 1985, Exp. 4), employing the same logic, preexposed goldfish to 10-sec presentations of red light for 160 trials over an 8-day period. In the test, a white light was used as the S^+, and a white light plus a red light was the S^-. The dependent variable was the number of shuttling responses. Again, there was no evidence of differences between the preexposed and nonpreexposed groups, with the exception that the preexposed group showed a general reduction in shuttling (to both S^+ and S^-), a finding that appears to run through all of these experiments.

The problem with these procedures is that, in test, the preexposed stimulus is presented as part of a compound with a novel stimulus. As has been noted elsewhere, the introduction of a novel stimulus or context at time of test may well have the effect of attenuating the amount of latent inhibition. Therefore, these designs must be considered as weak tests of latent inhibition effects.

However, one of these experiments (Shishimi, 1985, Exp. 3) did use a more typical, simple, and straightforward design to search for latent inhibition. Goldfish were given 0, 20, 40, 80, or 160 preexposures (10 sec each) to red or green lights. For half the groups, the test, classically conditioned shuttle responding, was to the preexposed color, and for half it was to the nonpreexposed color. Once again, there were no reliable main effects nor interactions for either type of preexposure or number of preexposures. Here, then, the evidence for the absence of latent inhibition is more convincing than in the previously described studies.

To be comprehensive, one must also take note of the last experiment in that series (Shishimi, 1985, Exp. 5). In an unusual design, animals of the preexposed group were placed in a shuttle box for four daily 40-min sessions. The two ends of the box were differentially illuminated with green and blue lights. The goldfish of the nonpreexposed group remained in their home tanks during this period, and thus received neither preexposure to the colored lights nor preexposure to the apparatus. The test phase consisted of 60 training sessions of 40 min each in which 10 programmed white-light–shock pairings were presented. However, the scheduled white light was turned on only when the animal was in one of the predesignated colored compartments. For half of the goldfish it was the green compartment, and for half the blue compartment. However, irrespective of the compartment in which the subject found itself, the shock was delivered at the scheduled time. In other words, the goldfish could "choose" to receive a predicted shock or a nonpredicted shock. The subjects of both groups came to prefer the signaled shock over the nonsignaled shock, but the discrimination was acquired much more rapidly by the group that was preexposed to the colors. Although this facilitation effect would seem to be contrary to what one would expect if there were latent inhibition, it should be noted that the design was confounded in that the stimulus-preexposed group also had apparatus preexposure that the control group did not. Other studies, with other species and tasks, have indicated that apparatus preexposure by itself may promote rapid learning.

Pigeon

All of the latent inhibition studies with pigeons, with one exception, used the autoshaping procedure. Those studies were reviewed in the section on autoshaping (chapter 2), where it was concluded that the evidence for latent inhibition in the pigeon autoshaping paradigm was, at best, weak. The one pigeon latent inhibition study that did not use an autoshaping procedure employed conditioned heart-rate response. Cohen and MacDonald (1971) gave *restrained* 0, 10, or 40 preexposures to 6-sec changes in illumination (from complete darkness) or to unpaired presentations of the light and shock. In the test, the light was paired with shock, and change in heart rate was used as the index of conditioning. Only the group that experienced 40 *unpaired* light and shock episodes showed a decrement in conditionability compared with the zero-preexposure group. There were no differences between the groups preexposed 0, 10, and 40 times to light only.

A conservative evaluation of the latent inhibition studies using autoshaping and heart-rate conditioning, taking into account the technical and interpretive problems of the former (see chapter 2) and the limited number of experiments on the latter, suggests that there is not yet sufficient evidence to conclude that the pigeon is capable of exhibiting latent inhibition.

Subhuman mammals

Just as the studies of the nonmammalian species indicate either absence of a latent inhibition effect or, at best, equivocal evidence for an effect, the evidence from a variety of mammalian species is quite clear in identifying the presence of latent inhibition. In addition to latent inhibition having been demonstrated in the rat under a variety of conditions (e.g., see the sections on conditioned avoidance, conditioned suppression, and conditioned taste aversion in chapter 2), it has also been demonstrated in the mouse (Alleva et al., 1983; Bignami et al., 1985; Oliverio, 1968; Schnur & Lubow, 1976), rabbit (see the section on EB and NMR conditioning in chapter 2), goat and sheep (Lubow, 1965; Lubow & Moore, 1959), cat (Wickens et al., 1983; cf. McDaniel & White, 1966). There has also been one study with dogs (Herendeen & Shapiro, 1976) and one with monkeys (Mineka & Cook, 1986). However, those experiments were uninterpretable, the former for reasons cited in chapter 2, and the latter because it did not use the basic latent inhibition paradigm. Nevertheless, the overall pattern is quite clear. Subhuman mammals display a robust latent inhibition effect, and this occurs in a wide variety of testing procedures.

Humans

As indicated earlier, the literature on latent inhibition in humans, though not extensive, is complex and warrants a separate chapter (chapter 5). However, to complete this section, a short summary of that work is required. Very briefly, then, latent inhibition has been demonstrated in young children (e.g., Kaniel & Lubow, 1986; Lubow, Caspy, & Schnur, 1982; Lubow, Rifkin, & Alek, 1976) and in older children and adults. These latter studies have employed a variety of testing procedures, including eyelid conditioning (e.g., Hulstijn, 1978; Schnur & Ksir, 1969), electrodermal conditioning (e.g., Siddle, Remington, & Churchill, 1985), simple instrumental learning (e.g., Ginton et al., 1975; Lubow, Weiner, Schlossberg, & Baruch, 1987), and discrimination learning (e.g., Lubow, Caspy, & Schnur, 1982). With the exception of the studies using electrodermal conditioning, it appears that demonstration of latent inhibition in older children and adults requires the use of a task, during the preexposure period, that diverts the subject's attention from the preexposed stimulus that will later, in the test, be the target stimulus. The interaction of age and masking will be the focus of the next chapter.

Summary

To recapitulate, there are several points to be made: (1) There is little or no evidence for a latent inhibition effect in invertebrate species; the absence of such an effect has been shown in a mollusc and in the honeybee. Similarly, there is no convincing evidence for the presence of latent inhibition in lower vertebrates, although one can be more confident of this conclusion with goldfish than with pigeons. Various mammalian species, on the other hand, all exhibit powerful latent inhibition effects, including mouse, rabbit, rat, sheep, goat, cat, and, among humans, children below the age of 6 or 7 years; older children and adults also show the effect, but only under preexposure conditions that mask the presentation of the to-be-tested stimulus.

5 Associative learning tests of the effects of stimulus preexposure in children and adults

Although the vast majority of experiments on latent inhibition have used animals as subjects, there has been increasing interest in the phenomenon as it occurs in humans. Because the normal procedures for producing latent inhibition are effective with children, but not with adults, who require a masking task, it is convenient to examine separately these two populations of subjects.

Children

The study of stimulus preexposure effects, using children as subjects, has not received much attention in the literature. Indeed, the work can be assigned almost *in toto* to the activities of three different laboratories. The earliest research, that of Cantor and his associates, concentrated on the effects of stimulus familiarity on *reaction time* (Cantor, 1969a,b). These efforts were continued by Kraut and Smothergill (e.g., Kraut, 1976; Kraut & Smothergill, 1980; Smothergill & Kraut, 1980), who started from a cognitive theoretical base (Posner, 1978; Posner & Boies, 1971). On the other hand, Lubow and his colleagues (e.g., Lubow, Alek, & Arzy, 1975; Kaniel & Lubow, 1986) have explored stimulus familiarity primarily within the context of its effects on subsequent *learning*. Because there is general agreement that the stimulus preexposure effect that is termed "latent inhibition" involves a learning deficit, the reaction-time studies will be omitted from this review.

Whereas the stimulus familiarization effect is based on reaction-time studies, stimulus preexposure also has decremental effects on later learning when the previously familiarized stimulus is employed as a conditioned or a discriminative stimulus, one to which new associations must be acquired. This paradigm, unlike reaction time, and more closely allied to those used in animal learning experiments, has only recently been used with human subjects. In this section, the four such studies that have used children as subjects will be reviewed (Kaniel & Lubow, 1986; Lubow, Rifkin, & Alek, 1976; Lubow, Caspy, & Schnur, 1982; Lucas, 1984).

Lubow, Rifkin, and Alek (1976): Effects of stimulus preexposure and context

Lubow, Rifkin, and Alek (1976) familiarized two groups of 6.5-year-old children to five "randomly" shaped, multicolored plaster-of-paris objects. Each group was given experience with a different set of objects, and then half of each group was tested with the familiar objects, and half with the other set. The test consisted of associating one of the five objects with reinforcement, a marble. The groups that were preexposed and tested on the same stimuli made more errors, took more trials to reach a learning criterion, and had longer response times than did the groups preexposed and tested on different stimuli. However, this was true only when the preexposure environment (e.g., the table on which stimulus objects were presented, which also varied across groups in color and shape) was the same as the test environment (S_oE_o). In fact, when the preexposure and test environments were different (S_oE_n), stimulus preexposure had a salutary effect on subsequent learning. The very strong influence of context changes on stimulus preexposure effects [such that either latent inhibition ($S_oE_o < S_oE_n$) or perceptual learning, i.e., facilitated learning ($S_oE_n > S_nE_n$), can be obtained] was discussed earlier. It will be recalled that context change also disrupts latent inhibition in animals in a variety of testing procedures (see chapter 3, "Role of Context").

Kaniel and Lubow (1986): Age and social class affect latent inhibition

Kaniel and Lubow (1986), in the only human developmental study of latent inhibition, used simpler, more easily specified stimuli than those just described. Middle-class children from nursery school, kindergarten, first grade, and sixth grade were given 0, 50, or 100 preexposures to either a pair of black squares differing in size or a pair of white squares differing in size. The nonpreexposed control groups had an equivalent amount of time in the preexposure phase compared with the stimulus-preexposed groups, as they were both required to solve a simple concept-formation problem while the target stimuli were either presented passively or not presented at all. In the test phase the subjects had to learn a black–white discrimination in which size was irrelevant. The nursery and kindergarten groups each showed a profound latent inhibition effect, requiring an average of 48 trials to reach the learning criterion (there were no reliable differences between the 50- and 100-preexposure conditions), as compared with an average of 27 trials for the nonpreexposed groups. However, latent inhibition was obtained only when the preexposed stimulus was later used as the positive stimulus.

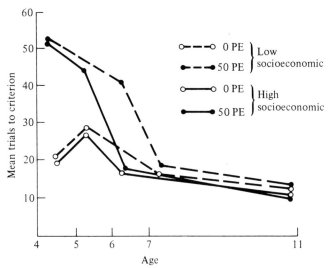

Figure 6. Mean number of trials to reach the learning criterion as a function of number of preexposures (0 or 50), socioeconomic status (low or high), and age. (Data from Kaniel & Lubow, 1986, Exp. 1 and 2.)

Of major interest was the finding that the large latent inhibition effect that was obtained with young children disappeared in the first grade. A parallel experiment with culturally deprived children indicated that latent inhibition was present in nursery, kindergarten, *and* first-grade children, disappearing only in the second grade. These two sets of data are illustrated in Figure 6.

Lubow, Caspy, and Schnur (1982): Latent inhibition is stimulus-specific; older children require masking

Lubow, Caspy, and Schnur (1982), in a series of experiments using 10–11-year-old children, demonstrated that latent inhibition in that age group could be obtained only when masking conditions were employed during the preexposure period. In their Experiments 3 and 4, subjects were presented with pairs of circle figures, each one of which was composed of four dimensions, each of two possible values. The dimensions were size of circle, color of circle, number of dots within the circle, and color of the dots. Whereas some groups were passively preexposed to repeated presentations of stimulus pairs by themselves, other groups received either no preexposure or stimulus preexposure in conjunction with a masking task. The masking task consisted of three centrally presented scrambled letters (anagram) that the subject was required to make into a word. The to-be-target stimuli flanked the masking task letters. Another group received only the masking task in the absence of the anagram material. Two types of tests were applied after the preexposure phase, one using the same stimulus materials that were preexposed (circles), and the other using different stimulus materials (squares) with the same test task (Exp. 3) or a completely different task (Raven matrices, Exp. 4). The results clearly confirmed that passive preexposure of the to-be-tested stimulus was *not* sufficient to produce latent inhibition in 10–11-year-old subjects. In order to obtain latent inhibition, the stimuli had to be presented *together* with a masking task.

Although the major finding of those experiments can be described in terms of the interaction between stimulus preexposure and masking, it was also shown that the latent inhibition effect was stimulus-specific; that is, there were no stimulus preexposure effects when the test stimulus differed from the preexposed stimulus. This stimulus specificity conforms to the findings of Lubow et al. (1976) with younger children, as well as to the findings in the animal literature.

Lucas (1984): Some contradictory findings

Finally, a recent study by Lucas (1984) used 6-year-old culturally deprived children from a kindergarten in Israel as subjects. Half of the children were preexposed to cards with one of three colored circular patches: blue, green, or yellow. There were 60 cards in all, so that each color was preexposed 20 times. Preexposure trials were given in blocks of three, with every color represented once in each block, and with the order varying between blocks. The other half of the subjects were preexposed to three letters (A, B, T) in the same manner as described for the patches. The letters may be considered as nonsense figures, because these Israeli subjects had no experience with a non-Hebrew alphabet. The test, conducted immediately after preexposure, was modeled after a standard paired-associate procedure. Half of each group was tested with the stimuli to which it was preexposed, and half with the stimuli to which the other group was preexposed. In both cases the preexposed material served as the stimulus terms in the paired-associate task. The response terms were the same for all groups; each stimulus had to be associated with the figure of a circle, bar, or triangle. The experimental design was a straightforward 2 × 2 factorial, with type of preexposed stimulus (color or letter) crossed with type of stimulus in test (color or letter). Thus, there were two conditions of test stimulus familiarity and two conditions of test stimulus novelty. Table 17 shows the mean numbers of correct responses that were achieved in 30 test trials.

As can be seen, the only significant difference was that between the familiar color group and the other three groups. However, the difference was in the opposite direction from what was expected. Instead of latent inhibition, the familiar color group showed superior learning as compared with the other groups. Thus, there were two findings in this exper-

Table 17. *Mean numbers of correct responses (and standard deviations) for the two familiar stimulus and two novel stimulus groups*

Test stimuli	Preexposed stimuli	
	Colors	Letters
Colors	22.44 (8.38)	14.00 (6.56)
Letters	12.00 (7.73)	12.67 (5.15)

Source: Lucas (1984).

iment that have to be explained: Why was there facilitation rather than retardation? Why was the facilitation found only for the color–color group and not also for the letter–letter group?

To begin with, it should be noted that the preexposure conditions of the Lucas (1984) experiment differed in a significant manner from the conditions in those studies that did obtain latent inhibition. Whereas all the successful latent inhibition studies preexposed a single simple stimulus, or, if complex, the complete stimulus was preexposed *simultaneously* (Lubow, Rifkin & Alek, 1976; Lubow, Caspy, & Schnur, 1982), in the Lucas study the preexposure was to either *three different* colors or *three different* letter shapes *successively*. One might argue that presenting different stimuli on successive trials, especially if the stimuli are from the same dimension (i.e., color), promotes stimulus differentiation. Stimulus differentiation, in turn, would facilitate paired-associate learning in the test phase. There is ample evidence for this contention in the human verbal paired-associate literature (Goss & Nodine, 1965). Moreover, it is known that active labeling of the preexposed stimuli also aids stimulus differentiation (Gibson, 1969). Indeed, children will spontaneously label colors (Dale, 1969). However, such labeling to the English letters, which for the Israeli children should be equivalent to nonsense shapes, would not appear likely to occur. Thus, the results of the Lucas (1984) experiment may be accounted for by two related factors: (1) stimulus differentiation induced by successive presentation of the stimuli and (2) the probable elicitation of a covert verbal response to the color. If these explanations are valid, then one should be able to produce latent inhibition with the Lucas procedure simply by masking the stimuli during the preexposure phase. Such a masking procedure should prevent both stimulus comparisons and labeling responses.

Adults

As with the stimulus preexposure research with children, there have been several studies with adult human subjects that have used the reaction-time paradigm as well as the more typical learning procedures. Once again, the reaction-time studies will be omitted from this review because of the less than direct relationship between measures of reaction time and measures of learning. In addition, the effects of stimulus preexposure on reaction time in adults are quite inconsistent (Cecil, Kraut, & Smothergill, 1984; Kraut & Smothergill, 1978; Kraut, Smothergill, & Farkas, 1981; Meyers & Joseph, 1968; Roelofs, 1982; Schnur & Lubow, 1987).

Studies of latent inhibition in adults with conventional learning tasks can be divided into the two traditional major categories of learning: classical conditioning and instrumental conditioning. Under the former rubric, studies of eyelid conditioning, Ivanov-Smolensky conditioning, and electrodermal conditioning will be examined. Within the latter category, studies of acquisition of simple instrumental responses and studies of more complex discrimination learning will be reviewed. In addition, several miscellaneous studies will be noted, including two on conditioned taste aversion, and one on the learning of verbal commands by retarded adolescents.

Classical conditioning of the eyeblink response

Of the seven studies using eyelid conditioning with adult humans in conjunction with nonreinforced preexposure of the to-be-conditioned stimulus (Allen, 1967; Grant et al., 1948, 1951; Hulstijn, 1978; Perlmuter, 1966; Schnur, 1967; Schnur & Ksir, 1969), only Hulstijn (1978) and Schnur and Ksir (1969) successfully demonstrated latent inhibition. Although the Grant et al. (1948, 1951) studies both employed fewer than the minimum requisite number of preexposures, Allen (1967) and Schnur (1967), who used considerably large numbers of stimulus preexposures, also failed to find a latent inhibition effect.

These early failures to obtain latent inhibition in adults, as compared with the ease of demonstrating such an effect in animals, prompted a search for an explanation of the differences and for an appropriate technique that might reveal the latent inhibition effect in human adults. Such a technique, masking, was first used successfully by Schnur and Ksir (1969). They modeled their procedure after that of Spence (1966), who used masking with human eyelid conditioning to retard the rate of extinction, and therefore to make the adult extinction curve similar to that obtained with animals. Like Spence (1966), Schnur and Ksir (1969) used a probability learning task as the mask. In this procedure, three lights face the subject. One serves as a signal or warning stimulus. When it is lit, the subject is required to guess which one of the two small lights, one to the right and the other to the left of the signal light, will be activated. The actual sequence of operation of the lights is prearranged and arbitrary. Instructions to the subjects are such that they are led to believe that the purpose of the experiment is to study problem solving (i.e., to discover the sequence of the lights), while the tone CS and airpuff US are merely meant to serve as distractors. It is important to note that these masking procedures were present throughout all phases of the experiment. Thus, in Spence's studies, the masking task was used during the acquisition *and* extinction phases, whereas in the Schnur and Ksir (1969) study it was employed during three phases: preexposure, acquisition, and extinction. This type of deployment of the masking task reflects the belief that in order to obtain either the slower extinction (Spence, 1966) or the latent inhibition (Schnur & Ksir, 1969) in humans that is characteristic of lower animals, it is necessary to prevent the adult subject from becoming aware of the *transition* from phase to phase. That the light guessing task successfully served this purpose, at least for the transition from acquisition to extinction, is attested to by the fact that when a post-experimental questionnaire was administered to subjects in a standard acquisition–extinction group, 76% reported that they recognized that there was a shift in conditions from acquisition to extinction. On the other hand, only 4% of the masked group reported any awareness of such a change (Spence, Homzie, & Rutledge, 1964).

It is, of course, well known that under these conditions of masking, Spence did show that extinction of human eyelid conditioning was severely retarded as compared with what occurred with absence of the masking conditions. Likewise, Schnur and Ksir (1969) demonstrated that, with masking, latent inhibition could be induced in adult humans. As in many experiments with animals, the latent inhibition effect was stimulus-specific.

The Schnur and Ksir (1969) data appear to confirm the hypothesis in regard to the masking task. However, there was a major problem with the design. Masking was not directly compared with absence of masking. Although cross-experiment comparisons suggest that the production of latent inhibition in adult humans does indeed require the masking task, a stronger case can be made by a direct within-experiment comparison of the two conditions. Such a design was used by Hulstijn (1978). After failing to obtain latent inhibition of the conditioned eyelid response with standard nonmasked stimulus preexposure and nonpreexposure groups, Hulstijn ran a second experiment in which he contrasted the masked and unmasked procedures. The former group was given a Spence-type masking task that continued from the preexposure phase through the test phase. He obtained significantly better conditioning in the nonmasked preexposure group than in the masked preexposure group. These results, although again strongly suggesting that masking is required to produce latent inhibition, still do not provide a completely adequate experimental design to answer the question without qualification. It is quite possible that masking, by itself, produced poorer conditioning in the acquisition phase, rather than the poorer conditioning being a result of stimulus preexposure, which is the implicit assumption of these types of studies.

The appropriate test requires a 2 × 2 factorial design with two levels of preexposure (zero and n) crossed with two levels of masking (present and absent). If masking is indeed a requirement for producing latent inhibition, then the group that receives stimulus preexposure plus masking should show the greatest retardation of conditioning relative to the other three groups. Unfortunately, such a design has not been executed with eyelid conditioning. However, studies by Lubow, Caspy, and Schnur (1982), with 11-year-old children, and Ginton et al. (1975) and Har-Evan (1977), with adults, confirm the expected interaction between stimulus preexposure and masking, albeit with instrumental learning tasks, as will be described later.

Ivanov-Smolensky conditioning

The basic characteristic of Ivanov-Smolensky conditioning is the procedure of using a verbal command as the US. Thus, for example, the CS, a tone, would precede the order "Squeeze bulb." Any anticipatory squeezing during the CS period, but before the command, would be considered a conditioned response. There have been few studies utilizing this technique, at least in the West. Indeed, there are only two Ivanov-Smolensky-type studies in conjunction with nonreinforced preexposure as a variable (Siebert, 1967; Sokolov & Paramanova, 1956). The latter, as reported by Sokolov (1963), indicated that preexposure to the to-be-conditioned stimulus resulted in a severe handicap to subsequent conditioning. In that study, in which subjects were adolescents, the CS, a tone, was paired with the US, the command "Raise your hand," while the CR was an electromyographic response during the tone presentation. Sokolov reported that most of the subjects who had the CS and US paired from the beginning were conditioned readily. Other subjects, who were preexposed to the CS before pairing the CS with the US, were conditioned either more slowly than the subjects without stimulus preexposure or not at all. Unfortunately, complete details of the study, particularly instructions, are not available.

The study by Siebert (1967)[1] employed college students. The CS was a moderate-intensity buzzer, and the US was the command "Close switch," to which the subject had to respond by raising one leg to close a microswitch. Anticipatory movements of the upper thigh muscles were recorded as the CRs. As can be seen in Figure 7, in spite of relatively low levels of conditioning, latent inhibition was demonstrated.

Of special interest is the finding of a latent inhibition effect without

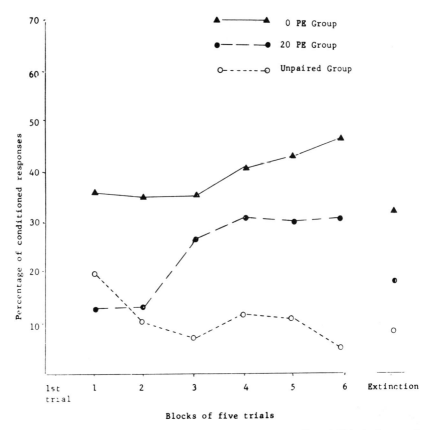

Figure 7. Percentages of conditioned responses over five trial blocks for acquisition and extinction for three groups: not preexposed; preexposed to 20 tones; and receiving unpaired CSs and USs during acquisition. (From Siebert, 1967.)

the use of a masking task. However, the instructions to the subjects were such as to promote masking-like conditions: "... Now you will be of the most use to us if you can keep your mind on anything except what is going on in here. Think about homework, a movie you have seen, anything at all."

It should also be noted that it is characteristic of these studies, as it also is for the galvanic skin response (GSR) studies to be reported in the next section, that the subject is unaware of the response that the experimenter is interested in measuring.

Classical conditioning of the GSR

Considering all of the work published on autonomic conditioning and the orienting response (OR), there is a surprising paucity of information in this area concerning the effects of preexposure of the to-be-conditioned stimulus. The first mention of such a procedure was made by Sokolov (1963, p. 249), who briefly referred to the necessity to reevoke the GSR (as one member of the OR family) after nonreinforced trials before conditioning can occur. He noted that it takes more trials to reinstate the GSR after nonreinforced stimulus preexposure than without it.

Maltzman, Raskin, and Wolff (1979, also reported in Wolff & Maltzman, 1968) specifically tested the hypothesis "that the greater the amount of habituation of cues prior to the onset of conditioning, the more difficult it is to establish a CR to one of these cues." Three groups received 0, 20, or 40 nonreinforced preexposures to different words. This was followed by the conditioning phase, in which the word "plant" (not present during the preexposure phase) was interspersed between filler words and followed by the US, a blast of noise. The zero-preexposure group reached a peak GSR–CR at the fourth CS, the 20-preexposure group at the seventh CS, and subjects with 40 preexposures at the sixth CS. Similarly, the magnitude of the peak GSR was a function of the number of preexposures. Thus, poorer conditioning was manifested by the word-preexposed groups, as compared with the nonpreexposed group, even though the word used in the conditioning phase was neither preexposed nor related to the preexposed word.

Surwit and Poser (1974) employed a more standard latent inhibition design. They presented groups of adult subjects with either 50 or 100 nonreinforced preexposures of the to-be-conditioned tone stimulus or with an irrelevant stimulus. The test phase was conducted at various times after preexposure: immediately, 1 hr, or 24 hr. The 50- and 100-tone-preexposure groups exhibited fewer short-latency GSR–CRs than did the groups exposed to the irrelevant stimulus. For long-latency responses, only the first trial block (5 trials) showed a latent inhibition effect. Amplitude, magnitude, and latency of the GSR were also analyzed, but showed no significant differences between groups. There were no reliable differences in the amount of latent inhibition between the zero-delay, 1-hr-delay, and 24-hr-delay groups.

In addition to the studies mentioned earlier, there have been two studies employing adult humans that have obtained either no differences in conditioning after stimulus preexposure (Zeiner, 1970) or a facilitory effect (Silver, 1973). Zeiner (1970) preexposed half of his subjects eight times each to both a red light and a green light, which then served as either a CS+ or a CS- in a differential conditioning procedure. The other half of the subjects were given conditioning without preexposure. All subjects received 20 acquisition trials to each CS. In the case of the CS+, 14 of these stimulus presentations were followed by shock, and the remaining six served as unreinforced test trials. The color that was reinforced was counterbalanced. Immediately following conditioning, all subjects were given six extinction trials to each CS. Stimulus preexposure decreased the magnitude of the GSR to both the CS^+ and the CS^- for the first few conditioning trials, but no lasting effects were noted.

Zeiner's failure to obtain latent inhibition may be attributable to a variety of causes, each one by itself sufficiently compelling to account for the results. First of all, the number of preexposures, eight to each stimulus, was considerably below the number that typically has been successful in producing the effect. Indeed, those GSR studies that did report latent inhibition used many more stimulus preexposures than did Zeiner: Brandeis (1974) 40; Maltzman, Raskin, and Wolff (1979), 20 and 40; Siddle et al. (1985), 20; Surwit and Poser (1974), 50 and 100.

The role of number of stimulus preexposures was specifically investigated by Brandeis (1974). Subjects were preexposed to 0, 6, 12, 18, 24, 30, 36, or 42 presentations of the to-be-conditioned stimulus. Although the probability measure approached statistical significance, only the magnitude measure proved to be reliably sensitive to the effects of number of preexposures. For short-latency magnitude responses, a significant latent inhibition effect was obtained after a minimum of 30 preexposures, whereas for the long-latency response the effect was present only after 42 preexposures.

A second possible reason for Zeiner's failure (1970) to obtain latent inhibition relates to the fact that he also preexposed both the to-be-conditioned S^+ and S^-. As mentioned in relation to the Lucas (1984) study, serial presentations of a number of stimuli, all of which will be used in the test phase, may promote stimulus differentiation. This, in turn, would serve to obscure any latent inhibition effect, particularly with a small number of stimulus preexposures. Indeed, a recent study by Siddle et al. (1985), which also preexposed the to-be-conditioned S^+ and S^-, did obtain a latent inhibition effect, but they used a total of 40 stimulus preexposures, as will be described later.

Silver (1973) also investigated the effects of prior CS presentations on the acquisition of classically conditioned GSR. Six groups of subjects each received 1, 4, or 16 preexposure trials to tone stimuli. Each of the groups was then divided into two during acquisition, one receiving unpaired CS–US presentations, and the other receiving paired presentations. The results, scored only in terms of magnitude of the short-latency response, showed that the paired groups made significantly larger CRs across acquisition trials than did the unpaired groups. Mean CR differences between the paired and unpaired groups increased directly as a function of number of preexposure trials, and although attenuation of the CR occurred for all groups, subjects who received 16 habituation trials, in the paired CS–US condition, showed the *least* amount of response decrement. Silver concluded that increased numbers of prior CS-alone trials led to stronger conditioning. This facilitation was, of course, opposite to the usual latent inhibition effect.

Although the number of preexposures, 16, would, according to past accounts, be borderline for producing latent inhibition, this does not explain why Silver (1973) found a facilitory effect. However, it should be noted that Silver excluded from the analysis subjects who fell asleep during the experiment. He did not report the number of subjects lost nor their group assignments. It may well be that the subjects who were eliminated were just those who showed the fewest CRs because of the effects of latent inhibition.

In summary, Lubow (1973a), in his review paper, concluded that the picture regarding the presence of latent inhibition for GSR conditioning in humans was unclear. There were not sufficient numbers of studies, nor were the data from the studies strong enough to reach a firm conclusion. Indeed, it was suggested that, as with the case of human eyelid conditioning, a masking task may be necessary to uncover the latent inhibition effect (Lubow, 1973a,b). Now, more than 10 years later, with the addition of the parametric study of Brandeis (1974), the ambiguities in the literature appear to be resolved. One can obtain latent inhibition of conditioned GSR in humans without an explicit masking task. However, the effects of preexposure should be looked for primarily in the short-latency responses, and unlike in other systems, it would appear that at least 30 preexposures are required.

These conclusions have recently been confirmed by Siddle et al. (1985). They used a differential GSR conditioning procedure to ensure that they would be measuring associative effects of stimulus preexposure, and they also employed a control group with irrelevant stimulus preexposure to ensure that the measured effects would be stimulus-specific. In their first experiment, half the subjects were given 40 preexposure trials, 20 to a diamond shape, and 20 to a triangle shape. The remaining half of the subjects received the same treatment, but with the letters H and F. The preexposure phase was immediately followed by the acquisition phase. Here, half the subjects received the same stimuli as in preexposure, and half received the other stimuli. The stimuli used as CS^+ and CS^- were counterbalanced. In this first experiment, the US was a tone to which the subjects were instructed to press a microswitch as quickly as possible. Twenty-four acquisition trials, 12 to CS^+ and 12 to CS^-, were administered, as well as 16 extinction trials. Several measures were recorded, including short-latency responses (1–4.5 sec) and long-latency responses (4.5–8 sec).

During preexposure, only short-latency responses occurred with sufficient frequency to be analyzed. In general, it was found that the visual stimuli elicited significantly greater numbers of such responses than were elicited in no-stimulus control periods, and the frequency of responding declined with number of stimulus repetitions. During acquisition, there was no difference between short-latency responses to the

Table 18. *Studies investigating the effects of stimulus preexposure on classically conditioned electrodermal responses*

Experiment	Number of PEs	Stimulus	Remarks
Brandeis (1974)	0, 6, 12, 18, 24, 30, 36, 42	Tone or light	LI: magnitude SLR[a] from 30 preexposures, and magnitude LLR[b] for 42 preexposures (other measures, no difference, see text)
Maltzman, Raskin, & Wolff (1979)	0, 20, 40	Word[c]	LI: trial of peak response and magnitude of peak response, both SLR
Siddle et al. (1985)			
Exp. 1	40[d]	Shapes, letters	LI: frequency of SLR
Exp. 2	40	Shapes, letters	LI: frequency of SLR; smaller effect for LLR
Silver (1973)	1, 4, 16	Tone	Facilitation: magnitude of SLR after 16 preexposures
Surwit & Poser (1974)	0, 50, 100	Tone	LI: response probability of SLR; no difference for amplitude, latency, or recruitment
Zeiner (1970)	0, 16[e]	Red or green light	No differences for magnitude of SLR or LLR

[a] SLR short-latency response.
[b] LLR long-latency respnse.
[c] The word CS in acquisition was different from the preexposed words.
[d] 20 to the to-be-CS⁺ and 20 to the to-be-CS⁻
[e] 8 to the to-be-CS⁺ and 8 to the to-be-CS⁻

CS⁺ and CS⁻ for the groups preexposed and tested with the same stimulus, but there was significantly more responding to the CS⁺ compared with the CS⁻ for the nonpreexposed group. No such effects were found for the long-latency responses. A similar pattern was exhibited during extinction. These data are in accord with those reported by Maltzman, Raskin, and Wolff (1979), but the latter also found such effects for responses following US omission, maximum CS⁺ response, and the trial in which the maximum response occurred. Siddle et al. (1985) found no such differences between groups.

In a second experiment, Siddle et al. (1985, Exp. 2) used the same procedures described for Experiment 1, but with a more conventional US in the acquisition phase: an intense burst of noise instead of the voluntary response to tone. In general, the results of the second experiment replicated those found in the first, with the exception that preexposure affected long-latency responding as well as short-latency responding during the first six trials of acquisition, with the preexposed group showing reduced responding as compared with the nonpreexposed group.[2] Siddle et al. (1985) concluded that their data clearly demonstrated a latent inhibition effect in human subjects, and furthermore the data suggest that there is a relationship between the orienting response and the conditioning process.

Whereas the first part of that statement conforms to the conclusions that we reached earlier, the last part would seem to be somewhat premature. One must remember that in the Siddle experiments the OR and the CR were one and the same response. To support the type of OR–latent-inhibition relationship that is suggested by Siddle, correspondences between an OR measure and a CR measure that are defined independent of each other are required.

Table 18 summarizes the various experiments that have sought to investigate the effects of preexposure of the to-be-conditioned stimulus on the acquisition of a classically conditioned electrodermal response. As already indicated, there is considerable evidence for a latent-inhibition-like effect, particularly for short-latency responses. This effect, it should be noted, is produced without the aid of a masking task. What is not clear is whether the effect reflects a decrement in stimulus associability or a waning of the orienting response that may be confounded with the conditioned response.

Instrumental learning

This section will look at those latent inhibition studies that were based on instrumental as opposed to classical conditioning techniques. In particular, five studies, all using the same basic preexposure and testing procedures, will be examined. Those studies (Baruch, Hemsley, & Gray, 1988a,b; Ginton et al., 1975; Har-Evan, 1977; Lubow et al., 1987) all employed preexposure to an auditory stimulus, with masking, and a subsequent test in which the subject had to learn to associate the preexposed stimulus with a new event. Because all the procedures followed those developed by Ginton et al. (1975), that first study of latent inhibition with an instrumental learning procedure will be examined in detail.

Ginton et al. (1975), after reviewing the pertinent studies on latent inhibition in adult humans and the effects of masking, concluded that it was the absence of attention to the CS at the beginning of the test phase that was responsible for producing the latent inhibition effect. Furthermore, they proposed that this lack of attention be regarded as an acquired inattention that is developed in the preexposure phase (cf. Lubow, Weiner, & Schnur, 1981). Thus, the preexposure phase typically starts with the presence of an attentional response to the to-be-conditioned stimulus, but this response diminishes as a function of the number of preexposures. They asked, however, what would happen if the attentional response to the to-be-conditioned stimulus were *not* elicited at the beginning of the preexposure phase? If, with the typical masking task, it is assumed that the subject is initially aware of the presence of the CS, but *learns* to ignore it, what, then, is the result of having that same stimulus presented under conditions of *initial* inattentiveness? In addition to assessing the effect of the initial attentional level to the to-be-learned stimulus on subsequent learning, Ginton et al. (1975) simply wanted to demonstrate the latent inhibition effect in adult humans. It will be remembered that, at that time, such latent inhibition had been an elusive phenomenon.

To accomplish those objectives, attention was manipulated during the preexposure phase by the use of a dichotic listening task. The whitenoises CS was presented either to the ear that the subject was monitoring, but for other information (the masking task), or to the ear that was not being monitored for that information. The first group was intended to represent a condition of stimulus preexposure with relatively high attention, as compared with the second group. Two other groups received no preexposure to the to-be-learned stimulus and were considered to have the highest level of attention at the beginning of the learning phase.

The masking materials consisted of a list of 40 nonsense-syllable pairs that was repeated, without a break, several times. The list was presented to one ear while the same list in reverse order was presented to the second ear. A white-noise stimulus (the to-be-conditioned stimulus) of 1.25 sec average duration was superimposed on one channel, so that there were from four to six stimulus presentations for each series of 40 nonsense-syllable pairs.

The subjects in all groups were told to attend to one ear and to count the number of times the syllable list was repeated. The ear to which a subject was told to attend and the presentation to that ear varied according to which one of four groups the subject was assigned. Two of the groups were preexposed to a list that included the to-be-conditioned stimulus, and two of the groups were preexposed to a list that did not include the to-be-conditioned stimulus. Of the two CS-preexposed groups, one received the to-be-conditioned stimulus in the ear that was monitored for the nonsense-syllable repetition (PE/M, preexposed-monitored group). The other CS-preexposed group received the to-be-conditioned stimulus in the ear opposite to the monitored one (PE/NM, preexposed-nonmonitored group). Both groups were subsequently tested with the conditioned stimulus presented to the *same* ear in which it had been preexposed. Of the two groups not preexposed to the to-be-conditioned stimulus, one was tested with the critical stimulus presented to the ear that had been monitored during the preexposure phase (NPE/M, nonpreexposed-monitored group). The other group was tested

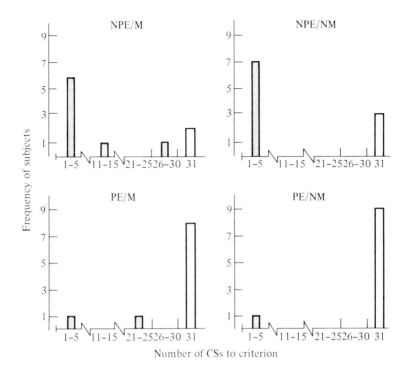

Figure 8. Distribution of subjects according to the number of CS presentations to reach the learning criterion. Empty bars indicate failure to reach the criterion; NPE/M, nonpreexposed to stimulus/monitored ear; NPE/NM, nonpreexposed to stimulus/nonmonitored ear; PE/M, preexposed to stimulus/monitored ear; PE/NM, preexposed to stimulus/nonmonitored ear. (From Ginton, Urca, & Lubow, 1975.)

with the conditioned stimulus presented to the ear that had not been monitored during the preexposure phase (NPE/NM, nonpreexposed-nonmonitored group).

For half the subjects in each group, the monitored ear was the right ear, and for half it was the left. The subjects were exposed to five repetitions of the 40-item list. For the stimulus-preexposed group, this resulted in 26 presentations of the CS.

After termination of the preexposure phase, all subjects were given the following instructions:

We are now starting with a new test. In this test all the stimulus material that you will hear on any track (in any ear) may be relevant. During presentation of the recording you will see the experimenter raising points on a score board; the raising of points is not arbitrary but is guided by a rule connected directly to the information you will be hearing on the recording. The rule is a 100 percent rule. Listen to the recording and watch the scores rise. The moment you think you have caught on to the rule push the button in front of you whenever you expect me to raise a point. If your decision was correct, a point will be subtracted instead of raised on the scoreboard. Your aim is to try and bring your score down to zero. Do not stop when you have reached zero, but go on pressing. I will tell you when to stop.

All groups were then presented with the same recording as that received by the stimulus-preexposure groups during the preexposure phase (i.e., nonsense syllables on both tracks, and the white-noise CS on one track). A point was added on the scoreboard each time the CS appeared and the subject did not push the button. The experiment ended either after the subject had correctly pressed the button five consecutive times or after 31 presentations of the CS.

The criterion for learning was the number of CSs to termination of experiment, not including the five correct presses to the CS if the subject had solved the problem. At the end of the test phase, all subjects in the stimulus preexposure groups were told what the CS was, and they were asked if they had heard it during the preexposure phase.

The results of the experiment were very clear and somewhat unexpected. First, as can be seen from Figure 8, most of the subjects who solved the problem did so within the first few CS presentations. These subjects were concentrated in the groups that had not been preexposed to the to-be-learned stimulus.

A comparison of the number of trials to reach the learning criterion for the subjects preexposed to the critical stimulus (Groups PE/M and PE/NM) and for the groups not preexposed to the critical stimulus (Groups NPE/M and NPE/NM) yielded highly significant differences in favor of the nonpreexposed groups. The difference between groups NPE/M and PE/M also was significant. These results show that there was a latent inhibition effect *irrespective* of whether preexposure of the to-be-conditioned stimulus was to the monitored ear or to the nonmonitored ear. In other words, stimulus preexposure with masking produced latent inhibition, and it did not matter whether the target stimulus was presented to the monitored ear (that attended to in order to solve the masking problem) or to the opposite ear.

In summary, this experiment clearly demonstrated a very powerful latent inhibition effect in human adults, but with a masking task. The procedure differed from the previously reviewed experiments with adult subjects in that it employed instrumental conditioning.

Baruch et al. (1988a,b) and Lubow et al. (1987) used a procedure similar to that of Ginton et al. (1975), but they modified the experimental design so as to eliminate the nonattended-ear conditions. Thus, there were two basic experimental groups, both of which received the masking task, but only one of which received the to-be-conditioned stimulus. The Lubow et al. (1987) study compared the effects of stimulus preexposure with masking versus the effects of no stimulus preexposure with masking for three classes of subjects: normal college students, hospitalized paranoid schizophrenics, and hospitalized nonparanoid schizophrenics. All three groups showed a marked latent inhibition effect. There were no significant differences between the populations nor interactions between the population types and the experimental treatments. However, Baruch et al. (1988a) did find, in addition to the normal latent inhibition effect, a sharp reduction in latent inhibition in acute schizophrenics as compared with chronic schizophrenics.

Baruch et al. (1988b) repeated the foregoing experiments, except that they applied the stimulus preexposure and nonpreexposure treatments to groups that differed in levels of psychoticism (Eysenck & Eysenck, 1976). In this case, in addition to the typical latent inhibition effect, there was a significant interaction between preexposure condition and psychoticism level. The high-psychoticism group displayed a smaller latent inhibition effect than did the low-psychoticism group.

Another study using the same procedures was performed by Har-Evan (1977). That experiment was noteworthy for two reasons: First, it directly compared the masking and nonmasking conditions and looked for an interaction between masking and stimulus preexposure. Second, it varied the intensity of the preexposed stimulus.

The basic experimental design was a 2×3 factorial, with two conditions of masking (present versus absent) and three conditions of stimulus preexposure (no preexposure, or preexposure to one of two not very different white-noise intensities, low and high). The masking task was similar to that used by Ginton et al. (1975).

As usual, the testing phase was the same for all groups. In this case, a discrimination task was employed, and the subject was required to respond differentially to the preexposed noise and to a second noise of a different intensity. Those subjects preexposed to the low-intensity noise had to discriminate it from the high-intensity noise. Those preexposed to the high noise had to discriminate it from the low noise. The subject's task was to press a button when the low-intensity noise was heard and to refrain from pressing the button when the high-intensity noise was presented.

Figure 9 shows the mean number of correct responses (hits plus correct rejections), over five blocks of eight trials each, as a function of the six conditions. The upper panel presents the effects of stimulus preexposure with masking, and the lower panel shows the effects of stimulus preexposure without masking. It is evident that the masking condition separated the three groups, so that only in the masking condition was the nonpreexposed group superior to the groups preexposed to the low- and high-intensity stimuli. In addition, there was a tendency for the high-intensity-preexposed group to perform more poorly than the low-intensity group. However, under the same conditions without masking

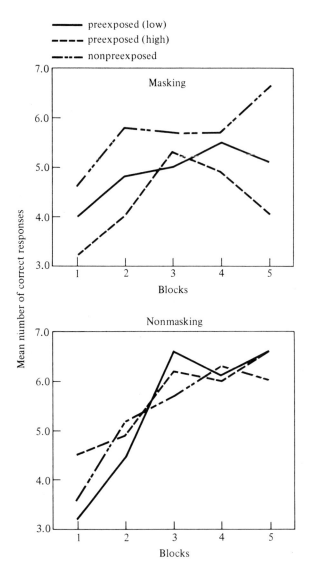

Figure 9. Mean numer of correct responses across five blocks for the six groups. Groups were preexposed to a low- or high-intensity auditory stimulus or were not preexposed. Each of these groups was preexposed either with or without a masking task. (From Har-Evan, 1977.)

(lower panel), the difference between groups were wiped out. The interaction is shown quite clearly in Figure 10.

One might argue about the meaning of the difference between the low- and high-intensity-stimulus preexposure groups, both because of the marginal statistical reliability and because there was a confounding

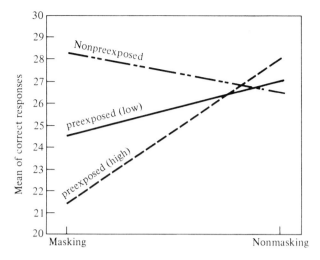

Figure 10. Interaction of type of preexposure (none, low intensity, high intensity) and masking; summary of Figure 9, without blocks. (From Har-Evan, 1977.)

of test response and preexposure intensity (it should be recalled that all groups had to respond with a button press to the low-intensity stimulus, irrespective of the condition of preexposure). Nevertheless, the major finding appears to be unequivocal: If latent inhibition is to be obtained, it requires that stimulus preexposure be conducted under conditions of masking.

This latter point requires some elaboration. It will be remembered that Spence's original reason for the use of masking was that it reduced the subjects' awareness of the *change* in contingencies from the acquisition phase to the extinction phase. Thus, he deployed the masking task so that it covered *both* phases, and therefore the transition. With the same logic, Schnur and Ksir (1969) and Hulstijn (1978) employed masking tasks that were continuous throughout the experiment, beginning with preexposure and extending through the test phase. Likewise, Baruch et al. (1988a,b), Ginton et al. (1975), and Lubow et al. (1987) presented the lists of nonsense syllables in both phases, although the subject was required specifically to attend to the list only during the preexposure phase. The Har-Evan (1977) study, however, was quite different. Masking was present *only* during preexposure. The fact that latent inhibition was obtained under these conditions suggests that the locus of the masking effect in inducing latent inhibition was not at the point of transition between preexposure and acquisition, but rather in the preexposure phase itself. Other studies that used masking only in the preexposure phase were those by Lubow, Caspy, and Schnur (1982) and Meiri (1984), both of which also obtained latent inhibition.

Two points summarize this section on latent inhibition in adults using a simple instrumental learning task. With the current paradigm, the production of latent inhibition in adults requires the use of a masking task; the locus of effect of such masking resides in the stimulus preexposure phase.

Other studies

In this section, three rather unusual and isolated studies of latent inhibition in adult humans will be described, dealing with the learning of verbal commands by retarded adolescents (Feldman, 1982) and conditioned taste aversion (Cannon, Best, Batson, & Feldman, 1983; Arwas, Rolnick, & Lubow, 1989).

In one of the very few latent inhibition studies that was application-oriented,[3] Feldman (1982) tested two severely retarded adolescents with verbal-command stimuli (e.g., "Close the window," "Stand up").[4] Using within-subject replications with counterbalancing of preexposed and nonpreexposed commands, he divided the experiment into three runs of 5 days each. During each run, two commands were presented. Following the establishment of a baseline response rate to the two commands, one of the commands was presented for 200 trials. This, in turn, was followed by the test–instruction phase, in which, for each day, both the preexposed and novel commands were alternately trained during two 20-trial blocks. Running order across days was randomized:

A training trial consisted of the trainer (a) establishing eye contact with the child, (b) saying the command, (c) waiting 10 seconds, (d) using physical assistance to guide a correct response if none occurred within 10 seconds, (e) praising and hugging the child for a correct response whether unassisted or assisted, (f) recording the child's response to the command as unassisted (i.e., correct response made within 10 seconds of the command–no physical assistance required) or assisted (i.e., no correct response made within 10 seconds of the command–physical assistance required). Two training sessions of each girl (a total of 160 trials) were videotaped behind a one-way mirror. The author scored each trial viewed as assisted or unassisted and later compared his scores to the trainer's record. Interrater reliability equaled 100%.

The subjects learned to respond to the preexposed commands much more slowly than to the nonpreexposed commands. The overall mean difference in compliance to the two command conditions was 36%.

The Feldman experiment was unique in several respects. It was the only latent inhibition study to use mentally handicapped humans and to use a socially meaningful situation, both for the stimulus (verbal commands) and for the response (compliance). Interestingly, not only was the latent inhibition effect obtained, but it occurred without the use of a masking task during preexposure. This is exactly what one would expect with severely retarded subjects; their behavior should be similar to that found with much younger subjects (i.e., latent inhibition in an associative test should be obtainable without the use of a masking task).

Cannon et al. (1983) and Arwas, Rolnick, and Lubow (1989) provided, yet one other type of study that used the latent inhibition paradigm with adult human subjects. Those experiments were of particular note because they employed the conditioned taste aversion procedure and replicated in humans what had been a common finding in rats and other animals.

The Cannon et al. (1983) study provides a potentially valuable application of latent inhibition procedures to solve a specific medical problem. When human cancer patients undergo chemotherapy, they develop severe gastrointestinal upset and nausea, and frequently they exhibit an aversion to food, to the degree of being anorexic (Bernstein & Borson, 1986). It has been suggested that this food aversion occurs as a direct result of the association of the drug-elicited stomach upset and food taste. Such taste aversion conditioning has been repeatedly demonstrated in lower animals. Similarly, it has also been demonstrated in animals that flavor preexposure protects that flavor against the development of an aversion when it is later paired with a noxious event. With this as a basis, Cannon et al. (1983) reasoned that preexposing adults to a particular taste before pairing it with a nausea-inducing drug should prevent the subject from acquiring aversion to that flavor. Indeed, they demonstrated that their volunteer subjects learned an aversion to a *relatively novel* cranberry flavor that had been paired with an apomorphine injection. However, as predicted, the aversions were reduced markedly when the subjects were given a pretreatment familiarization period with the flavor. Thus, Cannon et al. (1983) were able to produce latent inhibition of conditioned taste aversion with adults, a finding that has been repeated by Arwas, Rolnick, and Lubow (1989) using rotation-induced nausea as the US. It should be noted that no explicit masking task was used in either of these studies, although, of course, the subjects were unaware of which response was being conditioned or, indeed, that these were experiments in learning.

In summary, all three of the foregoing experiments produced a latent inhibition-like effect in adult humans. Of special interest is the fact that Feldman (1982) found the effect in severely retarded adolescents without masking, and Cannon et al. (1983) and Arwas, Rolnick, and Lubow (1989) found it with conditioned taste aversion without masking.

6 Neural substrates of latent inhibition

In the past, the extensive efforts to understand latent inhibition were directed at behavioral analyses of the phenomenon, the data from which we have discussed at some length in the previous chapters. More recently, however, a keen interest has developed in the neural substrates of latent inhibition. Although this certainly reflects a general trend in the experimental psychology of learning, the additional focus also can be partially explained by the relatively new interest in the area of attention among psychologists, as well as by the consensual opinion that latent inhibition reflects some aspect of attention (e.g., Lubow, Weiner, & Feldon, 1982). More specific manifestations of this direction can be seen in various attempts to evaluate the attentional deficits of schizophrenia by assessing impairments of latent inhibition (Baruch et al., 1988a,b; Lubow et al., 1987) and, at the same time, to explore the animal amphetamine model of schizophrenia by examining the effects of amphetamines as well as neuroleptics on latent inhibition in animals (e.g., Solomon et al., 1981; Solomon & Staton, 1982; Weiner et al., 1984; Weiner & Feldon, 1987; Weiner, Feldon, & Katz, 1987; Weiner, Lubow, & Feldon, 1981). In addition to these efforts, which will be described in the section on the effects of dopaminergic manipulations on latent inhibition, other brain systems have also been studied in this regard. These will be reviewed and discussed separately in sections on noradrenergic, serotonergic, cholinergic, septo-hippocampal, and opiate manipulations. In a final section, some conclusions will be presented in regard to which neural systems are involved in the development of latent inhibition.

Septo-hippocampal system

The first investigations of the physiological basis of latent inhibition were those concerned with the effects of hippocampectomy on latent inhibition (Ackil et al., 1969; Solomon & Moore, 1975). Those investigations were prompted by the leading theories of hippocampal function of that time, namely, those of Douglas and Pribram (1966), Douglas (1967, 1972), and Kimble (1968). According to these theories, the hippocampus is essential for registering and excluding from attention stimuli that have no significant outcomes. Since those original formulations, theories of hippocampal function have abounded (see *Physiological Psy-*

chology, 1980, Vol. 8; Gray, 1982; Schmajuk, 1984). Nevertheless, the view of the hippocampus as the register of incoming information, with an emphasis on stimuli correlated with nonreinforcement, has retained a central position in a number of theories. Thus, Gray, Feldon, Rawlins, Owen, and McNaughton (1978) described the hippocampus as a behavioral inhibition system that, among its other functions, inhibits organisms' responses in the face of nonreinforcement. Pribram (1986) proposed that the hippocampus is involved in learning to actively ignore nonreinforced events. Along more explicit attentional lines, Solomon and Moore (1975), Moore (1979), Solomon (1980), and Moore and Stickney (1980) argued that the hippocampus is responsible for tuning out irrelevant stimuli or for decreasing the salience/associability of stimuli. In the most recent version of this approach (Schmajuk & Moore, 1985), the hippocampus is said to control modifications in stimulus associability, as postulated in the attentional models of learning of Mackintosh (1975) and Pearce and Hall (1980). Hippocampectomized animals are incapable of decreasing the value of an attentional parameter, alpha, that determines the rate of association of the preexposed stimulus and reinforcement.

Clearly, the attentional theories would predict that hippocampal lesions should disrupt the development of latent inhibition. This prediction has been supported in several studies. Ackil et al. (1969) reported that aspiration lesions of the hippocampus abolished latent inhibition in a two-way active avoidance procedure. Similar results have been reported for dorsal hippocampal lesions for rabbit nictitating membrane response conditioning (Solomon & Moore, 1975), conditioned taste aversion (McFarland et al., 1978), and conditioned appetitive responding (Kaye & Pearce, 1987a,b).

Salafia and Allan (1980a,b, 1982) and Salafia (1987) reported that hippocampal stimulation during preexposure or during conditioning either attenuated or facilitated latent inhibition. This inconsistency is not surprising in view of the variable results often obtained in the literature on hippocampal stimulation (Gray, 1982). DeVietti et al. (1982) reported that, under certain conditions, latent inhibition was disrupted by mere placement of an electrode in the dorsal hippocampus.

There is abundant evidence that hippocampal and total septal lesions produce highly similar outcomes in a wide range of learning procedures (Gray & McNaughton, 1983). Not surprisingly, then, total septal lesions were found to disrupt latent inhibition in a two-way avoidance test (Weiss, Friedman, & McGregor, 1974). Likewise, lateral septal lesions were reported to disrupt latent inhibition in a successive discrimination, with tone serving as S^+ (Burton & Toga, 1982; Toga & Burton, 1979).

Noradrenergic manipulations

Experiments assessing the role of noradrenergic (NA) manipulations in latent inhibition have focused on central NA depletion, though using different lesioning techniques. The pioneering study that examined the effects of NA depletion, produced by 6-hydroxydopamine (6-OHDA) lesions of the dorsal noradrenergic bundle (DNAB), the principal noradrenergic projection from the nucleus locus coeruleus to cerebral cortex and hippocampus, was that of Mason and Lin (1980).[1] These authors chose to study the latent inhibition effect in order to test their "attention theory" of DNAB function, namely, that rats sustaining a lesion to the DNAB are impaired in selective attention.

In Mason and Lin's experiment (1980), rats were preexposed to tone–light pairings. Subsequently, the two stimuli, counterbalanced, were used as S+ and S− in a successive discrimination task. They found that nonlesioned, stimulus-preexposed animals acquired the discrimination more slowly than did their nonlesioned nonpreexposed counterparts. In contrast, the DNAB-lesioned preexposed and nonpreexposed animals did not differ in their discrimination performance (i.e., did not show latent inhibition). Consequently, those authors concluded that DNAB is involved in selective attention. Mason and Lin's results (1980) have been criticized on the basis of the fact that both lesioned preexposed and lesioned nonpreexposed rats showed a severe acquisition deficit, thus rendering interpretation of the results difficult (Robbins et al., 1982, 1985; Tsaltas, Preston, Rawlins, Winocur, & Gray, 1984). Indeed, Robbins et al. (1982) showed that when the results were not confounded with acquisition impairment, lesioned animals were not different from control animals in the amount of latent inhibition.

Although a general acquisition impairment provides the basis for a valid reservation, there is an additional, more important, problem in Mason and Lin's experiment stemming from their use of an unconventional preexposure procedure. Paired preexposure of stimuli (S_1–S_2) may produce quite different latent inhibition effects than S_1-alone preexposure (see chapter 3). Indeed, when Robbins et al. (1982) separately preexposed the two to-be-conditioned stimuli, tone and light, DNAB lesions did not affect latent inhibition.

Another study that can be interpreted along the preceding lines is that by Lorden, Rickert, and Berry (1983). In that experiment, two groups of animals, nonlesioned and DNAB-lesioned, were preexposed to a single stimulus, light, whereas two other groups were preexposed to tone–light pairs. All groups, including nonpreexposed, were subsequently tested for conditioned bar-press suppression to *light*. DNAB-lesioned light-preexposed animals conditioned more poorly than did

DNAB-lesioned nonpreexposed animals. Thus, NA depletion did not disrupt latent inhibition. In the tone–light paired condition, however, the DNAB-lesioned stimulus-preexposed group showed attenuated latent inhibition, in comparison with nonpreexposed lesioned controls. The conclusion that latent inhibition was present following single-stimulus preexposure in DNAB rats was perfectly justified and has been supported by additional experiments, as described later. However, though it is evident that multiple-stimulus preexposure leads to an observable deficit in DNAB-lesioned animals, this outcome may not be relevant to latent inhibition as such. First, as noted earlier, S_1–S_2 preexposure is not comparable to S_1 preexposure. Second, and more critically, Lorden et al. (1983) did not adhere to the paradigmatic S_1–S_2 preexposure procedure of latent inhibition. In order to make any meaningful comparisons regarding latent inhibition between the S_1 and S_1–S_2 preexposure procedures, the same stimulus must serve as S_1 in both the S_1-alone and the S_1–S_2 conditions, and latent inhibition to S_1, not to S_2, must be assessed.

There is considerable additional evidence that central NA depletion has no effect when conventional latent inhibition procedures are employed (i.e., single-stimulus preexposure, followed by conditioning to that stimulus). The absence of such an effect has been reported for conditioned suppression (Tsaltas et al., 1984), two-way active avoidance (Archer, 1982a), and conditioned taste aversion (Archer et al., 1983; Archer, Järbe, Mohammed, & Priedite, 1985). Moreover, using the taste aversion procedure, it has recently been shown that latent inhibition is not disrupted by central NA depletion produced by three different techniques: intracerebral injection of 6-OHDA to induce lesion of the DNAB; intraperitoneal injection of 6-OHDA into newborn rats; systemic administration of the neurotoxin DSP-4 in adult rats. Although these three methods produce differential patterns of NA depletion, in all cases latent inhibition was not affected (Mohammed, Callenholm, Järbe, Swedberg, Danysz, Robbins, & Archer, 1986).

Cholinergic manipulations

Central cholinergic (ACh) blockade by scopolamine leaves latent inhibition intact in conditioning of the nictitating membrane response in rabbits (Moore et al., 1976) and conditioned suppression in rats (Feldon, 1974). Using avoidance conditioning with mice, Oliverio (1968) found that scopolamine disrupted latent inhibition when given prior to the preexposure stage, but left latent inhibition intact when administered in both the preexposure and the conditioning stages. These results indicate that the disruption of latent inhibition was due to a state-dependency effect, that is, failure to transfer learned material from a drug state in preexposure to a nondrug state in conditioning.

Serotonergic manipulations

In similarity to experiments assessing the role of the NA system in latent inhibition, the work on the serotonergic (5-HT) system has focused on the effects of manipulations leading to central serotonergic depletion. Such depletion has been consistently found to abolish latent inhibition. Solomon, Kiney, and Scott (1978) employed systemic administration of parachlorophenylalanine (PCPA), which depletes whole-brain serotonin. PCPA-treated animals failed to develop latent inhibition in a two-way active avoidance test. Likewise, Lorden et al. (1983) showed that latent inhibition in a conditioned suppression test was disrupted following forebrain serotonergic depletion produced by 5,7-dihydroxytryptamine lesions of the medial and dorsal raphe nuclei. Solomon, Nichols, Kiernan, Kamer, and Kaplan (1980) examined the differential involvement of the mesostriatal versus the mesolimbic 5-HT system in latent inhibition. The mesostriatal and the mesolimbic 5-HT systems originate in the dorsal raphe (DR) nucleus and the medial raphe (MR) nucleus of the mesencephalic tegmentum, respectively. The DR provides the main serotonergic innervation of the striatum, whereas the MR provides the main serotonergic innervation of the medial septum and the hippocampus. They found that latent inhibition was disrupted in two-way active avoidance by electrolytic lesions of the medial, but not dorsal, raphe, pointing to the involvement of the mesolimbic, but not the mesostriatal, serotonergic system in the development of latent inhibition. Asin, Wirtshafter, and Kent (1980) also showed that electrolytic lesions of the medial raphe nucleus disrupted latent inhibition in a two-way active avoidance procedure. However, the same lesions did not produce this effect in the conditioned taste aversion procedure.

Dopaminergic manipulations

Investigations of dopaminergic (DA) involvement in latent inhibition have been carried out within the general framework of an animal amphetamine model of schizophrenia (e.g., Kokkinidis & Anisman, 1980; Robinson & Becker, 1986). This model is based on the findings that administration of amphetamine, which increases brain DA activity, produces in humans a syndrome that mimics paranoid schizophrenia and exacerbates symptoms in schizophrenic patients. Because schizophrenia is characterized, at least in part, by an attentional deficit (e.g., Anscombe, 1987; Bleuler, 1911/1966; Garmezy, 1977; Kraepelin, 1919/1971; Matthysse, Spring, & Sugarman, 1979; McGhie & Chapman, 1961; Shakow, 1962; Venables, 1964) or, more specifically, an inability to ignore irrelevant stimuli, Solomon et al. (1981) and Weiner et al. (1981, 1984) suggested that the latent inhibition paradigm might be suitable for demonstrating a schizophrenia-like attentional deficit in amphetamine-treated animals. More specifically, amphetamine was expected to abolish latent inhibition. This expectation was confirmed: Animals under amphetamine failed to develop latent inhibition (Solomon et al., 1981; Hellman, Crider, & Solomon, 1983; Weiner et al., 1984). However, differences between these studies emerged regarding the drug administration regimen necessary to produce the disruption of latent inhibition. Solomon et al. (1981) disrupted latent inhibition with five daily injections of d-amphetamine (4 mg/kg) in a conditioned avoidance procedure. In addition, they reported that latent inhibition was not disrupted by either acute administration of d-amphetamine (4 mg/kg) or five daily injections of 1 mg/kg. The latter result was also obtained by Hellman et al. (1983), who concluded that chronic administration and a high dosage of the drug were required for abolition of latent inhibition. In contrast, Weiner et al. (1984), using the conditioned suppression procedure, found that 15 daily injections of dl-amphetamine at 1.5 mg/kg abolished latent inhibition. Moreover, if the preexposure and the conditioning stages were separated by 24 hr, acute administration of dl-amphetamine (1.5 mg/kg), injected prior to preexposure and prior to conditioning, abolished latent inhibition. These results were obtained in the conditioned suppression and conditioned avoidance procedures (Weiner et al., 1984, 1988; Weiner, Israeli-Telerant, & Feldon, 1987). To complicate the matter further, Weiner, Israeli-Telerant, and Feldon (1987) showed that administration of dl-amphetamine at 6 mg/kg, either acute or chronic (8 days), left latent inhibition intact.

Weiner, Israeli-Telerant, and Feldon (1987) suggested that the critical difference in procedures between the two sets of studies involved the time interval separating drug injection and the beginning of the latent inhibition session. Solomon et al. (1981) administered one injection of the drug 50 min before the preexposure–conditioning session, whereas Weiner et al. (1984) and Weiner, Israeli-Telerant, and Feldon (1987) injected the drug 15 min before the start of preexposure and 15 min before the start of conditioning (because the two stages were separated by 24 hr). Brain levels of amphetamine rise substantially within the first 10 min following systemic administration, peak in 20–30 min, and then rapidly decline (Danielson & Boulton, 1976; Danielson, Petrali, & Wishart, 1976; Kuczenski, 1983; Kuhn & Schanberg, 1978). The levels of the drug in the brain 1 hr after a 5-mg/kg d-amphetamine injection

(Solomon et al. used 4 mg/kg) better approximate the maximal levels (30 min postinjection) of the drug following administration of 1 mg/kg than the maximal levels following 5 mg/kg (Danielson & Boulton, 1976; Danielson et al., 1976). These data suggest that Solomon et al. (1981) used a functionally low dose of amphetamine, which, according to Weiner, Israeli-Telerant, and Feldon (1987), would be expected to abolish latent inhibition. This possibility is particularly important when considering the conditions under which the conditioning sessions took place in the two experiments. Weiner et al. (1984) showed that abolition of latent inhibition by amphetamine at 1.5 mg/kg was critically dependent on administration of the drug in the conditioning stage. In the Weiner experiments (Weiner et al., 1984; Weiner, Israeli-Telerant, & Feldon, 1987), the conditioning stage commenced 15 min following drug administration and continued for an additional 15 min. In marked contrast, in the Solomon et al. (1981) experiment the conditioning (avoidance acquisition) stage started 65 min after drug administration. Because Solomon et al. (1981) did not specify the length of the intertrial interval, it is not possible to calculate the duration of their conditioning stage. However, on the basis of the comparative data presented earlier, it can be safely concluded that brain levels of amphetamine in the conditioning stage of the Solomon et al. (1981) study following d-amphetamine at 4 mg/kg were very similar to (and probably lower than) those in the Weiner experiments (Weiner et al., 1984; Weiner, Israeli-Telerant, & Feldon, 1987) following dl-amphetamine at 1.5 mg/kg. This conclusion also explains the fact that in the Solomon et al. (1981) study, the animals given d-amphetamine at 4 mg/kg acquired the avoidance response as effectively as did saline controls. Because Hellman et al. (1983) used a procedure almost identical with that of Solomon et al. (1981), but with d-amphetamine at 1 mg/kg, it is suggested that in their experiment the *functional* concentration of the drug 65 min following injection was too low to disrupt latent inhibition (Danielson & Boulton, 1976). Again, this possibility is supported by the fact that, in this experiment also, amphetamine-treated animals performed like saline controls (and even worse), although a dose of d-amphetamine of 1 mg/kg is know to enhance avoidance performance dramatically (e.g., Barrett, Leith, & Ray, 1972; Weiner, Israeli-Telerant, & Feldon, 1987). In summary, Weiner, Israeli-Telerant, and Feldon (1987) concluded that the discrepancy between their results and those of Solomon and his colleagues was due to differences between functionally effective dosages that were created by using different delay times between drug administration and testing. When this factor is taken into account, it appears that relatively low doses of amphetamine are more effective in abolishing latent inhibition than are relatively high doses.

Although amphetamine exerts a number of effects on the central nervous system (Groves & Rebec, 1976), several lines of evidence indicate that the effect of amphetamine on latent inhibition is mediated via the DA system. Thus, amphetamine-induced disruption of latent inhibition is prevented by concomitant administration of DA receptor blockers: chlorpromazine (1 mg/kg; Solomon et al., 1981) or haloperidol (0.1 mg/kg, I. Weiner & J. Feldon, unpublished data). In addition, Solomon et al. (1981) have shown that dopamine receptor supersensitivity produced by chronic haloperidol administration (21 days on the drug, followed by 7 days off) can combine with a low dose of amphetamine, which by itself is not sufficient to disrupt latent inhibition, to produce the latter effect.

Additional evidence for dopaminergic mediation of latent inhibition was obtained by Weiner and Feldon (1987) and Weiner, Feldon, and Katz (1987). These authors reported that the latent inhibition effect was enhanced by haloperidol (0.1 mg/kg) in both conditioned suppression and conditioned avoidance procedures. Moreover, haloperidol not only produced facilitation of latent inhibition under conditions in which normal animals developed latent inhibition, namely, following 40 reinforced tone preexposures, but also led to the development of latent inhibition under conditions in which normal animals did not show latent inhibition, following 10 reinforced tone preexposures. In addition, these experiments demonstrated that just as amphetamine must be administered in *both* preexposure and acquisition stages in order to reduce latent inhibition, similarly, to enhance latent inhibition, haloperidol must be present in both stages. When administered in the preexposure stage only, the drug yielded a normal, nonfacilitated latent inhibition effect.

These findings indicate that enhancement of DA transmission disrupts latent inhibition, whereas blockade of DA transmission facilitates this phenomenon. Moreover, several lines of evidence point to differential involvements of the mesolimbic DA system [which originates in the ventral tegmental area (VTA) and projects to the nucleus accumbens (NAcc)] and the mesostriatal DA system [which originates in the substantia nigra (SN) pars compacta and projects to the caudate putamen (CP)] in latent inhibition. Solomon and Staton (1982) showed that latent inhibition was disrupted by microinjections of d-amphetamine into the nucleus accumbens, but not into the caudate nucleus. This dissociation is in line with the findings described earlier that latent inhibition is disrupted by low, but not high, doses of amphetamine. These effects parallel those existing for the unconditioned effects of low and high doses of the drug and provide further support for an involvement of the mesolimbic DA mechanisms in latent inhibition. Low and high doses of amphetamine produce distinct unconditioned responses: The former produce locomotor hyperactivity, whereas the latter produce stereotyped behaviors (e.g., Groves & Rebec, 1976; Joyce & Iversen, 1984). There is abundant evidence that the locomotor enhancement is mediated by mesolimbic DA mechanisms, whereas stereotypy is principally mediated by striatal DA mechanisms (e.g., Creese & Iversen, 1975; Joyce & Iversen, 1984; Kelly, Seviour, & Iversen, 1975; Moore & Kelly, 1978; Pijnenburg, Honig, & van Rossum, 1975; Staton & Solomon, 1984). This suggests that both the locomotor enhancement and the abolition of latent inhibition that are produced by low doses of amphetamine are mediated by the mesolimbic DA system. In contrast, high doses of the drug that produce stereotypy via the striatal DA system do not affect latent inhibition.

Recently, Feldon, Ben-Shahar, and Weiner (unpublished data) showed that latent inhibition was enhanced by an atypical neuroleptic, sulpiride (100 mg/kg), which is believed to act preferentially on the mesolimbic DA system (White & Wang, 1983; Chiodo & Bunney, 1983). In similarity to haloperidol, sulpiride facilitated latent inhibition following 40 as well as 10 preexposures. These authors also showed that a low dose of this drug (8 mg/kg), a dose that blocked DA autoreceptors and thus led to enhanced DA transmission, disrupted latent inhibition.

Opiate manipulations

Bostock and Gallagher (1982) and Gallagher et al. (1987) employed heart-rate conditioning in the rabbit and reported that systemic or intraseptal (medial septum) administration of naloxone facilitated latent inhibition, whereas it was attenuated by intraseptal administration of levorphanol. However, these findings are difficult to interpret, because the experiments did not include proper nonpreexposed control groups.

Summary

Several conclusions may be reached in regard to the brain systems subserving the development of latent inhibition. First, it is clear that latent inhibition does not require an intact NA system. By the same token, it can be concluded that disruption of latent inhibition following hippocampal lesions is not mediated by the noradrenergic innervation of this structure. The same conclusion, although based on meager evidence, applies to the cholinergic innervation of the hippocampus.

In contrast to the NA and the ACh systems, the serotonergic and the dopaminergic systems are clearly involved in the development of latent inhibition. This conclusion is based on the following facts: (1) Latent inhibition is disrupted by an enhancement of dopaminergic transmission, either by administration of low doses of amphetamine or by low doses of neuroleptics (which block DA autoreceptors), as well as by forebrain serotonergic depletion following PCPA administration or raphe lesions. (2) Blockade of dopaminergic transmission by neuroleptics enhances latent inhibition. (3) Moreover, the existing data demonstrate that for both the DA and 5-HT systems, the mesolimbic rather than the mesostriatal portion is critical for the establishment of latent inhibition. Taken together with the fact that latent inhibition is disrupted by septal and hippocampal lesions, these findings implicate the limbic circuitry in the mediation of latent inhibition. Indeed, both mesolimbic systems (DA and 5-HT) have intimate connections with the septo-hippocampal system, as well as between themselves.

There is a well-documented hippocampal projection to the NAcc (Chronister, Sikes & White, 1976; Groenewegen, Vermeulen–Van Der Zee, de Kortschot, & Witter, 1987; Kelley & Domesick, 1982; Nauta & Domesick, 1984; Raisman, Gowan, & Powell, 1966; Swanson & Cowan, 1977; Takaori, Sasa, Akaike, & Fujimoto, 1982), and it has been suggested that at least some effects of hippocampal lesions, such as hyperresponsiveness and hyperactivity, may be due to lesion-induced changes in the NAcc (Isaacson, Springer, & Ryan, 1986; Totterdell & Smith, 1986). The disruption of latent inhibition following hippocampal lesions may also stem from disruption of the hippocampal input to the NAcc. Such lesions could lead to enhanced activation of the NAcc, thus disrupting latent inhibition.[2] If this supposition is correct, then manipulations that block NAcc activity, such as administration of high doses of amphetamine, or enhancement of serotonergic transmission, should prevent disruption of latent inhibition in hippocampal animals.

As for the 5-HT system, the medial raphe provides the main serotonergic innervation of the septo-hippocampal system (Soubrie, Reisine, & Glowinski, 1984; Azmitia, 1978; Azmitia & Segal, 1978; Geyer et al., 1976; Moore & Halaris, 1975; Wyss, Swanson, & Cowan, 1979; Kohler & Steinbush, 1982). Indeed, the most striking and consistent neurochemical correlate of medial raphe lesions is the almost total disappearance of hippocampal serotonin (e.g., Jacobs, Wise, & Taylor, 1974; Gately, Poon, Segal, & Geyer, 1985; Kostowski, Giacalone, Garattini, & Valzelli, 1968; Lorens & Guldberg, 1974; Lorens, Guldberg, Hole, Kohler, & Srebro, 1976; Rommelschpacher & Strauss, 1980; Williams & Azmitia, 1981). It has been suggested that the serotonergic projection to the hippocampus is responsible for "stop" signals to this structure in response to unimportant stimuli (Deakin, 1983). It is possible that the removal of such stop signals alters the hippocampal input to the NAcc, again leading to overactivation of the NAcc and latent inhibition disruption. In addition, the effects of medial raphe lesions on latent inhibition could be mediated by the well-documented antagonistic 5-HT–DA interaction in the NAcc (Costall, Naylor, Marsden, & Pycock, 1976; Costall, Hui, & Naylor, 1979; Carter & Pycock, 1978a,b; Fink & Oelssner, 1981; Geyer et al., 1976; Jones, Mogenson, & Wu, 1981; Lyness & Moore, 1981).

In summary, the existing data indicate that the neural substrates of latent inhibition include the mesolimbic DA system, the mesolimbic 5-HT system, and the hippocampus. The results of DA manipulations show that the disruption of latent inhibition is subserved by enhanced activity of the NAcc; and it is suggested that such activation is the final common path via which hippocampal lesions and 5-HT depletion abolish latent inhibition.

7 Theories and explanations of latent inhibition in animals

How can one explain the latent inhibition phenomenon? Why does nonreinforced stimulus preexposure of the to-be-conditioned stimulus result in a decrement in associability to that stimulus as compared with another stimulus that has not been preexposed? Thus far, this book has provided a description of the means whereby one can produce latent inhibition, attenuate latent inhibition, and even obliterate latent inhibition. If these conditions can be divided into those that are necessary and those that are sufficient, perhaps that is explanation enough. However, researchers have not always allowed themselves the comfort of such readily attained descriptions, but have sought other explanations for this phenomenon, usually either in neurophysiology, hypothetical (Hebb, 1955) or real, or in behavior, where "explanation" means "related to other behavioral concepts and/or empirical laws." Thus, for conceptual nervous system type explanations, writers interested in latent inhibition have appealed to habituation, particularly of the orienting response (Maltzman & Raskin, 1965; Wolff & Maltzman, 1968), and filter-type attention mechanisms (Ackil et al., 1969; Carlton & Vogel, 1967; Siegel, 1969a). More recently, as we saw in the last chapter, there has been a considerable amount of work on the real nervous system in regard to latent inhibition, with accompanying theoretical considerations.

Those inclined to find explanations within behavior theory have inspected the possibilities of conditioned inhibition (Reiss & Wagner, 1972; Rescorla, 1971) and competing or complementary responses (Lubow & Moore, 1959; Lubow et al., 1968). Others, unable to verify specific predictions based on a selective filter, conditioned inhibition, or specific competing or complementary response models, have turned to explanations of latent inhibition that invoke stimulus salience reduction (e.g., Domjan, 1971; Reiss & Wagner, 1972; Rescorla, 1971; Schnur, 1971).

Finally, several theories have incorporated salience or attention reduction with more general models of learning or information processing (e.g., Frey & Sears, 1978; Pearce & Hall, 1980; Mackintosh, 1975; Wagner, 1976), and one theory based on the conditioning of inattention has been developed specifically to account for latent inhibition in animals (Lubow, Weiner, & Schnur, 1981). Each of these explanations and theories will be discussed in the following sections, and a separate chapter will be devoted to a revised version of the Lubow, Weiner, and Schnur (1981) conditioned attention theory, which also takes into account latent inhibition data obtained from human subjects.

Conditioned inhibition

Conditioned inhibition has been defined by Rescorla (1971) as a "stimulus which comes through learning to specifically interfere with excitation" (p. 77). He proposed the use of two criteria for identifying a stimulus as a conditioned inhibitor: its summation with a stimulus that is a known excitor, and retardation of the development of a conditioned response to the stimulus. Furthermore, both criteria must be fulfilled in order to identify a conditioned inhibitor, because either criterion by itself can be reached by stimuli with other than conditioned inhibitory properties. After reviewing a variety of techniques that might result in meeting the two criteria, Rescorla (1969) concluded that conditioned inhibition is produced when the CS is negatively correlated with the US; that is, it serves as a signal for nonappearance of the US.

He suggested that the latent inhibition procedure (i.e., a nonreinforced preexposure of the to-be-conditioned stimulus) does not involve a negative correlation between CS and US and therefore should not result in conditioned inhibition even though this procedure "has been described as producing learned inhibition" (Rescorla, 1971, p. 77). Before examining the evidence for a relationship between conditioned inhibition and latent inhibition, some of Rescorla's initial assumptions should be questioned. It is by no means clear that presenting the CS without the US can be used as an example of a *not* negatively correlated CS-US. It was only the fact that subsequent experiments indicated an absence of conditioned inhibition that allowed Rescorla to state that nonreinforced stimulus preexposure was not an example of a negatively correlated CS-US procedure. Had conditioned inhibition been obtained, one could have argued equally well that the stimulus preexposure procedure did result in a negatively correlated CS and US. Certainly within the temporal period of such preexposure there is no correlation, but once the conditioning period begins, if there are any memorial effects at all, then immediately a negative correlation starts to accrue. It would, however, be reasonable to assume that such a negative correlation, as a psychological process rather than a statistical procedure, accrues at a faster rate under conditions of relatively successive contrasts than under the conditions of large block comparisons that are inherent in the latent inhibition paradigm.

However, the equivalence of latent inhibition and conditioned inhibition should be judged not on the basis of whether or not latent inhibition procedures violate the conditions hypothesized as being necessary to generate conditioned inhibition but rather on the basis of whether or not they meet the operationally defined behavioral criteria for identifying conditioned inhibition.

As will be seen, if one accepts the operational tests for conditioned inhibition to be summation with a known excitor and retardation of the development of a CR, then one must agree with Rescorla that the procedure of nonreinforced preexposure to the to-be-conditioned stimulus, although it produces a decrement in subsequent performance, is not related to conditioned inhibition.

Clearly, in regard to performance as an excitatory stimulus, a latently inhibited stimulus shares with a conditionally inhibited stimulus the property of entering into a new association more slowly than a neutral stimulus. Indeed, this retardation of learning is a major behavioral defining property of latent inhibition, together with stimulus specificity – also shared with conditioned inhibition. It follows, then, that any substantive difference between conditioned inhibition and latent inhibition must be sought in data from summation tests. This approach was recognized very early, and two independent investigations were undertaken to determine if the simple nonreinforced stimulus preexposure that characterizes latent inhibition procedures produces summation deficits as well as the known retardation effects (Reiss & Wagner, 1972; Rescorla, 1971).

In Rescorla's first experiment (1971), one group of rats received 76 preexposures to a tone, and a second group did not receive any stimulus preexposure. Tone paired with shock was then superimposed on a previously trained variable-interval (VI) 2-min food reinforcement schedule. As would be expected, the acquisition of suppression was slower for the preexposed group than for the nonpreexposed group. For the summation test, there also were two groups of rats, as before. However, in the second stage of the experiment, all subjects were given *light*–shock pairings superimposed on the VI 2-min food reinforcement schedule. In the test phase, subjects received both light alone and the tone–light compound while bar pressing (now without reinforcement). If tone preexposure results in conditioned inhibition, the presence of the tone should interfere with excitation in the summation test procedure; suppression to the light–tone compound should be greater than suppression to the light-alone condition for the tone-preexposed group as compared with the nonpreexposed group. These differences were not found. Even within the tone-preexposed group there was no evidence for superiority of the light–tone over the light.

In a second experiment in that study, Rescorla (1971) argued that if conditioned inhibition does accrue to the preexposed stimulus, then such preexposure should interfere with subsequent learning to that stimulus, but only in excitatory conditioning. Inhibitory conditioning, on the other hand, should be facilitated. The study tested this possibility by transferring from a simple stimulus preexposure procedure to a conditioned inhibition procedure. One group of rats received 80 tone preexposures, and the other no preexposure. All subjects had light–shock pairings superimposed on a previously trained VI 2-min food reinforcement schedule. Discrimination training, designed to establish the tone as a conditioned inhibitor of fear, was then initiated. Subjects continued to receive light, followed by shock, but also received the light–tone compound, not followed by shock. The tone-preexposed group was not able to use the light–tone as a signal that there would be no shock as well as was the nonpreexposed group. In other words, the tone-preexposed group learned to associate the inhibition of a fear response to a nonshock contingent stimulus more poorly than did the nonpreexposed group. This finding is incompatible with the notion that stimulus preexposure produces a stimulus that is a conditioned inhibitor.

In a related study, Reiss and Wagner (1971), using rabbits, demonstrated that 1,380 stimulus preexposures, as compared with 12 preexposures, did not significantly interfere with a conditioned eyelid response when that stimulus was compounded with another stimulus to which a CR had already been developed. In fact, conditioning scores for the 1,380-preexposure group were significantly higher than for the 12-preexposure group, a result quite inconsistent with a conditioned inhibition explanation. Other failures to find summation effects after simple nonreinforced stimulus preexposure include studies by Solomon, Brennan, and Moore (1974) and Solomon, Lohr, and Moore (1974), using rabbit nictitating membrane conditioning, and Solomon et al., (1978) using two-way active avoidance with rats.

In addition to summation and retardation tests for conditioned inhibition, there are other procedures that can be used to identify a conditioned inhibitor. A preexposed stimulus, if it has become a conditioned inhibitor, should *enhance* subsequent discrimination learning when it is used as the negative discriminative element. Thus, if stimulus X is preexposed, and if X has become a conditioned inhibitor, the discrimination A^+–X^- should be more easily trained than if X was not preexposed. Several experiments found exactly the opposite effect, namely, poorer discrimination learning as a result of such preexposure (e.g., Halgren, 1974; Hearst, 1972; Mellgren & Ost, 1971). In those studies, preexposure of the stimulus that was later to become the stimulus for inhibiting responding resulted in *increased* responding to S^-, as compared with a nonpreexposed group. A conditioned inhibition explanation of latent inhibition calls for decreased responding. Related data have been reported by Best (1975, Exp. 4) using conditioned taste aversion procedures.

Similarly, when X is preexposed and then used in a compound to designate the unreinforced stimulus, i.e., $A^+ AX^-$ – the procedure often used specifically to develop conditioned inhibition – if X has become a conditioned inhibitor as a result of preexposure, this discrimination should form more rapidly in the X-preexposed group than in a nonpreexposed group. Again, the evidence is quite to the contrary (e.g., Rescorla, 1971, Exp. 2).

Finally, process differences between latent inhibition and conditioned inhibition are suggested by data that indicate differential effects of physiological manipulations. In this regard, Lorden et al. (1983) reported that depletions of forebrain norepinephrine and serotonin affected latent inhibition and conditioned inhibition in dissimilar ways.

Altogether, the experiments cited earlier provide convincing evidence that latent inhibition is not a form of inhibitory conditioning. Subjects receiving simple stimulus preexposure were retarded in the subsequent development of *both* excitatory and inhibitory conditioning and showed no evidence of conditioned inhibition in a summation test.

The conclusion that latent inhibition and conditioned inhibition are not a single phenomenon is congruent with the Rescorla-Wagner model of classical conditioning (Rescorla & Wagner, 1972; Wagner & Rescorla, 1972). Thus, rather than treating the effects of nonreinforced preexposure as examples of conditioned inhibition, Rescorla (1971, p. 77) suggested that such procedures may "modify the salience of the stimulus or lead the organism to cease attending to that stimulus." Nevertheless, learning hypotheses other than conditioned inhibition are still tenable. Indeed, the information processing theories described in a later section, as well as conditioned attention theory, all propose that the rules for affecting the stimulus in the manner Rescorla described are the same rules as those for generating associative learning, and as he suggests, "they are also consistent with any learning propositions which ascribe to different stimuli different saliences and which furthermore permit these saliences to change with experience" (Rescorla, 1971, p. 81).

Competing responses

A difference approach to finding an explanation for latent inhibition is to look at overt responses occurring during the preexposure period, responses that are elicited by the to-be-conditioned stimulus and that may either facilitate or interfere with subsequent conditioning performance. The search for such an observable response was initiated by Lubow and Moore (1959). In that first demonstration of latent inhibition, it was suggested, from Exp. 1, that the retardation of learning was a result of learning a response to the preexposed stimulus that was incompatible with the later to-be-conditioned leg flexion.

Lubow and Moore (1959) argued that the decremental effects of stimulus preexposure might be accounted for in terms of an S–R contiguity paradigm. Because only one relevant stimulus was reliably preexposed, and it was the same as the test stimulus, negative transfer on the basis of a competing S–S association was difficult to assume. Whereas today plausible arguments might be offered in regard to stimulus–context associations, 30 years ago it was "natural" to infer that some competing response had become associated with the stimulus. An analysis of the responses in the preexposure and reinforcement phases of the Lubow and Moore (1959) study suggested that the characteristic response to the preexposed stimulus was a reflexive turning of the head in the direction of the stimulus (cf. Pavlov's investigatory reflex, 1927). In the first experiment, the preexposed stimulus, and, later the CS and US, were located directly to the right of the subject, and right head turns were in fact frequently observed. Sherrington's description (1906) of the various reflex figures indicates that when the head of an animal is turned, there is an increase in extensor tonus in the limb on the side toward which the head is pointed. There was thus some inferential evidence for the pairing of the preexposed stimulus with a reliable response. Furthermore, this extensor response was clearly in competition with the response that was to be conditioned during the reinforcement phase, flexion of the right foreleg.

The hypothesis that conditioning to the preexposed stimulus was decremented as a result of the association, during the preexposure period, of an extension of the right foreleg with the nonreinforced stimulus that was incompatible with the CR flexion required in the test had a readily deducible test. If, instead of having both the nonreinforced stimulus of Phase 1 and the US of Phase 2 delivered to the same side of the subject, these stimuli should be presented to opposite sides of the subject, one would expect facilitation of subsequent learning, rather than interference. In placing the preexposed CS and the US on opposite sides, the overt orienting response during stimulus preexposure should result in the subject turning away from the side to which the US will later be presented. This should lead to an increase in flexor tonus on that side for which flexion later will become, during the shock-reinforced phase, the conditioned response. In this case, then, with the preexposed stimulus and the US on the same side of the animal, facilitation was predicted. However, again, it was found that nonreinforced preexposure to the to-be-conditioned stimulus resulted in a decrement in performance. These results clearly did not enable Lubow and Moore (1959) to identify a competing mediating response. Although that analysis predicted and found a slower rate of learning to the preexposed CS when the CS and the US were on the same side of the animal, an equally valid prediction of facilitated learning, when the CS and the US were on opposite sides, was not supported.

Nevertheless, the inability to identify a mediating response does not nullify an S–R interpretation of the latent inhibition phenomenon, but rather eliminates one specific S–R hypothesis. The burden of proving such a hypothesis is borne by the investigator who proposes some other possible competing response.

Habituation theories

It has been commented, on occasion, usually at colloquia, that latent inhibition is "just habituation," with the implication that by subsuming a new set of data under the heading of a familiar concept, an explanation for those data is thereby provided. Nevertheless, one can understand the reasons for identifying latent inhibition with habituation, as there are apparent similarities in procedures and results.[1] Both latent inhibition and habituation procedures center on the repeated presentation of a single stimulus, and both take as evidence of the presence of the phenomenon in question some reduction in the amount of recorded behavior. However, these similarities are gross, and they overlook critical differences that obviate any notion of a one-to-one correspondence between latent inhibition and habituation.

Although habituation studies, like those of latent inhibition, expose subjects to a series of stimulus repetitions, the dependent variable in the former is always some component of the unconditioned response (i.e., some aspect of that response to the stimulus that is elicited on the first presentation). It follows from this that in the typical habituation procedure the dependent variable is measured contemporaneously with the stimulus preexposure phase of the experiment. Indeed, the usual experiment involves only a single phase, during which stimulus preexposure occurs and in which the decremented response is measured on each successive stimulus presentation.

Latent inhibition, to the contrary, uses a two-stage procedure in which stimulus preexposure precedes measurement of the dependent variable and in which the response explicitly does *not* represent an unconditioned response, but rather reflects *associative* strength. As discussed in several of the preceding chapters, a major methodological concern is, in fact, to keep the two measures – unconditioned response level and conditioned response level – independent of each other. To the extent that this can be accomplished, it eliminates the almost trivial explanation of latent inhibition – a reduction or habituation of the unconditioned response to the preexposed stimulus.

Habituation of the alpha response

The original unconditioned response to the purportedly neutral target stimulus, as described earlier, is sometimes referred to as the alpha response (Kimble, 1961). As a result of a study by Lubow et al. (1968, Exp. 1), which showed that the decremental effect of nonreinforced stimulus preexposure appeared from the first trial of conditioning and dissipated rather quickly over subsequent trials, it was hypothesized that latent inhibition may have been due to the habituation of the alpha response in the preexposed groups, in this case, a pinna response elicited by the to-be-conditioned tone stimulus. Although it should be possible to separate alpha responses from true CRs by means of differences in latency or form, as has been done successfully in eyelid conditioning, in the study by Lubow et al. (1968, Exp. 1) all pinna responses occurring during the CS were counted as CRs, thus allowing for the possibility that the total number of CRs included alpha responses. Because the frequency and magnitude of the alpha response decrease as a function of the number of presentations of the eliciting stimulus (habituation), subjects preexposed to the to-be-conditioned stimulus should exhibit fewer alpha responses during the beginning of acquisition than should those who had not been preexposed to the stimulus. The nonpreexposed subjects would thereby start acquisition with the full complement of alpha responses and thus achieve a higher total CR score. For the stimulus-preexposed group and the nonpreexposed group, then,

$$CR_m = (\alpha R - \alpha R_h) + CR_t$$

where CR_m is the frequency of measured CRs (in the same manner as recorded in Exp. 1), αR is the normal expected frequency of alpha responses, αR_h is the frequency of alpha responses elicited during the preexposure period, and CR_t is the frequency of the true (not alpha) CRs. As can be seen, αR_h will be high for the preexposed group, therefore lowering CR_m. For the nonpreexposed group, αR_h will be low, thus increasing the value of CR_m.

The test of this relationship and its value in explaining the latent inhibition phenomenon is straightforward. It was accomplished by manipulating the number of αRs elicited during the preexposure period. The logic of the experimental design, presented later, rests on the reasonable assumption that the number of alpha responses is proportional to the intensity of the eliciting stimulus.

A 2 × 4 factorial design with two levels of preexposure (0 and 20) orthogonal to 4 levels of stimulus intensity was employed. Each of the four 20-stimulus-preexposed groups received different intensity CSs during preexposure. During the acquisition phase, immediately following preexposure, the four nonpreexposed groups each received one of the four CS intensities paired with shock to the pinna. The stimulus-preexposed groups received the same-intensity CS to which they had been preexposed. If latent inhibition of the pinna response of the rabbit is a result of habituation of the alpha response in the preexposed group, as compared with absence of such habituation in the nonpreexposed group, such that alpha responses are included in the recording of CRs for the nonpreexposed groups, but not included for stimulus-preexposed groups, then one would predict that the group with the highest CS intensity should exhibit the most latent inhibition (where latent inhibition is defined as number of CRs for the nonpreexposed group minus the number of CRs for the stimulus-preexposed group), and the group with the lowest CS intensity should exhibit the least latent inhibition. The results indicated that the effect of stimulus preexposure was to reduce the number of CRs uniformly, irrespective of stimulus intensity. These data indicate that the alpha response habituation hypothesis is not applicable to explaining the latent inhibition phenomenon, at least not for this particular preparation.

The foregoing results do not deny that there is habituation of alpha responses, but rather confirm that such habituation is independent of the latent inhibition effect. It is quite apparent, then, that repeated nonreinforced stimulus presentation may give rise to a number of quite different and independently specific effects. Indeed, there are additional examples of this point. Russo, Reiter, and Ison (1975) demonstrated that the response decrement to the habituated stimulus was independent of the ability of the stimulus to evoke interreflexive inhibition. Similarly, Krauter (1973)[2] presented acoustic startle stimuli prior to their use as conditioned stimuli in a conditioned suppression procedure and found that the decrement in the ability of the stimulus to enter an associative relationship proceeded independent of its ability to elicit a startle reflex.

Additional examples of the dissociation between habituation of unconditioned responding and latent inhibition derive from studies by Lubow and Siebert (1969) and Domjan and Siegel (1971). Lubow and Siebert (1969) showed that the effects of repeated stimulus presentation on the acquisition of conditioned suppression to that stimulus were independent of the effects that such presentations had on unconditioned suppression to the stimulus. Similarly, in an experiment by Domjan and Siegel (1971), 5 tone presentations resulted in elimination of unconditioned response to the stimulus, and 25 presentations were needed to obtain an associability decrement to the stimulus in the subsequent test.

Note that the foregoing studies all examined the relationship between the habituation of unconditioned responses to the CS and the development of conditioned responses to the CS, and they found that, at least for some responses, there were independent habituation and latent inhibition effects. However, this is not meant to suggest that latent inhibition is independent of the habituation of *every* response. As will be seen, that may not be the case for the relationship between latent inhibition and the orienting response.

Latent inhibition and long-term habituation

The foregoing experiments all confirm that during repeated nonreinforced stimulus presentation, different behavioral changes take place simultaneously. As already noted, one of the factors that determines the particular effect that is obtained is the type of test. Even within those studies that examine the decline of an unconditioned response to a given stimulus over trials, there is a distinction to be made between contemporaneous response measurement (short-term habituation) and delayed measurement (long-term habituation) (Davis, 1970; Davis & Wagner, 1968): "It is important that students of habituation acknowledge the separate influences of the conditions under which habituation is produced and the conditions under which habituation is evaluated, and that generalizations made concerning the effects of various parameters upon habituation are not potentially misleading, by virtue of ignoring these separate influences" (Davis & Wagner, 1968). Indeed, studies of habituation using *delayed* testing come considerably closer to the latent inhibition procedures, and at least one theory (Wagner, 1976) has proposed a common theoretical mechanism whereby such habituation and latent inhibition are governed by the same principles.

The major empirical difference between short-term and long-term habituation appears to be that for the former there is an inverse relationship between certain variables (e.g., stimulus intensity, intertrial interval) and the amount of habituation (i.e., as intensity or interval increases, the amount of habituation decreases). In contrast, for long-term habituation these variables produce a direct relationship with the amount of habituation (i.e., as intensity or intertrial interval increases, habituation increases).[3] Because it is this latter function that also characterizes latent inhibition, this suggests that latent inhibition may be related to long-term habituation. Some of the empirical evidence for the distinction between the two types of habituation and the relationship between long-term habituation and latent inhibition is provided next.

Davis and Wagner (1968) exposed animals to different stimulus intensities and subsequently tested with a common intensity. They obtained a direct relationship between stimulus intensity and rate of habituation, in contrast to the generally described inverse relationship for contemporaneous, short-term presentations and test (Thompson & Spencer, 1966; Groves & Thompson, 1970; Petrinovitch, 1973). Similarly, Davis

(1970), manipulating the interstimulus interval (ISI) used during stimulus presentation and test, found more habituation with long ISIs, a finding contrary to the common observation of greater habituation with short ISIs (Thompson & Spencer, 1966; Groves & Thompson, 1970; Petrinovitch, 1973). The inverse relationships were obtained when there was no delay between stimulus preexposure and behavioral recording (i.e., when the contemporaneous procedure was used). On this basis, Davis and his colleagues differentiated between the two processes: short-term habituation, measured in intervals up to 1 min, and *inversely* related to stimulus intensity and ISI length, and long-term habituation, measured after 24 hr, and *directly* related to stimulus intensity and ISI length. Thus, with different test conditions, separable short-term and long-term effects of repeated nonreinforced stimulus presentation become evident. These effects would also seem to reflect different processes operative during nonreinforced stimulus presentation, as they bear different relations to two central variables affecting habituation; stimulus intensity and frequency.

Latent inhibition studies, which use yet another type of postpreexposure test (acquisition of a *new* response to the previously preexposed stimulus), demonstrate an additional long-term effect of repeated nonreinforced stimulus presentation: a decrement in the associability of the habituated stimulus. Like the long-term response decrement obtained in habituation studies, there is some evidence that an associative decrement to the habituated stimuli in latent inhibition experiments bears a direct relation to stimulus intensity and ISI duration (e.g., Lantz, 1973; Schnur & Lubow, 1976; Spurr, 1969).

The inverse relationship between stimulus intensity or frequency and rate of habituation that characterizes short-term habituation has been described as encompassing those parametric features that differentiate habituation from conventional conditioning (e.g., Petrinovitch, 1973). Similarly, habituation models employing conditioning principles are criticized on the grounds that such models require the direct stimulus intensity and frequency functions (Groves & Thompson, 1970). The direct stimulus intensity and frequency functions obtained in the experiments of Davis and associates suggest that for long-term habituation effects, there is indeed some resemblance among habituation, conditioning, and latent inhibition. More recent evidence for this contention has been provided by Wagner and his students; for review, see Wagner (1976, 1978, 1979).

Latent inhibition and general habituation theories

Models of long-term habituation and latent inhibition effects require the operation of a mediator that enables the transfer of information or state from stimulus preexposure to the subsequently assessed performance. The correspondence between the variables affecting long-term habituation and those affecting conditioning suggests that the two phenomena may share a common mediator. As will be seen in chapter 8, Lubow's conditioned attention theory (CAT) employs both conditioning and mediator constructs to explain the effects of repeated nonreinforced stimulus presentation. CAT proposes that during repeated nonreinforced stimulus presentation, a classically conditioned central mediating state of inattention to the presented stimulus is established. Conditioning of inattention to the presented stimulus is postulated to take place simultaneously with and independent of changes in the unconditioned response to the stimulus. In other words, CAT postulates that repeated stimulus presentation results in elicitation of at least two independent processes: First, a decrement in stimulus-elicited responses (habituation); second, conditioning of inattention to the stimulus. The latter process is proposed as a mechanism underlying the long-term response and associability decrements that result from repetitive stimulation.

Wagner (1976) also has proposed that habituation and latent inhibition can be accounted for by a single set of processes: priming of short-term memory. Here, too, the basic explanatory variable is stated in terms of learned associations, in this case between the preexposed stimulus and the contextual stimuli that accompany it. Because the burden of explaining latent inhibition in Wagner's priming theory and in Lubow's CAT resides more with conditioning than with habituation, both of these theories will be examined in more detail in later sections.

Perhaps the most influential theory of habituation has been that of Thompson (e.g., Groves & Thompson, 1970; Thompson, Groves, Teyler, & Roemer, 1973). Although there has been no direct attempt to relate their dual-process theory of habituation to latent inhibition, there are several points of intersection worth noting. The theory states that "every stimulus that evokes a behavioral response has two properties; it elicits a response and influences the 'state of the organism' " (Groves & Thompson, 1970, p. 440). The response is elicited in the S–R pathway, which "is the most-direct route through the central nervous system from stimulus to discrete motor response. State is the general level of excitation, arousal, activation, tendency to respond, etc., of the organism" (Groves & Thompson, 1970, p. 440). Repeated presentation of an effective stimulus results in decremental processes in the S–R pathway, termed habituation, and an incremental process in the state of excitation, termed sensitization. The two processes are assumed to be independent, and each process can occur in relative isolation. However, they interact to yield the final, common behavioral outcome. Groves and Thompson stated that when a strong stimulus is presented regularly at a relatively slow rate, temporal conditioning may occur. Temporal conditioning is a phenomenon of state and does not occur in the S–R pathways. The effect on state is to produce a temporally conditioned tendency towards increased state, a condition fundamental to the development of conditioned responses. Thus, an alteration in central state, subject to classical conditioning, is postulated to take place during repeated nonreinforced stimulus presentation.

Groves and Thompson do not suggest a direct measure of state, or sensitization, as opposed to habituation. The two processes are difficult to differentiate because they are yoked in their dependence on stimulus repetition. As will be seen in the next chapter, CAT proposes that it is the conditioning of inattention to a repeatedly presented nonreinforced stimulus that provides the mechanism by which the Groves-Thompson central mediating state is established. Furthermore, CAT suggests that the degree of state conditioning is, at least in part, reflected in a subsequent associability decrement of the habituated stimulus. Thus, CAT provides the mechanism for "state" alteration during repeated stimulus presentation, and the latent inhibition paradigm provides a sensitive measure of this alteration.

One can note other resemblances between the dual-process habituation theory and CAT. In the former, the theoretical curves for habituation and sensitization are similar to the theoretical curves for attentional response magnitude for S_1 and S_1-S_2, respectively (see Figure 12, p. 192). However, the similarities are more apparent than real. Although the Thompson-Groves theoretical curves are generated by the same single-stimulus input, they must be summated to get a behavioral prediction. The CAT curves are generated by independent input conditions – S_1 and S_1-S_2 repetitions – and predict, separately, subsequent associability. Nevertheless, it would not be surprising if the empirical S_1 habituation curve, using the S_1-S_2 procedure, followed the same general course as described for the theoretical attentional response and subsequent associability (i.e., an initial increase, followed by a decline), although the peaks and slopes would certainly be quite different.

Indeed, it might be the case that for repeated exposure to high-intensity S_1-alone presentations, there is an initial increase in subsequent associability, followed by a decrease. The well-known "learned helplessness" effect that follows preexposure to shock would illustrate the decremental effect (Maier & Seligman, 1976). What is less well known, and hardly alluded to in the literature,[4] is the large number of studies showing that unsignaled shock preexposure can *enhance* subsequent learning with shock as the US. However, this occurs *only when there is a small number of such shock preexposures* – up to 10 (e.g., Anderson et al., 1968; Anderson, Tyson, & Williams, 1966; Anisman &

Waller, 1971a,b; Baum, 1969; Kurtz & Pearl, 1960; Kurtz & Walters, 1962; Pearl, Walters, & Anderson, 1964). A recent study reported by Bignami et al. (1985) suggested a similar effect when varying the number of preexposures to the to-be-conditioned stimulus preceding avoidance learning. A low number of such preexposures produced facilitation, and a high number retardation, of subsequent conditioning.

This facilitation effect, then, might be viewed as confirmation of a fundamental similarity between the Groves-Thompson dual-process theory of habituation and a process affecting associability. However, it would first be necessary, in the foregoing studies, to rule out the effects of the small numbers of shock preexposures on sensitization, because escape-avoidance scores may confound activity with measures of associability. Just as it has been found that frequent, long-duration shock may result in a decrease in activity, which, when transferred to an active escape-avoidance test, is expressed in relatively poor performance (e.g., Anisman, de Catanzaro, & Remington, 1978; Glazer & Weiss, 1976a,b), a small number of shock exposures might well increase general activity and be reflected in relatively superior performance in an escape-avoidance test. In both cases, response habituation can be differentiated from associability by employing associative tests in which the effects of the two processes will be expressed in opposite directions. Thus, in a passive avoidance test, increased general reactivity will result in poor test performance, whereas increased associability will produce good performance. Indeed, some of the previously cited studies employed passive avoidance procedures and did find superior performance for shock-preexposed groups as compared with nonpreexposed groups. Other types of tests that are capable of separating associative and nonassociative effects include those developed for learned helplessness (e.g., Jackson, Alexander, & Maier, 1980).

Habituation of the orienting response

Although general theories of habituation may implicitly be relevant to latent inhibition, habituation of certain specific responses *may* be directly useful in gaining an understanding of the latent inhibition phenomenon. An examination of the effects of nonreinforced preexposure to a stimulus other than the associative deficit that characterizes latent inhibition itself may, indeed, be a fruitful tactic in this approach. And, of course, one of the simplest of these effects is that of habituation, in which the probability of eliciting an unconditioned response to a given stimulus decreases as a function of the number of times that stimulus is presented. The alpha response, which provides one example of such a possibly relevant specific response habituation, was examined earlier in this chapter. However, there is yet another kind of response that also habituates during nonreinforced stimulus preexposure: the orienting response (OR).[5]

Habituation of the OR, unlike habituation of the unconditioned alpha response to the preexposed stimulus, is a candidate for explaining latent inhibition because of its nonspecific character. That is, because preexposure to a given stimulus interferes with the associability of that stimulus with *any other* stimulus or response, a *specific* response component that may become habituated cannot account for the *general* decline in associability.

The section on conditioning of orienting responses in chapter 2 describes, in some detail, the various components of the orienting response. It is sufficient to note here that a review of the pertinent literature, which included GSR conditioning, cardiac conditioning, and observing response conditioning, indicated that the relationship between stimulus preexposure and the acquisition of a subsequent orienting response was anything but clear. However, this circumstance, by itself, does not relate to the critical theoretical question: Does stimulus preexposure produce reduced associability to that stimulus because the OR to the stimulus has been habituated (i.e., reduced over trials) *during preexposure*?

An OR theory of latent inhibition must be based on the primary assumption that the acquisition of a CR is facilitated by the presence of the OR. Support for this statement can be found in both the Russian literature (e.g., Sokolov, 1963; Voronin, Leontiev, Luria, Sokolov, & Vinogradova, 1965) and the Western literature (Maltzman, 1977; Maltzman, Weissbluth, & Wolff, 1978). If learning is facilitated by the presence of the OR, then any procedure that reduces the magnitude of the OR should also decrease the effectiveness of learning. However, as already noted, the OR is not a unitary dependent variable (Lynn, 1966). In addition, Sokolov (1963) distinguishes between two types of ORs, as well as between ORs and defensive reactions, which share many, but not all, of the component changes. Unfortunately, most investigators usually choose only one dependent variable, GSR, and therefore it is not clear whether they are measuring differences in the defensive reflex, the orienting reflex, or one of several other states that are correlated with changes in GSR. At best, only by examining the pattern of change in several of the response systems can one differentiate between the various processes.

Because that has not been done, the most that can be said is that there are several responses that undergo changes with repeated nonreinforced stimulation and that these changes may be correlated with an interference with subsequent learning. More specifically, in regard to the relationship between OR habituation and poor subsequent learning, it has been shown that habituation of the GSR to a list of words prior to conditioning reduces the level of GSR conditioning to a related word (Maltzman et al., 1978).[6] Here, again, there is a confounding of the OR and CR, and therefore these data are not directly relevant to the argument. Nevertheless, on this basis, Schnur (1971) argued that Sokolov's habituation model of the OR might provide a plausible explanation for the latent inhibition phenomenon. Once more, because elicitation of the OR is presumed to facilitate conditioning, then extinction of the OR, as might occur during nonreinforced stimulus preexposure, should hinder conditioning. Furthermore, if latent inhibition is the result of extinguishing the OR, any condition that reactivates the OR should restore normal conditionability.

Schnur (1971) tested this deduction. Two groups received preexposure to the to-be-conditioned stimulus. For one of the preexposed groups (Group 1) the stimulus was subsequently paired with shock; for the other (Group 2), a new stimulus was superimposed on the previously preexposed stimulus, and both, together, were paired with shock. A third group received no preexposure, followed by the conditioning phase, in which the same primary stimulus as used for Groups 1 and 2 was paired with shock. The strength of the stimulus–US association was assessed with a lick-rate-suppression test. The usual latent inhibition effect was found. However, there were no differences between Groups 1 and 2. Both groups showed the same amount of latent inhibition, indicating either that the addition of a novel stimulus during CS–US pairing, intended to reinstate the OR, was not effective in performing that function (there was no independent measure of OR to confirm or reject this possibility) or that the reinstatement of the OR was not effective in overcoming the decrement produced by stimulus preexposure. Schnur noted that certain ceiling effects may have been operating during the CS–US pairing stage and that these may have obscured any attenuation of latent inhibition resulting from the compounding procedure. In a second experiment, he attempted to reduce the strength of the CS–US association by introducing unpaired USs. The basic design and logic remained the same, and so did the results. Adding a novel stimulus did not restore conditionability. Opposite results, however, have been reported by Lantz (1973) and by Hall and Pearce (1982). The latter study is of particular interest because it used the omission of an expected weak shock as the means to restore associability.

These conflicting data should not be too surprising. The current value of the OR as an aid in understanding other behaviors is reduced because of failure to find reliable relationships between the OR dependent measures (e.g., Allen, 1971; Furedy, 1968; Maltzman, Gould, Barnett, Raskin, & Wolff, 1979) and failure to establish sufficient and necessary stimulus conditions to elicit the OR (e.g., McCubbin & Katkin, 1971; Siddle & Heron, 1975). Furthermore, Sokolov's statements concerning the relationships between OR and conditioning are neither precise nor clear enough to allow the OR concept to be used to explain yet another set of data. For example, Sokolov suggests that too strong an OR may interfere with, rather than facilitate, learning. But there are no independent criteria for deciding when an OR is too strong, too weak, or optimum.[7]

Nevertheless, more recent work has presented some convincing evidence that certain components of the galvanic skin response (GSR),[8] particularly first-interval anticipatory responses (FARs), "show virtually all of the important characteristics of an OR. It declines over trials, recovers to stimulus change, shows dishabituation on the trial following a novel stimulus, shows spontaneous recovery with rest and potentiation of habituation with repeated training series, is larger with irregular than with regular interval, and increases with stimulus intensity" (Ohman, 1983, p. 333). Similar statements can be made about P300[9] evoked brain waves (e.g., Roth, 1983). However, once again, even though the same stimulus conditions can be used to elicit the two different OR indicators (FAR and P300), there is very little covariation between the two response systems (e.g., Becker & Shapiro, 1980; Roth, Blowers, Doyle, & Kopell, 1982). Blowers et al. (1986, p. 101) attempted to resolve this discrepancy by suggesting that "perhaps the OR is best viewed, as Kahneman (1973) noted, as reflecting an effort to analyze the alerting stimulus, and as a complex pattern of preparation for future stimuli and responses. While the SCR (GSR) may be indicative of all these processes, it would appear . . . that the P300 is associated more with preparation for future stimulus processing."

Another deficiency of the OR explanation of latent inhibition is revealed in an examination of the logic of Schnur's study (1971). Schnur argued that adding a novel stimulus to the preexposed stimulus during the acquisition phase should re-evoke the habituated OR. True. However, is it reasonable to expect that the OR increment provided by a novel tone or light is required *in addition to* the OR increment produced by the novel shock on the first CS–US pairing? One would think that the novel shock would be sufficient to produce the OR increase that is necessary for conditioning.

At this point, then, there is little evidence to support an OR habituation explanation of the latent inhibition phenomenon, primarily because of the virtual absence of studies specifically designed to test the hypothesis. And although it might be reasonable to ask what such studies might look like, first the OR itself must be firmly grounded on both sides of the functional equation relating independent variables to dependent variables, lest we end up building a theoretical house of cards from a house of theoretical cards.

This problem can be avoided, in part, by designing studies in which the OR and the to-be-learned response are indexed independent of each other. As already noted, the GSR studies of Maltzman and others, where the GSR was the measure both of the OR and of the conditioned association, do not meet this criterion, nor do studies such as Schnur's and Lantz's, which simply *infer* a change of OR from the results of the association test.

Recent experiments in England, however, have measured ORs and the conditioned response independent of each other. This has been accomplished for standard Pavlovian conditioning (Kaye & Pearce, 1984a), inhibitory conditioning (Pearce & Kaye, 1985), blocking (Kaye & Pearce, 1984b), and latent inhibition (Hall & Channell, 1985b; Kaye & Pearce, 1987a). In each study the OR was defined as a behavioral response: rearing in front of, or touching (with paw or snout), a 10-sec jewel light on a panel that also included the source of appetitive reinforcement. In general, the data from these studies indicate a high degree of relationship between the OR to a stimulus and the associability of that stimulus. Conditions that provide for reliable elicitation of the OR to a stimulus are also conducive to rapid associative learning to that stimulus. Thus, in regard to latent inhibition, Kaye and Pearce (1984a, Exp. 5) showed that the greater the number of nonreinforced stimulus preexposures, the greater the habituation of the OR, and the poorer the subsequent learning in regard to that stimulus. The latter finding, though graphically large, was not statistically reliable. There is, however, additional confirmatory evidence for a relationship between OR habituation and latent inhibition. Kaye and Pearce (1987a), in a recent paper that extended their efforts to make a case for the OR as an index of associability, reported two important findings: (1) A decline of OR during stimulus preexposure foreshadowed subsequently poorer conditioning to that stimulus, as compared with a group for which such an OR decline was absent. (2) Dorsal hippocampal lesions had similar effects on OR and conditioning.[10] The fact that both of these points were made in a simple appetitive conditioning experiment *and* in an appetitive *discrimination* experiment, where the preexposed stimulus was an S-, provides evidence for the general validity of their findings. Together with other arguments that they present, these two studies also provide convincing evidence that the two measures, while reflecting a common underlying process, are in fact independent of each other, *in terms of measurement.*

A positive relationship between OR habituation and latent inhibition was also found by Hall and Channell (1985b). However, in addition, they reported one finding that is disturbing to the OR–latent inhibition hypothesis. A context change that was not sufficient to produce dishabituation of the OR nevertheless did interfere with the display of the latent inhibition effect. These data, then, contrary to the data presented earlier, suggest that latent inhibition and OR habituation can be dissociated, the former being context-specific, the latter not. However, this conclusion must be accepted cautiously, for several reasons. First, it is in contradiction to the data from other manipulations; Thus, stimulus preexposure and hippocampal lesions do affect OR habituation and latent inhibition in the same manner. Second, the Hall and Channell (1985b) conclusion was based on a single cross-experiment comparison (their Exp. 2 and Exp. 3). Third, there is a logical problem: Hall and Channell found that good, new learning (relatively high performance of the group preexposed to the stimulus in environment A but tested in a familiar environment B) did not depend on restoration of the OR. However, it does not logically follow from the preceding that latent inhibition will not occur in the presence of a high OR, a finding that would be needed to support the notion that there is a complete dissociation between latent inhibition and habituation. (It is, of course, possible to have an asymmetrical dissociation rather than a symmetrical, double one; the Hall and Channell data are consistent with such an interpretation.) For all of the foregoing reasons, it seems prudent, at least until additional evidence to the contrary is presented, to assume that the OR, at least as indexed by an overt behavioral observing response, does provide an index of associability, and as such is a good predictor of latent inhibition effects.

The attractiveness of this type of OR habituation explanation of latent inhibition lies in its apparent simplicity. However, the explanation, as it is, addresses merely this question: *What* is learned during stimulus preexposure? And the answer that is provided, at least provisionally, is "a reduction in OR." Second, by implication, the negative transfer from preexposure to test is accounted for by requiring a minimum level of OR activity to support the development of new associations.

What is lacking, up to this point, is a statement of the rules that govern the reduction of the OR during stimulus preexposure and its redevelopment during the new-learning test phase. The OR redevelopment requirement is demanded by the implicit assumption, noted earlier, that after habituation of the OR, some minimum level is required for new learning to proceed.

The foregoing descriptions of the empirical characteristics of the OR and its relationship to latent inhibition are compatible with contemporary major theories of latent inhibition (Lubow, Weiner, & Schnur, 1981; Mackintosh, 1975; Pearce & Hall, 1980; Wagner, 1976). These theories are differentiated one from the other primarily on the basis of how they formulate the rules for the reduction and redevelopment of the OR, at least to the extent that OR can be equated to alpha (α) or attention.

Selective attention

Although selective attention theories might easily be placed in the section on information processing, we have chosen, for several reasons, not to do so. In general, these theories, at least the ones that relate to latent inhibition, are products of the 1960s; unlike the newer theories discussed in the next section, they do not rely on the Rescorla-Wagner (1972) model of classical conditioning. In addition, the selective attention theories place a heavy emphasis on competition for attentional resources (i.e., the inverse hypothesis), whereas the newer theories inadvertently accept the inverse rule (Rescorla & Wagner, 1972), reject it (Mackintosh, 1975), or ignore it (Lubow, Weiner, & Schnur, 1981).

As noted, one of the major characteristics of most selective attention theories is the inverse principle. According to this principle, attention is considered to be a fixed quantity, so that within any given situation there is competition for attentional resources: More attention to A means less attention to B. This principle was explicitly stated in those early selective attention theories that were designed to account for animal discrimination learning (e.g., Lovejoy, 1968; Mackintosh, 1965; Sutherland, 1964; Zeaman & House, 1963). As an example, Sutherland and Mackintosh (1971) wrote about a series of selective filters that analyze incoming stimulation. Each filter encodes a different dimension or feature. A particular aspect of the stimulus is paid attention to in proportion to the extent to which a particular analyzer or filter is "tuned in." "The more strongly one analyzer is switched in, the less strongly are others switched in" (Sutherland, 1964, p. 149).

In attempting to apply this type of selective attention approach to latent inhibition, one need only add that nonreinforced preexposure of a stimulus should reduce the strength with which the analyzer is switched in, at least relative to a novel stimulus. However, Sutherland and Mackintosh (1971) provided rules only for the strengthening of the analyzer relative to a novel stimulus, not for reducing analyzer strength below that of a novel stimulus. It is this last requirement that is necessary to account for the latent inhibition effect. Indeed, Sutherland and Mackintosh were puzzled about how to handle the latent inhibition data:

When different outputs are not consistently followed by events of importance to the animal, analyzer strengths revert to their base level. The theory as it stands, does not however, cope with all instances of habituation. Consider, for example, some experiments conducted by Lubow and Moore (1959) and Lubow (1965).... The more preexposure trials given, the more slowly leg flexion developed as a CR to the stimulus. Now according to our model this result could only be explained if the strength of the analyzer detecting the light or rotor was weakened relatively to the other analyzers operative in the situation, since by rule 2 analyzer strengths sum to a constant amount. Since neither the light nor the background stimuli were followed by events of any consequence to the animal there is no reason to expect any differential weakening of the analyzers for either set of stimuli. It may be that when a stimulus is presented for a brief time against a constant background, its novelty first gives it control of behavior but that with repetition it ceases to be novel and hence the output yielded from its analyzer becomes weaker. There is in fact no postulate in our theory that accounts for such habituation. [Sutherland & Mackintosh, 1971, p. 68]

As such, then, because of the absence of rules for degrading the filter below the value for the novel stimulus, the theory has only limited value for the explanation of latent inhibition. In addition, the critical inverse rule would seem applicable only to compound-stimulus preexposure studies (unless specific provisions are made for contextual stimuli).

Nevertheless, one study attempted to evaluate the inverse rule as it might affect compound preexposure (Schnur, 1971). The prediction was that stimulus preexposure to one element of a later to-be-compounded stimulus should enhance conditioning to the other element relative to a compound stimulus where both elements are novel. This follows from the assumption that the total filter capacity is a constant and that, as one reduces the potential of one filter (in this case through stimulus preexposure), then one should increase the potential in the other filters. The implicit premise that underlies this approach relating latent inhibition to selective attention must be that stimulus preexposure does not result in retarded learning to that element when placed in a reinforced compound, as compared with the same stimulus when it is novel, but rather that the nonpreexposed element of the compound should show enhanced learning compared with the same element in a novel compound. The two experiments reported by Schnur (1971) indicate that preexposing one CS element of a compound does not enhance conditioning to the other element as would be predicted by the selective filter theory. Similar results were obtained by Lubow, Wagner, and Weiner (1982, Exp. 2).

The OR habituation theory and the selective filter theory seek some change in the internal state of the stimulus-preexposed subject for their explanation of latent inhibition. In the first case, decrement in selected responses are presumed to reflect some hypothetical internal condition that must be at a certain level for learning to occur. Stimulus preexposure has the effect of reducing this level to less than optimum. In the second case, the stimulation is said to affect the states of hypothetical analyzers. Analyzers are postulated to be strengthened when their outputs make consistently correct predictions about future events that are important to the organism. However, nothing is postulated as to the importance of predicting no change, which is the essence of the latent inhibition paradigm.

Information processing theories of latent inhibition

Contemporary explanations of latent inhibition may, for several reasons, be loosely described as information processing theories. First (as shared with the OR habituation and selective filter theories), the explanatory burden is carried primarily by inferred processes that occur *within* the organism. Second, the terminology has been borrowed, at least in part, from computer analogies. Most important, however, as in analyses of learning in general, and conditioning in particular, there is a central role for CS–US contingencies. Concomitantly, the role of contiguity as a necessary condition for the establishment of associations is depreciated. Because there is a consensus that latent inhibition is a form of learning, it is inescapable that contemporary theories of the phenomenon should also rely heavily on explanatory processes that focus on some aspect of contingency analysis.

By the early 1970s there had been sufficient numbers of published latent inhibition studies to establish it as a reliable effect that must be considered by learning theories. To a large extent that recognition was initiated by Rescorla and Wagner's successful attempts to distinguish between conditioned inhibition and latent inhibition (Reiss & Wagner, 1972; Rescorla, 1971), a distinction that was necessary to preserve the symmetrical relationship between conditioned excitation and conditioned inhibition, as proposed in their model of conditioning (Rescorla & Wagner, 1972; Wagner & Rescorla, 1972).

Indeed, the Rescorla-Wagner treatment of the early latent inhibition studies illustrates how a novel empirical observation may affect an extant theory. First, it raises the question of how the new phenomenon relates to existing theories. Here, several orders of fit can be identified. Most comforting is the situation where the new observation has been predicted by the theory. At the other extreme, the theory may be formulated in such a manner as to exclude the possibility of the new fact, in which case, of course, either modification or rejection of the theory is required. More often than not, however, the theory, existing as it does prior to the new discovery, does not address itself to the new observation, but is either completely silent or somewhat vague in regard to the relevance of the discovery. Regardless of the differences noted earlier, one general approach is to attempt to integrate the new data with the old theory. That was the direction taken by Rescorla and Wagner and by those theories that built on the Rescorla-Wagner model of conditioning. The other direction, exemplified by Lubow, was to start with the basic new finding, seek out those variables that modulate and limit the effect, and construct a new theory specifically designed to account for the phenomenon.

Needless to say, there is no correct choice between these two styles of theorizing, only a question of the degree of success within each approach. Even if one takes as criteria for successful theorizing such traditional standards as parsimony, integration and organization of extant data, generation of new experiments, and prediction of new data, it is not a simple task to assign differential grades to the theories. The very generality of the criteria, and even more important the absence of weights within and between categories (e.g., What is more important for a good theory – integration or prediction?), makes theory grading a very subjective exercise. In the final analysis, whereas it is sometimes easy to know when to reject an explanation, it requires considerable effort, and perhaps a fool's courage, to rank those theories that have yet to be discarded. These difficulties are, in no small part, a consequence of the fact that the requirements for the falsification of a theory are not always apparent. The conditioned inhibition explanation of latent inhibition nicely illustrates the case in which such falsification criteria are clearly available. The explanation states quite simply that latent inhibition is an example of conditioned inhibition. Critically, a conditioned inhibitor is explicitly identified as a stimulus that exhibits operationally defined properties of retardation and summation. Consequently, the experimentally derived data, which show that the procedure for producing latent inhibition does not result in a summation effect, are sufficient to conclude unequivocally that latent inhibition is not an example of conditioned inhibition. Some of the other explanations of latent inhibition are not as easily falsifiable, either because of the absence of strong, formal statements or because the consequences of the theory are tested only within the limited framework from which the theory itself was derived.

An instructive example was provided by Wagner and Rescorla, as well as others, when they concluded that nonreinforced stimulus preexposure produced the subsequent learning deficit because preexposure resulted in reduced stimulus salience or reduced attention to the stimulus. Basically, this conclusion was derived from the following facts: (1) Latent inhibition has been shown *not* to be an example of conditioned inhibition, (2) *nor* was it explainable by reference to competing responses, complementary responses, and so forth, and (3) the salience/attention decrement explanation was not contradicted by other available data. Thus, the salience/attention reduction explanation of latent inhibition can be said to have been arrived at by default. The Rescorla-Wagner model of conditioning relates this presumed loss of stimulus salience to the CS learning rate parameter, α. The model is discussed later, both because it offers an early attempt to integrate latent inhibition into a more general theoretical framework of classical conditioning and because it provides the basis for the development of other influential models that relate to latent inhibition (e.g., Mackintosh, 1975; Pearce & Hall, 1980).

Rescorla-Wagner model[11]

A major theme in contemporary theories of classical conditioning, one that relates to contingency analysis, is that learning occurs only to the degree that the US is "surprising" (e.g., Wagner, 1976). Evidence supporting this view has been obtained in a variety of experimental paradigms, but it was Kamin's demonstration (1969) of blocking that first focused attention on the heuristic value of this proposition. In the blocking paradigm, conditioning to one element, X, of a compound status AX, is severely limited if AX–US pairings are preceded by A–US pairings. Prior conditioning to A is said to block subsequent conditioning to X. To account for these findings, Kamin (1969) proposed that blocking occurs because the US is an ineffective reinforcer during compound conditioning trials; for conditioning to occur, the US must be surprising or unpredicted. In the blocking paradigm, the US is "expected" (predicted) on compound conditioning trials, AX–US, by virtue of the A–US pairings during prior conditioning. Because the US is not surprising, the associative strength of X is not incremented during compound conditioning. The notion that US effectiveness (and therefore learning) depends on the degree to which the US is surprising has been expressed formally by Rescorla and Wagner (1972) in their model of classical conditioning. According to that model, the US is effective only to the extent that it is not predicted by the total aggregate of cues that preceded it. Specifically, changes in the associative strength of a given stimulus, A, are governed by the following rule:

$$\Delta V_A = \alpha\beta(\lambda - \bar{V}) \tag{1}$$

In this equation, V represents the associative strength of stimulus A, α is a learning rate parameter representing stimulus salience, β is a learning rate parameter for the US, λ represents the limit of associative strength supportable by the US, and $\bar{V}$ represents the aggregate associative strength of all stimuli present on a given trial.

The Rescorla-Wagner model defines surprising and expected USs in terms of the value of $\lambda - \bar{V}$. The US is expected when the value of $\lambda - \bar{V}$ is small, and the US is surprising when the value of $\lambda - \bar{V}$ is large. Moreover, the model states explicitly that learning is promoted to the *degree* that the US is surprising: A change in associative strength on a given trial is a direct function of the difference between the associative strength already accrued to all stimuli present on that trial, $\bar{V}$, and the total associative strength supportable by the US, λ. In the blocking paradigm, acquisition training to one stimulus, A, increases the associative strength of that stimulus, V_A. On the compound trials, AX–US of the blocking procedure, $\lambda - \overline{V}$ is relatively small owing to the contribution of V_A to $\overline{V}$. The US is therefore, by definition, expected, and further increments in associative strength to A or X are limited.

Rescorla and Wagner (1972; Wagner & Rescorla, 1972) have shown that this model is capable of explaining an impressive variety of results from classical conditioning experiments when changes in associative strength can be reasonably attributed to variations in the effectiveness of reinforcement and nonreinforcement. However, because the nonreinforced preexposed stimulus in the latent inhibition paradigm is retarded in both excitatory and inhibitory conditioning, latent inhibition cannot be explained in terms of changes in V in the Rescorla-Wagner formula. An alternative approach suggested by Rescorla and Wagner (1972) is to account for latent inhibition by assuming that preexposure reduces the salience (α) of the CS. But as Wagner and Rescorla (1972) acknowledge, their conditioning model does not provide rules governing modifications of CS salience: "Apparently, the present model will require modifications in order to accommodate such data, perhaps along the 'attentional' lines suggested by Sutherland and Mackintosh (1971)" (Wagner & Rescorla, 1972, p. 329). Thus, a model that has provided a coherent explanatory framework for a number of empirical phenomena, while scrupulously eschewing an attentional construct, requires modification "along attentional lines" to account for latent inhibition.

Indeed, several modifications to the Rescorla-Wagner model have been proposed, as, for example, those by Frey and Sears (1978), Mackintosh (1975), Wagner (1978), and, more recently, Pearce and Hall (1980). Some of these theories directly address the problem of rules for changing salience, often substituting terms such as attention (Mackintosh, 1975) or associability (Pearce & Hall, 1980) for the salience-related α of Rescorla and Wagner.

Mackintosh's theory of attention[12]

The Rescorla-Wagner model emphasizes the US, which is effective in promoting new learning only to the extent that it is not already predicted by the total aggregate of cues that preceded it. If the US is *otherwise* unpredicted, rehearsal or processing is initiated, and associative strength accrues to the CS that immediately and reliably precedes the US.

Mackintosh's theory (1975), on the other hand, emphasizes the CS and is primarily concerned with changes in attention ($\Delta\alpha_A$), which in turn affect changes in associative strength, ΔV_A. The Mackintosh α (attention to CS) is related to the Rescorla-Wagner α (CS salience), with the important exception that for Rescorla-Wagner, the α magnitude is fixed by the physical characteristics of the stimulus (e.g., intensity), whereas for Mackintosh, α changes with experience. Indeed, the rules for the changes in α are what lie at the heart of Mackintosh's theory. For Mackintosh (1975), attention to A increases when A is a more accurate predictor of reinforcement or nonreinforcement than are other available stimuli. Similarly, attention to A decreases when A is a less accurate predictor of reinforcement or nonreinforcement than are other available stimuli. These statements are expressed formally in equations (2) and (3):[13]

$$\Delta\alpha_A \text{ is positive if } |\lambda - V_A| < |\lambda - V_X| \qquad (2)$$

$$\Delta\alpha_A \text{ is negative if } |\lambda - V_A| \geq |\lambda - V_X| \qquad (3)$$

Here, V_X represents the associative strength of all stimuli present on a given trial (i.e., the $\bar{V}$ of Rescorla and Wagner).

Although the specification of rules for changing α was prompted, in part, by the phenomenon of latent inhibition, their application to latent inhibition is problematic, because it necessitates the introduction of an asymmetry in the foregoing rules. Latent inhibition is predicted by assuming that the preexposed stimulus (A) and the context (X) are equally good predictors of trial outcomes (nonreinforcement). Mackintosh (1975) states that "it is only the phenomenon of latent inhibition that necessitates the otherwise rather unhappy assumption that α declines even when $\lambda - V_A$ is equal to $\lambda - V_X$" (p. 289). Even with the specification of the rules for changes in α, Mackintosh raises the possibility that the "phenomenon of latent inhibition would lie outside the scope of the theory – to be explained, perhaps, in terms of some simpler mechanism of habituation" (p. 290).

Rescorla and Wagner (1972), then, in regard to latent inhibition, would say that nonreinforced stimulus preexposure reduces stimulus salience, and this is reflected in a reduced α value. Mackintosh (1975) significantly expanded on this notion by providing the rules by which α changes. Interestingly, these rules are quite similar in form to the linear operator model employed by Rescorla and Wagner for describing changes in associative strength ΔV_A; see equation (1). Furthermore, because Mackintosh basically accepts this model for changing associative strength, but adds a similar equation for the learning rate parameter α, the new equation represents a two-factor theory of acquisition of associative strength with an attentional (learned) process and an associative strength growth process. As such, it is similar to the older two-stage selective attention models of discrimination learning (e.g., Zeaman & House, 1963; or its later revision by Fisher & Zeaman, 1973; Lovejoy, 1968; Sutherland & Mackintosh, 1971). The major difference is that Mackintosh (1975) forgoes the inverse hypothesis, explicitly stated in the foregoing theories, and implicitly in Rescorla and Wagner

(1972), and formalizes the conditions for changes in α (attention), as expressed in equations (2) and (3). Thus, to repeat, attention to a stimulus increases when that stimulus "predicts an otherwise unexpected reinforcer," and attention to a stimulus decreases when that stimulus "signals no change in reinforcement from the level expected on the basis of other events" (Mackintosh, 1975, p. 287).

Like the theory or Rescorla and Wagner (1972), the Mackintosh theory makes explicitly strong predictions for conditions in which there is a competition for the acquisition of associative strength (i.e., for conditions in which there are compound stimuli, such as in blocking and overshadowing). As already noted, these predictions (as opposed to the theoretical mechanisms from which they are derived) differ from those of Rescorla and Wagner primarily because of the abandonment of the inverse hypothesis. The gaining of associative strength to stimulus A in an AB compound is not accomplished at the expense of the associative strength to B. This inverse linkage is broken by Mackintosh by the assumption that λ, like α, is stimulus-specific. Lambda becomes the total amount of associative strength supportable by a given US *for a particular stimulus*. Thus, in the expression $(\lambda - V)$, the λ for US_A is different for CS_A and CS_B, and the different lambdas retain their independence. It can therefore be said that λ for Mackintosh is both US-specific and CS-specific. For Rescorla-Wagner, lambda is, of course, US-specific, but *not* CS-specific. It is this lack of CS specificity that provides the basis for the implicit inverse hypothesis of their model, the same inverse hypothesis found in most selective attention theories. Ironically, it is Mackintosh's theory that emphasizes attention, yet discards the inverse rule. Although this is not the place to evaluate the empirical evidence for the inverse hypothesis, the difference between the two theories is noted because of the implications for latent inhibition. In the section on selective attention theories, the evidence for the inverse rule in regard to latent inhibition was examined and found wanting. Clearly, however, this conclusion should not devalue either the Rescorla-Wagner theory or Mackintosh theory, because neither makes strong claims in regard to latent inhibition, and both confess to an uncertainty in the treatment of the latent inhibition phenomenon.

This uncertainty is further reflected in the Mackintosh model by the following problem. It is postulated that α may vary from 0 to +1, the value on any given trial being dependent on the history of the stimulus, as described earlier. Equations (2) and (3) provide the mechanism by which α may increase or decrease. Now consider the prototypical latent inhibition experiment. For the group that is preexposed to stimulus A, on the first stimulus presentation, α_A is zero because stimulus A is no better at predicting the absence of a US than is the context stimuli, C (i.e., $\overline{V}$, all of the other stimuli present at the same time), then $\lambda - V =$

$\lambda_C \bar{V}$,[14] thus producing a minus value for $\Delta\alpha_A$. Subtracting $\Delta\alpha_A$ from α_A, the latter of which is zero, produces a negative value for α_A. Now if α_A is negative (not to be confused with a negative $\Delta\alpha_A$) or even zero, this means that the subject is not paying attention to the stimulus on that trial. If the subject is not paying attention to stimulus A, it follows that there should be no change in α_A as a result of that trial. Consequently, when beginning the acquisition–test session, the α level of the stimulus-A-preexposed group should be identical with that of the nonpreexposed group for which A is novel. No attention, no processing, no change in α.

This problem may be overcome by reverting to the traditional assumption, as expressed by the α of Rescorla and Wagner (1972) or the stimulus intensity dynamism construct of Hull (1952), as well as others, that the physical characteristics of the stimulus also influence attention to the stimulus, as, for example, stimulus intensity or the contrast of the stimulus with the background. Indeed, Mackintosh (1975) recognized this possibility: "A comprehensive theory must specify the limits within which α may change and the conditions determining the starting value of α_X" (p. 290). He noted that the effects of stimulus salience do not allow α simply to vary from 0 to 1 on the basis of the history of reinforcement of that stimulus, and he suggested that in addition to the α value, as determined by equations (2) and/or (3), a fixed value reflecting the physical characteristics of the stimulus might be required. However, this still does not solve the latent inhibition problem. To the extent that the physical characteristics of the stimulus provide a fixed value, impervious to the effects of experience, at the onset of acquisition the stimulus-preexposed group begins with this minimum, residual α level. The nonpreexposed group likewise begins at the same point. Thus, once again, the learning rate parameter α is the same for both groups, and there should be no latent inhibition effect. The phenomenon of latent inhibition would seem to require that α be allowed to vary from -1 to 1, with the starting point for a nonpreexposed stimulus being some small, stimulus-specific, experience-free, positive value of α. In regard to the question how any learning occurs when attention is a negative value, one solution may take the following form: Discard the notion of attention when using α as a learning rate parameter, and merely substitute the word "associability." Granted that this smacks of semantic legerdemain, nevertheless, because attention and associability are interchangeable in Mackintosh's theory, at least one can obviate the psychological difficulties of dealing with negative attention and the practical difficulties of figuring out how learning ever gets started when a condition of negative attention is attached to the to-be-learned-about stimulus.

Pearce-Hall model

Pearce and Hall (1980) compared the Rescorla-Wagner and Mackintosh models in their ability to explain blocking and related phenomena, and they concluded in favor of the latter. With that as a starting point, they proceeded to build their own theory based on the major premise of Mackintosh that change in associative strength is primarily determined by modification of CS associability (as opposed to Rescorla-Wagner's emphasis on the US), which in turn is modulated by the degree to which the CS predicts reinforcement [equations (2) and (3)]. However, they stated that "there is reason to doubt the adequacy of the rules that Mackintosh suggests should govern changes in α. The inadequacy becomes apparent when we consider the application of the model to latent inhibition" (Pearce & Hall, 1980, p. 537).

In this regard, Pearce and Hall (1980) claimed that Mackintosh had no problem in dealing with the *prototypical* latent inhibition phenomenon, which is adequately covered by equation (2). Because for the stimulus-preexposed group λ, V_A and V_X are all zero, α will decline, and thus at the beginning of acquisition, α for this group will have a smaller value than for the nonpreexposed group.

Pearce and Hall deviated from Mackintosh's theory because of the latter's inability to handle the Hall-Pearce variation of the S_1–S_2 effect. Briefly, Hall and Pearce (1979) found that a series of CS–US pairings, in which the US was a weak shock, produced negative transfer when the shock intensity was raised, as compared with a group that had not been preexposed to the CS–weak-shock pairing. This effect was demonstrated with a conditioned suppression procedure and has since been replicated on a number of occasions (Ayres et al., 1984; Hall & Pearce, 1982; Kasprow et al., 1985; also see Dickinson, 1976; Pearce et al., 1982), although not with nictitating membrane conditioning in the rabbit (Ayres et al., 1984). These data were interpreted by Pearce and Hall (1980) as illustrating that associability of a stimulus can decline at the same time that associative strength of that stimulus is increasing. Mackintosh's theory (1975) cannot accommodate this effect. Because the CS in the first stage is a better predictor of the weak shock than is any other stimulus, then, according to equation (2), the α of that stimulus should be incremented, placing it at a higher value than that for the control group that did not experience the CS–weak-shock pairing. Consequently, Mackintosh would predict positive transfer to the CS–strong-shock phase, rather than the negative transfer that was actually obtained. In light of these results, Pearce and Hall (1980), while still retaining the emphasis on CS effectiveness that characterizes Mackintosh's model (1975) of conditioning, completely rejected his *rules* for governing changes of such effectiveness (i.e., the rules for incrementing and decrementing α). Rather than $α_A$ increasing in value when A predicts an otherwise unexpected reinforcer, and decreasing when it predicts no change in reinforcement "from the level expected on the basis of other events" as expressed formally in equations (2) and (3), Pearce and Hall (1980) offered a new principle for governing changes in α. In addition, they completely rejected the notion that there is *any* change in US effectiveness during the course of conditioning.

The development of their theory starts with some basic information processing concepts used by Wagner (1978). The model assumes a limited-capacity central processor in which the work of associating the CS and US is accomplished during the time that they are jointly present there. Entry into this limited-capacity processor is determined independently for CSs and USs, with USs "always likely to gain access to the the processor," whereas CS access is limited

so that only those needed for learning, gain entry. Stimuli that fully predict their consequences will be denied access to the processor, whereas stimuli that have recently been followed by surprising or unexpected events will receive processing. In other words (and in direct contrast to Mackintosh's model), we suggest that a stimulus is likely to be processed to the extent that it is *not* an accurate predictor of its consequences. [Pearce & Hall, 1980, p. 538; emphasis added]

Formally, the preceding ideas are expressed as follows:

$$α_A^n = |λ^{n-1} - V_A^{n-1}| \tag{4}$$

where $α_A^n$ represents the associability (i.e., amount of processing) of stimulus A on trial n, $λ$ is the intensity of the US on the previous trial $(n - 1)$, and V_A is the associative strength of the stimulus on the previous trial $(n - 1)$. Both α and λ may vary from 0 to 1.

Pearce, Kaye, and Hall (1983) have recognized that equation (4) provides an all too rapid reduction of associability (α). Indeed, only one nonreinforced presentation of the stimulus reduces α to zero. To overcome this problem, which violates both common sense and the empirical data, Pearce et al. (1983) replaced equation (4) with equation (5):

$$α^n = γ|λ^{n-1} - V^{n-1}| + (1 - γ)α^{n-1} \tag{5}$$

which is designed to decrease the rate of loss of α by taking into account not only the immediately preceding trial but also an "exponentially weighted moving average of all its preceding values (Killeen, 1981)." Gamma, which can vary from 0 to 1, determines the relative contributions of trial $n - 1$ and the other trials to the value of α.

Associability (α) is reflected in associative strength by way of the equation

$$ΔV_A = S_A α_A λ \tag{6}$$

in which an increment in associative strength is governed by the values of CS intensity (S_A), US intensity (λ), and, of course, associability (α). Equation (6) explicitly acknowledges that there is a contribution of CS and US intensity to associative strength that is independent of the contribution to associability.

The net result of Pearce and Hall's model is that as conditioning proceeds (at least once beyond a problematic first trial), associability of the CS will decrease. At the onset of conditioning, α will be large, close to the value set by λ, but as trials progress, the associative strength of A increases, and therefore the value $(\lambda - V_A)$ decreases, and α declines. In this manner, Pearce and Hall accounted for the poor learning of the CS–weak-shock preexposure group as compared with the nonpreexposed group (e.g., Hall & Pearce, 1979).

The more basic latent inhibition effect is treated as follows:

When a novel stimulus is presented in the absence of a reenforcer, λ will be 0, and, since the associative strength of the stimulus is also 0 the associability of the stimulus will decline. When the stimulus is subsequently paired with a US, no learning will occur on the first conditioning trial, and in this way a difference will be established between the associative strength acquired by a preexposed stimulus and that acquired by a novel stimulus. [Pearce & Hall, 1980, p. 539]

In summary, latent inhibition is produced by a decline in α that occurs as a result of the CS *fully predicting* its consequences. A series of trials of CS alone or CS–US pairings fully predict what follow the CS, in one case nothing, and in the other, for example, shock. What was not clearly accounted for in the CS–weak-shock studies was the fact that although the CS–weak-shock manipulation produced latent inhibition, it produced *less* latent inhibition than in the CS-alone group. This important result was found in the studies of Hall and Pearce (1979), Kasprow et al. (1985), Pearce et al. (1982), and Schachtman et al. (1987), we well as in a study by Dickinson (1976) in which there was transfer from a CS–food, CS–alone, or nonpreexposure stage to a CS–shock stage. To account for this ordering of data with the Pearce-Hall model, one would have to assume that a series of CS-alone trials makes the CS a better predictor of "no consequence" than does the same number of CS–weak-shock (or any CS_1–S_2) pairings make the CS a predictor of weak shock. Intuitively it would seem to be quite the opposite; namely, learning a negation should require more trials than learning an affirmation. (Certainly the data on the feature positive effect,[15] which in a sense pits negation against affirmation, indicate that an organism more readily uses the former than the latter.)

Nevertheless, Pearce and Hall offer some interesting experimental designs to test their predictions. If, as they posit, associability (α) decreases as predictability increases, it should be possible to restore associability to a CS that has lost that associability, as a result of either the traditional CS-alone or the CS–weak-shock procedure, by introducing a trial(s) on which the CS is followed by a surprising (i.e., nonpredicted) event. Indeed, Lantz (1973) reported an attenuation of normal latent inhibition when, after a series of CS-alone trials, the CS was followed by a loud, unexpected noise. Hall and Pearce (1982, Exp. 1 and 2) reported a similar finding for the CS–weak-shock preparation, when the weak shock was unexpectedly omitted on two trials that immediately preceded the CS–strong-shock acquisition phase.

These data would seem to offer strong support for the Pearce-Hall (1980, 1985) model, as opposed to the Wagner (1976, 1978) model, because in the former it is the prediction of consequences of the CS that is central, and in the latter it is how well the CS is predicted, as, for example, by the context. Nevertheless, Hall and Pearce (1982) recognized that their data might be interpreted in such a way that the omission of the weak shock would be considered to be a change in context, and consequently associability would be restored to the CS because the context would no longer prime the CS in short-term memory. These two interpretations, however, can be differentiated on the basis that the Pearce-Hall model is CS-specific, whereas the Wagner model is context-specific. Consider the following: In the CS–weak-shock procedure, two CSs, *A* and *B*, are each, separately, paired with the weak shock. Then, just before the test, half of the group are given a few *A* trials with the US omitted, and half are given a few *B* trials with the US omitted. In the test phase, half of each group are conditioned with the same stimulus for which the US was omitted, and half are conditioned with the stimulus that was different from the one that was omitted. The predictions are clear. Pearce and Hall would opt for restoration of associability only to that stimulus for which the US was omitted. On the other hand, Wagner would predict that the change in context should restore conditionability to both stimuli (if he were indeed to accept the omission of weak shock as a change in context). Hall and Pearce (1982, Exp. 3), in fact, ran such a study and reported data that supported their position. However, data from the foregoing experimental design can provide only evidence that the stimulus-specific effect is larger than the "context change" effect. The design allows nothing to be said about whether or not there is *also* a context change effect. To assess the latter, it would be necessary to employ comparison groups in which there was no context change. As reviewed in the empirical chapters, there is abundant evidence that context change from preexposure to conditioning *does* attenuate latent inhibition (i.e., restores associability). That Hall and Pearce (1982) demonstrated a stronger effect of a CS–predicting manipulation than of a predicting–CS manipulation (and the latter within a very limited definition of context) should not obscure the fact that their theory does not encompass the very real context effects that have been extensively documented in earlier chapters.

The CS–weak-shock effect is a core prediction for the Pearce-Hall model. Therefore, the logic of the designs that have been used to produce supporting data deserves close scrutiny. In the first paper that demonstrated the Hall-Pearce effect (Hall & Pearce, 1979), a three-group design was employed; (1) no preexposure, (2) CS preexposure, and (3) CS–weak-shock preexposure. In terms of test performance, the groups ranked 1 > 3 > 2, which demonstrated normal latent inhibition (1 > 2) and, according to the authors, latent inhibition for Group 3 (1 > 3), thus apparently confirming the theoretical prediction that associability can decrease even as associative strength increases. But conclusions drawn from comparing Groups 1 and 3 must take into consideration that Group 3 differed from Group 1 on two counts: In Group 3, as compared with Group 1, the CS predicted weak shock, *and* the CS was more familiar. Consequently, differences between the two groups can be attributed to either prediction or familiarity. On the other hand, it will be remembered that Group 3 showed less negative transfer than Group 2. Here, however, note that both groups had the same amount of CS familiarity, and one might argue that the CS *equally* predicted its consequences in both groups: absence of consequences in Group 2, and weak shock in Group 3. That, in spite of no differences in either CS familiarity or predictability of CS consequences, the two groups were conditioned at different rates (CS–weak-shock faster than CS alone) suggests that the theoretical meaning of the CS–weak-shock manipulation must be reexamined.

Indeed, it is possible to design experiments that will directly compare the contrasting predictions derived from familiarity with those derived from the Pearce-Hall conditioning model – namely, that a CS loses its associability to the extent that it comes to fully predict its consequences. To test the different predictions, a set of experiments can be constructed that in Phase 1, across groups, will vary CS duration (i.e., familiarity) and keep constant the number of trials (predictability), or will vary the number of trials beyond learning asymptote (CS familiarity varied, but predictability constant). Although these designs might appear to be straightforward tests of the positions, there simplicity is misleading. For the first example, finding a familiarity effect would not irreparably damage the Hall-Pearce theory. Why should CS associability not also be determined by temporal factors between the CS and the shock that it predicts, or by CS duration? As to the second example, the manipulation is weak because it is not clear, theoretically, that asymptotic conditioning performance is equivalent to full predictability.

There are stronger manipulations that could be made. For instance, one could compare the effects of Stage 1 S_1–S_2 presentations where S_2 is always the same (quality, intensity, duration) and the effects of S_1–S_2 presentations where S_2 is varied. Clearly, in the preceding example, S_1

familiarity is constant, but predictability is not. Although one could predict, as a result of these two conditions, learning rate differences in the subsequent S_1–US acquisition stage, the direction of the differences might be dependent on the number of Stage 1 trials. Indeed, the basic Hall-Pearce effect might be modulated by number of Stage 1 pairings before asymptotic learning. One should entertain the very reasonable possibility that associability first increases and then decreases, with perhaps the state of full predictability being arrived at considerably before reaching asymptotic performance measures of associative strength.

Frey-Sears model

Frey and Sears (1978), like Mackintosh (1975) before them, recognized that the inability of the Rescorla-Wagner theory to handle latent inhibition is a result of the learning rate parameter, α, being a fixed quantity. They extended Rescorla-Wagner with an idea developed by Kirk (1974), by which α_A increases with reinforcement of stimulus A and decreases with nonreinforcement of A. This notion, they pointed out, is consistent with Mackintosh's treatment of α, and similarly it avoids the assumption of an inverse hypothesis. Frey and Sears (1978) stated that a computer simulation of Kirk's model was successful in "encompassing" latent inhibition. Nevertheless, they proposed a modification of Kirk's "dynamic attention" rule, one that retains most of the positive features of the Rescorla-Wagner model. Starting, then, with the basic Rescorla-Wagner equation, $\Delta V_i = \alpha\beta(\lambda - \overline{V})$, Frey and Sears proposed the following for the modulation of α:

> The central notion of this rule is that α should represent the information value of a stimulus as reflected by its recent associative strength. On a reinforced trial, the change in α is represented formally as
>
> $$\Delta\alpha_i = \varphi \cdot (E_r - \alpha_i) \qquad (7)$$
>
> where α_i is the stimulus-specific learning rate parameter for stimulus i, φ is a rate parameter for changes in attention, and E_r is equal to the associative strength of stimulus i, when V_i is greater than E_0 and is equal to E_0 otherwise. The term E_0 represents a minimum target value for α_i on reinforced trials and is always positive (i.e., $0 < E_0 < \lambda$).
>
> On nonreinforced trials, the change in α is represented as
>
> $$\Delta\alpha_i = \varphi \cdot (E_n - \alpha_i) \qquad (8)$$
>
> where α_i and φ are defined as above, and E_n is equal to the associative strength of stimulus i when V_i is positive and is equal to zero otherwise. [Frey & Sears, 1978, p. 323]

Associability (or attention), then, according to Frey and Sears (1978), is incremented, as in other theories of this type, according to a discrepancy rule. Here, however, the discrepancy is between the associative strength of the CS as acquired from previous trials and α itself. According to Frey and Sears, this formulation (together with the catastrophe rule[16] for converting associative strength to performance) has been used successfully in simulation experiments to produce a number of learning phenomena, including latent inhibition. Mackintosh (1983) interpreted the Frey and Sears (1978) position as being similar to that of Lubow, Schnur, and Rifkin (1976): "if a stimulus is presented alone without reinforcement, animals will learn that it signals no event of consequence [and] a signal that has previously signalled nothing will only with difficulty be established as a signal for reinforcement" (p. 230).

Wagner's priming theory of conditioning and habituation

All of the cognitive theories of conditioning and latent inhibition derive from the notion that information processing is some function of the current discrepancy between presented event(s) and expected event(s). Although the specific formulations are expressed in a variety of equations, with gross qualitative differences in terms of which events are important and even the direction of the relationship between discrepancy and associability (inverse or direct), they all trace a path back to the Rescorla-Wagner formulation [equation (1)]. Therefore, it was only appropriate for Wagner himself to confront the problem of latent inhibition, one that essentially was not addressed in the original model.

Indeed, Wagner's priming theory (1976, 1978) was designed to cover considerably more than just latent inhibition, in particular, habituation and classical conditioning phenomena. In its most general form, the theory begins by accepting the Rescorla-Wagner linear operator description (or some variation of it; see Wagner, 1978, p. 206) for incrementing associative strength. Thus, the amount of stimulus processing or conditioning is a direct function of the degree of surprise, where the expectancy–surprise continuum is defined in terms of "the degree to which the US can be assumed to be predicted by the total aggregation of cues which precede it" (Wagner, 1978, p. 183). However, in the priming theory, the naked Rescorla-Wagner equation is dressed in the information procesing clothing used so extensively by cognitive psychologists. Thus, the information processing schema consists of four units: a sensory register, short-term memory (STM), long-term memory (LTM), and response generator. The important operations, those for generating a discrepancy signal, are controlled by the two memories. LTM is conceptualized as "a number of representative elements [so-called gnostic units, idea, logogens, images, etc.] interconnected via an associative network" that has unlimited capacity (Wagner, 1978, p. 178). STM is defined as the set of elements that "are currently active." As usual, STM is considered to have a limited capacity, and representation in STM is unstable. Representations in STM may decay with time and/or be displaced by new representations. Within STM it is possible to maintain the activity of a particular representation. This maintenance is called rehearsal, and it has a numer of consequences, the most important of which is to reduce access to STM by other incoming stimuli and to promote processing of the rehearsed stimulus. Finally, "permanent associative connections are developed between representative elements only to the degree that the elements are *jointly* active in STM" (Wagner, 1978, p. 180; emphasis added). As in Sokolov, a critical stimulus comparator-like process determines the subsequent fate of that stimulus. For Wagner, the comparison process occurs in STM. A stimulus that is inputted to STM is matched against an output from LTM. A mismatch results in rehearsal (i.e., further processing); the absence of a mismatch produces no further processing elaboration. In addition to this particular circuit, labeled *retrieval-generated priming* because the source of the comparison is from LTM, there is a different circuit that is called *self-generated priming*. Here the comparison stimulus is drawn from stimuli that were immediately preceding in time and are retained in STM. In both cases, whether dealing with self-generated priming or retrieval-generated priming, a mismatch between incoming and stored information results in a discrepancy signal that initiates further processing.[17] Only stimuli that are unexpected receive further processing.

Within Wagner's theory, an important role is advocated for contextual stimuli. When a repeated stimulus is presented within a particular context, an association is formed between that stimulus and the context stimuli, such that when the context alone is presented, a representation of the target stimulus will be primed (via retrieval from LTM) in STM, and thus, of course, the representation will provide the basis for comparison with the next incoming target stimulus.

Considering the foregoing formulations, Wagner accounted for latent inhibition in the following manner: When a subject is preexposed to a stimulus (S_1), an association develops between that stimulus and the context stimuli provided by the apparatus. At the time of the acquisition–test, the subject is placed in the same apparatus, and the context stimuli prime (from LTM) a representation of S_1 into STM. As a result of this priming, when the S_1–US trial occurs, S_1 is already in STM and, by definition, is expected (predicted). Therefore, S_1 does not receive any further processing; that is, S_1 and US are not conjointly rehearsed, or at least are only weakly rehearsed. Consequently, the association S_1–US is incremented more slowly in the stimulus–preexposed group than in the nonpreexposed group, in which there was no priming of S_1 by context.[18]

Wagner's priming explanation of latent inhibition produces several rather direct deductions. In particular, (1) latent inhibition should be context-specific, and (2) extinction of the presumed stimulus–context association should attenuate latent inhibition. Furthermore, because the context–S_1 association is an instance of associations in general, the same variables that are proposed to affect DS–US associations should also affect context–S_1 associations. Although Wagner did not explicitly make such a statement,[19] logically it would seem to be a necessary asumption. It follows from this assumption that the various procedures that have been used to support the Wagner model, as applied to CS–US associations, should have similar effcets when applied to context–S_1 associations. These procedures include (1) using distractor stimuli immediately after the CS–US pairing in order to interfere with rehearsal and (2) providing a better predictor of the US than the nominal CS, as in blocking, or, instead of this retrieval-generated priming of the US, reaching the same condition via self-generated priming. It should be quite clear that a number of these predictions are not easily translated into experimental manipulations because of obvious qualitative differences in the operations that specify the CS–US relationship, as compared with the context–S_1 relationship, where context, unlike CS, is not a punctate stimulus.

The strongest predictions, and the ones that have been subject to test, involve context change and context extinction. Let us examine the evidence.

Context change. Because the model requires a context–S_1 association during preexposure, such that at acquisition–test the *same* context will prime (from LTM) the S_1 and thus prevent conjoing rehearsal with the US, it follows directly that if the acquisition–test environment is *different* from the stimulus preexposure environment, there should be a potent attenuation of latent inhibition. Such context specificity of latent inhibition has been extensively documented in conditioned suppression, conditioned taste aversion and discrimination learning (see chapter 3). Furthermore, even when the acquisition–test environment is familiar, but the familiarity is achieved independent of the stimulus preexposure, there is still a sharp reduction or absence of latent inhibition (Channell & Hall, 1981; Dexter & Merrill, 1969; Hall & Channell, 1986; Hall & Minor, 1984; Kaye, Preston, Szabo, Druiff, & Mackintosh, 1987; Lovibond et al., 1984).[20]

Context extinction. If during stimulus preexposure a context–S_1 association is formed, then this association should be weakened by an extinction procedure in which the organism is subsequently placed in the context without S_1. Such manipulations have been made in several experiments. Wagner reported positive results, namely, that a context-alone phase that followed a context–S_1 phase restored the associability of S_1 in the subsequent CS–US pairing test. Wagner reported such data for both rabbit eyelid conditioning and conditioned suppression procedures (Wagner, Pfautz, & Donegan, 1977, reported in Wagner, 1979 p. 67). However, a conditioned suppression study by Baker and Mercier (1982) provided only weak support for this contention, with the effect appearing only under very restrictive conditions. Hall and Minor (1984), who also used conditioned suppression tests, failed to find any context extinction effect.[21] (For a fuller discussion of these experiments, see chapter 3.)

Relatedly, studies by Hall and Minor (1984, Exp. 6) and Hall and Channell (1985a, Exp. 1, 2, and 4) that preexposed the context prior to preexposing the stimulus in that same context indicated that that procedure potentiated the latent inhibition effect. Although Wagner did not address this type of manipulation directly, it would appear that, if anything, his theory would predict quite the opposite.

In summary, the two sets of context studies, context specificity and context extinction, provide ony partial support for Wagner's theory, with the predicted effects occurring reliably ony in the former. However, it should be noted that these studies addressed only one aspect of the Wagner priming theory – that of context–S_1 associations. These experiments did not address the core of the theory, which concerns priming itself. Nevertheless, the data suggest that even if the priming mechanism concept is viable the assumption that context–S_1 associations are sources of priming effects that can account for latent inhibition would appear to be unwarranted.

Priming tests. Tests of the efficacy of priming as an explanation of latent inhibition have been provided more directly by studies that have manipulated "distractor" stimuli. Again, the experimental logic begins with the assumption of a context–S_1 association. However, instead of disrupting that association by means of context changes or context extinction, the disruption is produced by punctate stimuli that presumably operate, according to Wagner, by interfering with rehearsal and thus mitigating subsequent priming. Indeed, a series of experiments by Wagner and his associates provided elegant support for the theory as it relates to classical conditioning and habituation (Primavera & Wagner, cited in Wagner, 1976, pp. 200–201; Terry, 1976; Wagner, Rudy, & Whitlow, 1973; Whitlow, 1975). These studies provide considerable evidence for the idea that distractors interfere with the STM representations that are the basis for associative learning.

In regard to latent inhibition, those studies that showed the S_1–S_2 effect (see chapter 3) can easily be interpreted to fit the Wagner model. Thus, when S_2 immediately follows S_1 in the preexposure phase, it should interfere with the rehearsal of context–S_1, and thus in the test the context should not prime S_1. Consequently, there should be less latent inhibition in the S_1–S_2 group than in a S_1-alone group. As detailed earlier, such data have been reported (e.g., Lubow, Schnur, & Rifkin 1976; Szakmary, 1977a,b), indeed extensively if one includes the Hall-Pearce effect (e.g., Hall & Pearce, 1979). However, a series of studies by Baker and Mercier (1984) and Mercier and Baker (1985) using a variety of S_1–S_2 procedures, including one in which S_1 was predicted by S_2 (S_2–S_1), all failed to show attenuation of latent inhibition, as compared with an S_1 unpaired group. It is important to note that those studies all used long-duration intervals between onset of S_1 and onset of S_2 (ISI). Clearly, within the framework of conditioned attention theory, this could be a possible source for the ineffectiveness of the manipulation (cf. Lubow et al., 1975). The theory states that S_2 maintains attention to S_1 as a result of a conditioning relationship between them; as such, the interstimulus interval is a critical parameter. Similarly, for priming theory, the interstimulus interval would be important. The fact that some of Baker and Mercier's conditions involved a simultaneously compounded S_1 and S_2, with still no attenuation of latent inhibition, might be considered more troublesome for Wagner, particularly because Baker and Mercier independently demonstrated an S_1–S_2 association, which should have subtracted from the amount of context–S_1 association. They claimed that this prediction followed from the characterization of STM as having limited capacity. However, without reference to the extent of the limitation, it would seem unfair to use these data to refute Wagner on this point. Indeed, it is rather arbitrary to hold that this limited-capacity memory can operate on two events (i.e., CS–US or context–S_1 associations), but not on three events.

Other considerations. In addition to the equivocal empirical support for the Wagner model, there are some apparent internal inconsistencies in the model, at least as it applies to latent inhibition. Specifically, there are two problems: The model can be questioned as an account of the loss of associability during preexposure, and as an account of the transfer of that loss to subsequent conditioning. With regard to the loss of associability during preexposure, how is a context–CS association learned if priming in STM prevents rehearsal? The ubiquitous context is always present and available to STM via self-generated priming. From this it follows that the context is already primed when the CS is preexposed, and conjoing of context–CS processing should be vitiated.

The mechanism for explaining the subsequent interference with conditioning during CS–US trials is also questionable. Why cannot a primed representation of the CS in STM be processed conjointly with the presented US? As Wagner (1978) has assumed, learning involves the development of associative connections between *representations* of stimulus elements that are jointly active in STM. That is, associations are always formed between representations of events, not the events themselves, so why whould it matter whether the event enter the central processor via sensory channels or via retrieval from LTM (cf. Estes, 1959)? Thus, on conditioning trials, the context retrieves the CS from LTM to prime a CS representation in STM, and the US representation enters STM via the usual sensory channels. In any case, CS and US representations should be processed concurrently in STM, and conditioning sould proceed apace. At best, the Wagner model, as applied to the conditioning phase of the latent inhbiition paradigm, would suggest that the presented CS is redundant. But if the CS is not processed via the usual sensory channels because it is *already represented* in STM via retrieval-generated priming, then conjoint CS-US rehearsal seems assured.

Evaluation of prediction theories

The major theories outlined in the previous section share a common denominator, namely, they agree that associative learning is governed by some predictive relationship between the CS and US. The theme of prediction appears in a number of apparently distinct variations. Mackintosh (1975, 1978) claimed that the effectiveness of a CS in entering into an association with a particular US or reinforcer is a positive function of how well that CS has predicted the US in the past. Associability increases with the predictive power of the CS. Pearce and Hall (1980) assumed exactly the opposite: Associability decreases with an increase in the ability of the CS to predict its consequences. Wagner (1976, 1978), on the other hand, asserted that CS effectiveness is a function of how well the CS has been predicted by the present context. Unlike either Mackintosh or Pearce and Hall, who emphasized, albeit in opposite directions, the earlier *consequences* of the CS, Wagner stressed the earlier *antecedents* of the CS. In the former cases, the focus is on how well the *CS predicts*, and in the latter, on how well the *CS is predicted*.

Although one may try to determine which of these formulations of prediction conforms most closely to the data, and clearly there are differential tests of the various positions, there remains a concern about the logic of the formulations, in regard to the use of the explanatory term "prediction." In its most general form, the phenomenon that is to be explained is that of the association of two events, A and B. To explain that association by referring to some statement about how well A does or does not predict B, or how well A is predicted by C, would seem to beg the question. Prediction, as a psychological phenomenon, itself depends on association. Is there any meaning to the sentence "A predicts B", without A also having entered into an association with B? To the extent that A predicts B (or the absence of B), A is associated with

B. To the extent that A does not predict B (nor the absence of B), A is not associated with B.

The metatheoretical problem that we encounter here is analogous to the one described by Johnston and Dark (1986) in their discussion of the differences between causal and effect theories of selective attention.

> This explanation introduces an infinite regress because the same questions that were asked about how individuals pay attention now have to be asked about how the attention director pays attention. In tracing the history of atomic theory, Heisenberg (1958) notes that "Democritus was well aware of the fact that if atoms should . . . *explain* the properties of matter – color, smell, taste – they cannot themselves have these properties" (p. 69). Likewise, if a psychological construct is to explain the intelligent and adaptive selection powers of the organism, then it cannot itself be imbued with those powers. [Johnston & Dark, 1986, p. 68]

In addition to a metatheoretical problem, there are empirical phenomena that are not congruent with any of the formulations that rely on predictability. The importance of stimulus intensity in simple conditioning, as well as in modulating effects in blocking, overshadowing and latent inhibition, though acknowledged in various ways as affecting learning rate parameters, is perhaps insufficiently emphasized in terms of its possible significance in understanding the process of association. Similarly, the facts of one-trial learning, as in conditioned taste avoidance and conditioned suppression, and one-trial latent inhibition, as in conditioned taste avoidance, are difficult to accommodate within the framework of a theory that emphasizes prediction (although not impossible). Likewise, the various context effects, and the fact that the duration of the preexposed stimulus can be exchanged for the number of preexposures in affecting latent inhibition, would seem to set limits on the generality of the prediction construct. Indeed, the foregoing emphasis on the physical characteristics of the stimuli, A and B, rather than on the contingency between them, is not to deny the importance of such a contingency in the development of an association. The evidence is clearly in favor of such a notion. Nevertheless it should remind us of two points: (1) The older laws of association, mechanistic or not, are not to be discarded because of the newer cognitive emphasis on prediction. (2) Predictability as an explanatory concept must be firmly anchored in experimental manipulations. If predictability is to help us understand how two events become associated, it must have an independent, operationally defined existence of its own, one that clearly differentiates between predictability as a causal variable and predictability as an effect.

The various equations in the preceding sections should not mislead the reader on this point. These equations define the *direction* of predictability (i.e., whether A is predicting B or B is being predicted by A) and the consequences of such predictability (i.e., increased or decreased associability), but the equations do not provide a set of operations for manipulating the degree of predictability. (Might it be useful to consider that only two states exist – predictable and not predictable?) This shortcoming is somewhat alleviated by the fact that there is sufficient definition within the system to allow for at least gross ordinal statements. Thus, within the blocking paradigm, the stage 1 A–US procedure sets A up as a better predictor of the US than B when it is added in the second stage, AB–US. But the question can be raised, once again, concerning the point in the stage 1 series of A–US pairings at which A comes to predict the US. Operationally, a single prior pairing makes A a better predictor of the US than the novel B in stage 2. Yet clearly one has to differentiate between predictability as *operationally* defined and predictability as an organismically *acquired* relationship. This same problem would appear to lurk in those accounts of latent inhibition that utilize predictability as an explanatory concept.

Three new hypotheses

The information processing theories in the previous sections focus on the *predictive* relationship between the preexposed stimulus and some consequence, placing latent inhibition within a larger framework of a general conditioning theory. Three recent theories, retrieval failure, trace conditioning, and inhibition of delay, are conceived in an entirely different fashion. One, trace conditioning, is a hypothesis specifically designed to explain latent inhibition. According to the theory, stimulus preexposure produces a *functional* trace conditioning procedure in acquisition. The second approach, retrieval failure, includes latent inhibition within the general context of a large variety of learning deficit paradigms, including overshadowing and blocking. The theory claims that these deficits are not due to reduced stimulus associability, but rather to an inability of the organism to retrieve the stored information. These two theories, as well as the inhibition-of-delay explanation, provide the most recent theoretical contributions to the latent inhibition literature, and are described next, together with relevant data and critical analyses.

Retrieval failure

Recent experiments by Miller and his associates have suggested that a variety of paradigms that purport to produce learning decrements can be best understood in terms of retrieval failure (Miller, Kasprow, & Schachtman, 1986). Thus, such phenomena as overshadowing (Kasprow, Cachiero, Balaz, & Miller, 1982), blocking (Balaz, Gustin,

Cacheiro, & Miller, 1982), and latent inhibition (Kasprow, Catterson, Schachtman, & Miller, 1984), traditionally treated as failures of the target stimulus to *acquire* associative strength, have been reinterpreted by these authors. The relatively poor test-phase performance is viewed as a function of the animal's inability to *retrieve* the information encoded during the acquisition phase, rather than a failure of association during the acquisition phase. The evidence for this contention is provided by a set of experiments indicating that the normal deficit produced by the experimental treatment is reversible. Reversibility, at least if it cannot be attributed to changes in motivation or to relearning, is the criterion for implicating retrieval failure, rather than acquisition failure, as the source of the test performance deficit.

For latent inhibition, the claim would be that nonreinforced preexposure to the target stimulus has no effect on the future associability of that stimulus (as would be claimed by the major theories discussed in the preceding sections), but rather that such preexposure proactively interferes with the subject's ability to retrieve the information (in the third, test stage) from the acquisition session (i.e., the CS–US association). Although no specific mechanism of interference is supplied, the logic for demanding a retrieval failure, as opposed to an acquisition failure, explanation of latent inhibition, at least for part of the data, is reasonable to the extent that there is evidence that latent inhibition is, indeed, reversible by a postacquisition treatment. There are two types of experiments that are relevant to the problem: the effects of reminder cues (Kasprow et al., 1984) and the effects of delay between acquisition and testing (Kraemer et al., 1988; Kraemer & Ossenkopp, 1986; Kraemer & Roberts, 1984).

Reminder treatments. Kasprow et al. (1984) used a four-stage procedure with conditioned suppression to demonstrate that a reminder treatment can reverse the latent inhibition effect. Stages 1 and 2 consisted of stimulus preexposure (or simply apparatus preexposure for the control group) and CS–US pairings, respectively. In stage 3, the reminder treatment, two presentations of the shock US were delivered in an apparatus different from that used in the other stages. The degree of suppression to the CS was evaluated in the fourth stage. The postacquisition reminder treatment completely abolished latent inhibition. These results would seem to suggest that associative strength to the target stimulus in the second stage is acquired normally, but cannot be displayed in the test phase because of a retrieval failure, and that failure can be overcome with an appropriate reminder treatment.

These data are certainly provocative and do not seem to be easily accommodated by other theories of latent inhibition. However, the interpretation of Kasprow et al. (1984) should be accepted with some caution. There appear to have been several problems in the logic of the experimental design (problems that may also be applicable to their studies of overshadowing and blocking). Because the reminder treatment was applied in stage 3, might it not be acting on stage 1 rather than stage 2, as Kasprow et al. (1984) assumed? Although it is difficult to envisage a mechanism by which this would occur, it remains a logical possibility. However, even if true, it would not necessarily invalidate the retrieval failure hypothesis.

More important, the theory explicitly recognizes that the reminder treatment must be an element of the presumed-to-be-acquired but difficult-to-retrieve association. Thus, in their previous studies of blocking and overshadowing, CS, US, and context elements were successfully used as reminder cues. In the 1984 experiment, these authors noted that they were unable to produce reminder effects with CS and context. This finding may be related to the fact that CS and context are common to both the preexposure and acquisition phases, whereas the US is present only in the acquisition session. Therefore, CS and context reminders may serve to retrieve either the memory of the CS-alone experience or that of the CS–US pairing, or some interaction between them. As an additional point, these data suggest that the effect of the US may *not* be specific to its role in stage 2 (i.e., any strong postacquisition stimulus may attenuate latent inhibition). Clearly, if that is the case, the reminder interpretation will not be applicable, although, again, a general retrieval failure interpretation could still be viable.

Effects of delay between acquisition and test. As indicated earlier, *any* reversal of latent inhibition that follows the CS–US acquisition phase may provide evidence for retrieval failure, as opposed to acquisition failure. If, for example, latent inhibition is found after a short delay between the acquisition stage and the test stage, but not after a long delay, this is presumptive evidence that with the short delay the CS–US association was present, but not manifest, and that something happened during the longer time period that allowed the originally encoded association to be retrieved. In such a situation, one need not specify what has happened in order to identify the short-delay performance decrement as being due to some type of retrieval failure.

What, then, is the evidence for attenuation of latent inhibition with an increase in acquisition–test delay? Three studies addressed this issue (Kraemer et al., 1988; Kraemer & Ossenkopp, 1986; Kraemer & Roberts, 1984), and all used the conditioned taste aversion procedure. Kraemer and Roberts (1984) reported a marked reduction in the amount of latent inhibition when the test phase followed the acquisition phase by 21 days, as compared with 1 day, at least for some preexposed stimuli. Similar results were reported by Kraemer et al. (1988) for

12-day-old rats when using a 10-day interval. On the other hand, Kraemer and Ossenkopp's (1986) data, with adult rats, showed that groups preexposed and conditioned to the same flavor exhibit the same amount of latent inhibition when tested 1 or 21 days after acquisition.

Clearly, it would be useful to have some parametric data on the effect of acquisition–test interval, particularly in some paradigm other than conditioned taste aversion. The data reported earlier, which were used to support the retrieval failure hypothesis of latent inhibition, specifically *increased* flavor aversion with lengthening of the acquisition–test interval, may reflect the reappearance of neophobia rather than reappearance of the CS–US association. Kraemer and Roberts (1984) did report some control procedures designed to rule out a neophobia reactivation interpretation. One set of groups, stimulus-preexposed and not preexposed, did not receive a CS–US pairing. Whereas one might expect that a group with a long delay between preexposure and test would show a recovery of neophobia, thus producing reduced intake of the preexposed flavor independent of any CS–US associative effect, such data were not obtained. However, the neophobia interpretation may still be valid. A control procedure in which the CS and US are presented unpaired will be necessary to check for the possibility that the US sensitized the neophobic response.

In addition to the relative paucity of evidence concerning the retrieval failure explanation of latent inhibition, there are several other problems that the theory faces. The theory is stated in such a way that it is difficult to test. Thus, whereas any postacquisition treatment that reduces latent inhibition would seem to support some version of retrieval failure, the absence of such an effect does not allow one to conclude that the retrieval failure hypothesis is faulty. This problem arises because there is no independent definition of the conditions for producing an effective reminder treatment, nor is there a proposed mechanism for describing how reminder treatments work in latent inhibition. The difficulty is highlighted by the findings of Miller and associates. On the one hand, CS, US, and context reminders are effective in overcoming overshadowing and blocking deficits (Balaz et al., 1982; Kasprow et al., 1982), but only US reminders are effective for reversing latent inhibition effects (Kasprow et al., 1984).

A second problem concerns the fact that the retrieval failure hypothesis, though easily applied to the analysis of three-stage latent inhibition procedures (preexposure, acquisition, test), where postacquisition treatments can be applied and *subsequently* measured, would not seem to be suitable to the analysis of two-stage latent inhibition procedures (preexposure, acquisition–test), in which acquisition performance and test performance for latent inhibition are one and the same. Typically, such experiments show degraded performances for the stimulus-preexposed group *from trial 1 onward*, a finding that suggests, almost by definition, reduced associability of the preexposed stimulus (i.e., a failure of acquisition itself, rather than a failure of retrieval from acquisition).

It is, of course, possible that both effects occur. Kasprow et al. (1984) recognized that when they stated that their retrieval hypothesis was not necessarily incompatible with other theories of latent inhibition, as, for example, conditioned attention theory. Thus, stimulus preexposure may result in reduced attention during the CS–US acquisition stage, and it is this lowered level of attention during acquisition that interferes with subsequent retrieval in the test, although not affecting associative strength during the CS–US acquisition phase. In this regard it should again be noted that, indeed, the retrieval failure hypothesis is, by itself, incomplete, in that it does not address itself to a mechanism by which stimulus preexposure affects performance. Effective reminder treatments remain undefined, and the effects of variables that are known to influence the degree of latent inhibition, such as number of preexposures, duration of stimulus preexposure, context, and so forth, remain unexplained. It is the attempt to carry this latter burden that requires, at a minimum, a marriage between the retrieval failure hypothesis and one of the theories that view latent inhibition primarily as an example of an associability deficit.

Trace conditioning hypothesis

DeVietti et al. (1987) have recently suggested that latent inhibition may be an instance of a trace conditioning effect. The main thesis is based on the following assumptions and reasoning: A "normal" (i.e., nonpreexposed) stimulus is functionally composed so that property X (which psychologically may be related to salience or attention, and may be indexed by heart-rate change or some other OR measure) is distributed more nearly equally across its duration than is property X across the duration of the preexposed stimulus. Stimulus preexposure serves to shorten the duration of property X, which, with repeated presentations, moves closer and closer to stimulus onset. The net effect is that for the stimulus-preexposed group, at the time of the acquisition stage, X is at some temporal interval from US onset, whereas X for the nonpreexposed group, which in acquisition receives the CS for the first time, is contiguous with (or at least closer to) US onset. As a result, the stimulus-preexposed group at the time of conditioning is "functionally" in a trace conditioning situation, whereas the control group is in a situation with more effective conditions for acquisition. In other words, latent inhibition occurs because stimulus preexposure creates a functional trace conditioning paradigm that is less effective for conditioning than is delayed conditioning.

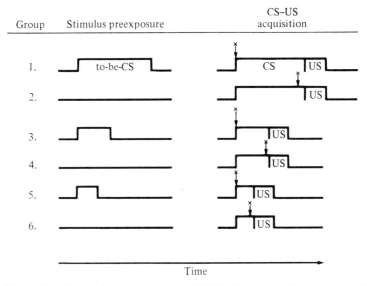

Figure 11. Schematic representation of DeVietti's trace conditioning hypothesis of latent inhibition. Groups 1, 3, and 5 were preexposed to stimuli of different durations. Groups 2, 4, and 6 were the respective control groups, not preexposed, but conditioned with same-duration CS as their matching stimulus-preexposed groups. Conditioning improved as the interval between X and the US decreased.

It follows from DeVietti's analysis that the amount of latent inhibition should be a positive function of the duration of the preexposed stimulus, at least when stimulus duration is the same in the preexposure and acquisition stages. As can be seen in Figure 11, at the start of the acquisition session the effective CS–US interval (i.e., the time between X and the US) increases for the stimulus-preexposed group as a function of stimulus duration, but does not change for the nonpreexposed groups.

Indeed, DeVietti has demonstrated that there is a strong effect of stimulus duration on latent inhibition, particularly as an interaction with number of stimulus preexposures. Thus, latent inhibition can be produced with very few preexposures, but only if the stimulus duration is quite long (DeVietti & Barrett, 1986a,b; DeVietti et al., 1987).

In a particularly critical experiment (DeVietti et al., 1987, Exp. 3) for the trace conditioning hypothesis, groups of rats received either zero or six 15-sec stimulus preexposures. In acquisition, the 15-sec stimulus was presented, with the US occurring 1, 3, 6, 9, 12, or 15 sec after CS onset. Latent inhibition was obtained for the longer interstimulus intervals, but not for the shorter ones.

Although these data are congruent with the trace conditioning hypothesis, they may also be interpreted in other ways. For example, suppose (1) the OR is elicited only by stimulus onset, (2) the OR declines as a function of number of stimulus preexposures, (3) a certain threshold amount of OR is required for normal conditioning to occur, (4) pairings of CS and US in the acquisition stage serve to re-evoke the OR, and (5) the longer the ISI (CS onset to US onset) in acquisition, the more difficult to rearouse the OR. Given all of the preceding assumptions, each one of which is supportable by empirical evidence, the data are explainable by a "traditional" model without recourse to the trace conditioning hypothesis.

Finally, if the trace conditioning hypotheses is to be considered as a general theory of latent inhibition, it will be required to integrate such phenomena as the attenuation of latent inhibition by context change, S_1–S_2 effects, and many others. It is not immediately apparent, at least at this stage of development, how this can be accomplished, or indeed how anything can be accounted for except stimulus duration effects.

Inhibition-of-delay hypothesis

Another recent explanation of the latent inhibition effect also focuses on the temporal aspects of the preexposed stimulus, in this case on what Pavlov (1927) called "inhibition of delay." The inhibition-of-delay phenomenon simply describes the fact that measures of conditioning indicate that the conditioning strength (e.g., number of responses, magnitude of the response) is greater toward the termination of the CS than toward the onset of the CS. This type of finding has been attributed by some (e.g., Hammond & Maser, 1970; Millenson & Dent, 1971) as being due to the encoding of stimulus duration, which then allows the organism to more accurately predict the arrival of the US.

On the basis of the foregoing reasoning, Schachtman et al. (1987) argued that stimulus preexposure might facilitate the stimulus duration discrimination and therefore enhance inhibition of delay in a stimulus-preexposed group as compared with a nonpreexposed group:

> Inhibition of delay would then occur rapidly and the overall magnitude of the conditioned response (CR) would be less than that seen in control subjects given no preexposure.... The [hypothesis advanced earlier] raises the possibility that latent inhibition may reflect superior learning of a temporal discrimination during the CS, rather than a retardation of conditioning. [Schachtman et al., 1987, p. 301]

The hypothesis, on the face of it, appears quite unreasonable in that it is difficult to envisage how it would apply to those procedures that have successfully demonstrated latent inhibition but that use only one or two CS–US pairings, as in conditioned taste aversion and very frequently in conditioned suppression. The scores of such studies have been duly noted in previous sections. A similar problem is encountered with those studies demonstrating latent inhibition with a very low number of preexposures. Again, in conditioned taste aversion, such a finding is common (see chapter 2). The effect can also be obtained with conditioned suppression with a long-duration stimulus (DeVietti & Barrett, 1986a,b; DeVietti et al., 1987).

Fortunately, the ample reasons to reject the hypothesis a priori are also supported by empirical evidence that Schachtman et al. (1987) themselves provided. Thus, the hypothesis requires that there be greater latent inhibition in a group preexposed to a constant-duration stimulus than in a group preexposed to a variable-duration stimulus. In fact, both groups exhibited similar latent inhibition effects. Even more damaging to the hypothesis was the finding that stimulus preexposure *slowed* the development of inhibition of delay (both in acquisition of conditioned suppression and postasymptotic performance). This retardation of the appearance of inhibition of delay is exactly opposite to what would be predicted from the theory.

8 Conditioned attention theory of latent inhibition

In general, conditioned attention theory, CAT[1] (Lubow, Schnur, & Rifkin, 1976; Lubow, Weiner, & Schnur, 1981), states that nonreinforced preexposure to a stimulus retards subsequent conditioning to that stimulus because during such preexposure the animal *learns not to attend* to it. The theory is based on the use of attention as a hypothetical construct, with the characteristics of a Pavlovian response, and on the specification of reinforcement conditions that modify attention.

The assumption that changes in attention to stimuli are a function of reinforcement conditions may be traced to Lashley (1929) and Krechevsky (1932). Likewise, Lawrence (1949) suggested that the "acquired distinctiveness of cues" might be a gradual learning process subject to traditional analysis. In more recent theorizing, changes in attention as a function of reinforcement conditions have been emphasized by Mackintosh (1975), Frey and Sears (1978), and Pearce and Hall (1980), as well as in the "selective attention" theories of Lovejoy (1968), Sutherland and Mackintosh (1971), Trabasso and Bower (1968), and Zeaman and House (1963).

In similarity to selective attention theories, CAT treats attention as a response, occurring on stimulus presentation, the probability of which is increased when it is followed by reinforcement and decreased when it is not reinforced. However, CAT differs from those theories in a number of important respects: the conditions specified for the changes in the attentional response; the mechanism postulated to govern such changes; and the course of these changes with repeated stimulus presentation.

More specifically, CAT makes four critical assumptions about changes in the attentional response, R_a. First, with repeated stimulus presentation the attentional response inevitably declines. Second, the decline in R_a is conditionable in the Pavlovian sense. Third, under certain conditions of repeated stimulus presentation, R_a may be temporarily increased; this increase of attention is also conditionable. Fourth, a certain minimum level of attention to a stimulus is a prerequisite for its entering into effective associations. With an increase in attention to a stimulus, the associability of that stimulus is enhanced; conversely, with a decrease in R_a, stimulus associability is reduced.

In similarity to most of the cognitive theories discussed in preceding sections, CAT postulates that when a stimulus, S_1, is repeatedly presented without being followed by reinforcement, there is a decline in the attentional response to that stimulus. However, CAT makes two additional explicit assumptions about this decline: First, the decline in attention is postulated to be a *conditioned* decrement in which the US supporting the conditioning of inattention is the consistent *absence* of any effective stimulus event following S_1. Consequently, the decline in attention is subject to the empirical laws of conditioning as they relate to such variables as stimulus intensity, intertrial interval, and so forth. Second, it is postulated that the decline in the attentional response is an *inevitable* consequence of repeated stimulus presentation. This assumption is critical for predicting the course of changes in attention to a stimulus when it *is* followed by another event.

According to CAT, when S_1 is repeatedly followed by the presence of a second event, S_2, the attentional response to S_1 will be incremented. The rationale for predicting an increment in R_a to S_1 with a small number of S_1–S_2 pairings is as follows: During early preexposures, S_1 by itself elicits R_a. This is consistent with the large number of studies showing that the magnitude of latent inhibition is a function of the number of preexposures. With a low number of stimulus preexposures, R_a is still present, and thus there is no latent inhibition. Likewise, S_2 would be expected to elicit R_a, at least during the early stages of preexposure. In normal classical conditioning, the S_2 as, for example, a mild acid solution in the mouth, would elicit unconditioned salivation. When paired with an S_1, as, for example, a tone, the tone would come to elicit salivation. Similarly, the R_a to S_2, after a number of pairings with S_1, should also come to be elicited by S_1. However, because S_1 by itself *also* elicits R_a, the total amount of R_a to S_1 is the summation of that produced by S_1 alone and that supplied by the S_1–S_2 conditioning of R_a to S_1. Thus, the total amount of R_a to S_1 after n S_1–S_2 trials is greater than that to an S_1 when it has not been preceded by S_1–S_2 pairings.

When R_a to S_1 is incremented over repeated trials as a result of its being paired with S_2, it is considered as a conditioned attentional response. For a conditioned increment in R_a to S_1 to occur, S_2 must be temporally contiguous to S_1 and must itself elicit an R_a. S_2 may be a different stimulus, a response, a conventional US, or, in general, any stimulus change or event that consistently follows S_1. Conditioning of attention will take place whether or not S_2 is able to support conditioning of an overt response to S_1 (i.e., the elaboration of a traditional CR to S_1). However, because S_2 is not followed by an S_3 over repeated trials, it, too, will eventually lose its power to evoke R_a and thus cease to maintain R_a to S_1. Consequently, the initial increase in R_a to S_1 will be followed by its decline. The magnitude and duration of the S_2-supported increment in R_a to S_1 will be functions of the strength of R_a elicited by S_2, which, as already noted, over repeated S_1–S_2 pairings will also inevitably decline.

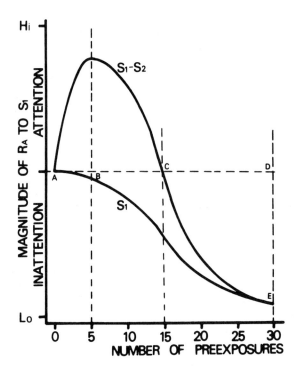

Figure 12. Magnitude of the theoretical attentional response (R_a) as a function of the number of preexposures of S_1 and the numer of preexposures of S_1–S_2. Latent inhibition occurs below line A–D, and facilitation occurs above line A–D. The origin (A) and the end point (E) for the S_1 and S_1–S_2 functions are the same. (From Lubow, Weiner, & Schnur, 1981.)

Figure 12 shows the theoretical course of R_a to S_1 for repeated paired presentations of a moderate-intensity S_1–S_2, as well as the course of R_a to S_1 alone. In regard to Figure 12, two points should be noted: First, no claims are made for the exact shape of the functions, only for the general prediction that for S_1 there will be some monotonic decline of R_a as a function of the number of preexposures, and for S_1–S_2 there will be an initial increase in R_a, followed by a subsequent decrease in R_a as a function of the number of preexposures; further, the increase in R_a will occur relatively early in preexposure. Second, Figure 12 serves to relate R_a more precisely to the concepts of attention and inattention.

Attention and inattention are meant to be on a continuum, not to refer to two separate processes. For purposes of exposition, however, levels of R_a that are higher than the level of R_a elicited on first presentation of S_1 will be called attentional, and levels that are lower than that initial level will be called inattentional. Finally, it is necessary to elaborate on the relationship between the theoretical curves for R_a to S_1 during preexposure and the predicted performance during a subsequent learning task where S_1 serves as a to-be-associated event. It is simply assumed that R_a must be elicited by any event that is to enter into an association *before* such an association can be made. Thus, in any transfer test, as, for instance, in going from the preexposure phase to the acquisition phase in the latent inhibition paradigm, the rapidity of conditioning to S_1 when followed by a US will be a direct function of the R_a level to S_1 as carried over from the last trial of preexposure.

Some further comparisons between CAT and current theories relating attention and learning are in order. Whereas in the formulations of Mackintosh (1975) and Frey and Sears (1978) changes in attention to relevant (reinforced) and irrelevant (nonreinforced) stimuli are assumed to be opposite and independent, increasing when a stimulus predicts a change in reinforcement and decreasing when it signals no such change, CAT assumes that decreases and increases in attention to a stimulus are on a single continuum and do not refer to two independent and separate processes. Repeated stimulus presentation is assumed to result in *eventual* loss of attention to that stimulus, whether or not it predicts reinforcement. A major function of reinforcement is to retard the decline in attention to the repeatedly presented stimulus.

This description of the course of changes in attention is critical for predicting transfer of training effects. Because reinforcement, according to CAT, slows the (inevitable) decline in attention to the repeatedly presented stimulus that precedes it, the only proper condition against which this function may be evaluated is a repeated presentation of the same stimulus in the absence of reinforcement. Thus, if a stimulus, S_1, is followed by another stimulus, S_2, and if the associability of S_1 is assessed in a subsequent transfer test in which S_1 is paired with a third stimulus, S_3, the results obtained in the transfer stage must be compared with results for a control group for which S_1 is preexposed *alone*, rather than with results for a group for which S_1 is novel. Because in the S_1–S_2 condition S_1 is preexposed, a loss of attention will accrue to S_1, and therefore it will have lower associability than a novel S_1. However, S_1 in the S_1–S_2 condition will have a higher associability than S_1 preexposed alone, because S_2 in this condition serves to attenuate the decline in attention to S_1. Thus, only by comparison with the S_1-alone presentation can the attention-maintaining property of S_2 become evident. By contrast, in the formulations of Mackintosh (1975) and Frey and Sears (1978), an increase in attention to a reinforced stimulus, and thus its subsequent extra-associability, is evaluated by comparison with a novel stimulus.

In addition to studies that indicate that pairing a stimulus with another stimulus attenuates latent inhibition, but does not increase its associability as compared with when it is novel, there is now evidence that even repeated pairing of a stimulus with reinforcement (such as weak shock) results in its loss of associability *as compared with a novel stimulus,* while at the same time increasing the associability of that stimulus *as compared with the condition of its repeated presentation alone* (Ayres et al., 1984; Hall & Pearce, 1979, 1982; Kasprow et al., 1985; Schachtman et al., 1987). It appears, then, that the changes of attention during repeated stimulus presentation, as formulated by CAT, provide a potent framework for predicting the subsequently observed effects of such presentations.

However, as stated earlier, CAT describes more than the general course of changes in attention. It specifies the mechanism whereby changes in attention take place, and thereby it is capable of predicting the variations in the course of transfer as a function of variables such as stimulus intensity, intertrial interval, and stimulus frequency. CAT uses empirical generalizations from data derived for classical conditioning experiments to account for changes in attention. In this sense, conditioning is viewed as a primary process for determining both the direction and the amount of attention. Here it should be emphasized that CAT, unlike the theories discussed in the previous sections, relies on the *operations and empirical laws* of classical conditioning, not on any particular theory of conditioning, to account for changes in attention.

The utility of the theory depends, in part, on being able to state the conditions of S_1 and S_2 that will elicit R_a. In a narrow sense this has already been specified. When either event is novel, it will elicit R_a. Furthermore, because R_a is a response that is subject to the laws of conditioning, all the variables affecting conditioning will similarly affect R_a (many of these are specified in the next section).

At this point, the problem of circularity must be addressed. How can we, on the one hand, use conditioning to explain changes in attention and then, on the other hand, use changes in attention to explain conditioning? This is justified on the basis that the empirical laws of conditioning and the theoretical statement regarding the rise and fall of attention are not in a one-to-one correspondence.

A dramatic illustration comes from the studies demonstrating that CS–US pairings will retard subsequent acquisitions using either the same CS or the same US, or both. As an example, Hall and Pearce (1979) showed that pairing a tone or light with a weak shock for 52 or more trials interfered with the acquisition of conditioned suppression to the tone or light when it was later paired with a strong shock – a case of a clear *lack of correspondence between the acquisition of associative strength and the acquisition of inattention, an effect that has been replicated a number of times* (see chapter 3, "Second-Stimulus Effects").

In general, it should be noted here that because of the rise and fall of R_a in the S_1–S_2 procedure, many transfer effects, whether they be from latent inhibition, sensory preconditioning, blocking, or overshadowing, should be sensitive to variations in the number of trials during the acquisition phase (i.e., during the phase for acquisition of blocking, overshadowing, etc.) and, in fact, should exhibit either positive or negative transfer depending on the number of such acquisition trials. This may be one way of reconciling the apparently discrepant results in regard to the effects of posttrial episodes (PTE), where, for example, Mackintosh and his colleagues have found that a surprising PTE may facilitate conditioning to an otherwise recalcitrant CS, as in unblocking (Dickinson & Mackintosh, 1979; Mackintosh, Bygrave, & Picton, 1977), whereas Wagner and others have found that a surprising PTE may interfere with conditioning (e.g., Donegan, Whitlow, & Wagner, 1977; Kremer, 1979).

In summary, then, although the same empirical laws of conditioning govern the acquisition of CR and R_a, the theoretical course of R_a is different from the empirical course of CR. These differences allow a number of significant predictions that cannot otherwise be generated. Thus, the apparent circularity of CAT is broken.

Recapitulating, CAT explains the effects of repeated stimulus presentation on subsequent learning by postulating changes in attention to the preexposed stimulus. These changes in attention are governed by empirical laws of conditioning, such that attention to the target stimulus at first increases (conditioned attention), if it is paired with a second stimulus that also elicits an attentional response, and then decreases (conditioned inattention) when the target stimulus is paired with a second stimulus that no longer elicits an attentional response, or if it is not paired with any stimulus. This formulation allows for a variety of specific predictions, many of which are not easily derivable from other theories of preexposure effects. In general, those variables that affect classical conditioning should also affect the conditioning of attention and inattention.

Predictions deriving from the theory can be divided into two classes: The first includes those that follow from the conditioning of inattention to S_1 when S_1 is repeatedly presented by itself. These predictions not only reflect the idea that such preexposures result in conditioning but also provide some counterintuitive notions, such as the prediction of more latent inhibition with a high-intensity preexposed S_1 than with a low-intensity S_1. The second class includes those predictions that concern the modulation of attention through the S_1–S_2 manipulation. Using conditioned attention theory, 16 predictions are generated. These predictions, together with the evidence relating to them, are presented next.

Predictions related to conditioning of inattention to S_1

1. Latent inhibition will be relatively long-lasting (i.e., distinguishable from sensory adaptation or fatigue-like effects). Siegel (1970), using rabbits and the conditioned eyelid response, found no significant differences in the amounts of latent inhibition as a function of delay (0 and 24 hr). A 24-hr delay was used by Lantz (1973) and many others in a conditioned suppression task. Similarly, a 48-hr delay was used by Carlton and Vogel (1967) and Lubow and Siebert (1969) to obtain latent inhibition. Crowell and Anderson (1972) demonstrated the effect with a 7-day delay between stimulus preexposure and testing.

Other conditioned suppression studies that reported latent inhibition with relatively long delays between stimulus preexposure and acquisition and/or test are cited in chapter 3. Even more striking, however, are the data from conditioned taste aversion, which were summarized in Table 4. As can be seen, it is possible to retain latent inhibition effects over periods of tens of days. Indeed, the standard procedure for conditioned taste aversion requires a minimum of 1 day for recovery after flavor–toxin pairing. Thus, the typical latent inhibition study with this paradigm, together with conditioned suppression, has provided scores of studies demonstrating that latent inhibition is a long-term effect and thus is distinguishable from sensory adaptation, fatigue, short-term habituation, and other transient phenomena.

2. Latent inhibition will be relatively stimulus-specific and will demonstrate a stimulus generalization gradient. Studies that have employed between-subject and within-subject designs indicate that the decremental effect of stimulus preexposure is stimulus-specific (Carlton & Vogel, 1967, Exp. 2; Lubow & Moore, 1959; Schnur, 1971; Schnur & Ksir, 1969; Siegel, 1969a). Furthermore, Siegel (1969b) has demonstrated that latent inhibition generalizes as a function of similarity between the acquisition-phase stimulus and the preexposed stimulus.

In addition to the foregoing, there is, again, the considerable amount of evidence for latent inhibition specificity and generalization from studies with conditioned taste aversion. The relevant experiments are reviewed in chapter 3.

3. Latent inhibition will be a positive function of number of stimulus preexposures. A previous review of the literature (Lubow, 1973a) strongly supported this prediction. Among those animal studies that used groups with 10 or fewer stimulus preexposures (Ackil & Mellgren, 1968; Domjan & Siegel, 1971; 5 and 2, 5 and 10, respectively), no latent inhibition was obtained. Those studies that used 20 or more preexposures did obtain latent inhibition (Ackil & Mellgren, 1968; Chacto & Lubow, 1967; Domjan & Siegel, 1971; Lubow, 1965; Lubow et al., 1968; Mellgren & Ost, 1971; Siegel, 1969a).

Since 1973, a number of additional experiments have been performed with the number of preexposures as a parameter. Lantz (1973) and Szakmary (1977b), using conditioned suppression in rats, and Clarke and Hupka (1974), using Pavlovian conditioning of the nictitating membrane response in rabbits, reported the same general results. Lubow, Wagner, and Weiner (1982) preexposed rats to a tone–light compound for 0, 20, 40, or 80 trials; a conditioned suppression test indicated greatest suppression for zero preexposures, and significantly less suppression for 20 preexposures, with no differences between 20 and 40 preexposures or between 40 and 80 preexposures. Weiner (1983) examined the curve for acquisition of latent inhibition more precisely. That study used single-stimulus preexposure, with conditioned suppression, and 0, 5, 10, 15, 20, 35, 45, 60, 75, or 90 preexposures. The data indicated quite clearly that latent inhibition was a negatively accelerating function of number of preexposures. The exact shape of the function, no doubt, is dependent on the species, testing paradigm, and stimuli that are employed.

4. Latent inhibition will be a positive function of the time between stimulus presentations [i.e., the intertrial interval (ITI) during the preexposure session]. Crowell and Anderson (1972) provided evidence that spaced stimulus presentations during preexposure produced more latent inhibition than did massed presentations. Lantz (1973) explicitly manipulated the ITI during preexposure, using 2-, 10-, 30- and 150-sec ITIs between stimulus presentations. She also found that longer ITIs during preexposure produced greater decrements in subsequent learning. Similar results were obtained by Schnur and Lubow (1976).

5. Latent inhibition will be a positive function of stimulus intensity. A positive relationship between the amount of learning decrement and the intensity of the preexposed stimulus was obtained by Crowell and Anderson (1972) and by Schnur and Lubow (1976), both of whom used a conditioned suppression procedure. Similarly, preexposure of a high-salience stimulus produces a greater latent inhibition effect than does preexposure of a low-salience stimulus (Midgley, Wilkie, & Tees, 1988); see the section on stimulus intensity in chapter 3.

6. Latent inhibition will be subject to external inhibition (i.e., will be disrupted by the introduction of an extraneous event following the S_1-alone presentations). The event may be a different stimulus, a response, an environmental change, or some other event. In terms of CAT, the presentation of the extraneous event will serve to restore attention to the preexposed stimulus. Lantz (1973) presented a novel light stimulus following 60 tone presentations. Subsequently, the tone was paired with a shock US in a conditioned suppression paradigm. She found that the interpolated light disrupted latent inhibition, as compared with a group in which no light stimulus was presented prior to

Table 19. *Experimental design for demonstrating blocking of conditioned inattention*

Group	Stage 1 Preexposure	2 Paired preexposure	3 Acquisition-test
Blocking	A	AX	X
Control		AX	X

conditioning. Lantz suggested that the light stimulus might serve an attentional function, thus reinstating the effectiveness of the repeated stimulus. Rudy et al. (1977), using taste aversion learning, demonstrated disruption of latent inhibition by exposing the animals to novel environmental stimulation prior to conditioning. Those authors concluded that the effect of the novel external stimulus was to reverse the prior habituation. A similar reversal of the associability deficit was accomplished by Hall and Pearce (1982), who used unexpected removal of the US in the CS–weak-shock preexposure paradigm.

The numerous studies using conditioned suppression and conditioned taste aversion that demonstrated that latent inhibition was disrupted when the context was changed from stimulus preexposure to acquisition (see the later section "Context Effects Considerations") may well belong in the category of external inhibition effects. However, other treatments of context effects are, of course, viable and will be presented later in this chapter.

7. Latent inhibition will be subject to blocking. If a stimulus to which conditioned inattention already has been established is then paired with a second stimulus, conditioning of inattention should develop more slowly to the second stimulus than to the stimulus to which conditioned inattention has accrued. The design, which parallels that used in blocking, is illustrated in Table 19. Rudy et al. (1976) preexposed animals to light alone and then to a light–noise compound. The animals were subsequently conditioned to the noise stimulus by pairing it with shock in a conditioned suppression situation. Light preexposure prior to compound preexposure interfered with development of latent inhibition to the noise. Although that finding was contrary to what was expected by those authors, that is precisely the blocking effect that CAT predicts. Prior conditioning of inattention to the light during its nonreinforced presentations *blocked* conditioning of inattention to the noise.

8. Latent inhibition will be subject to overshadowing. According to CAT, if a compound stimulus made up of two elements of differential salience is preexposed, more latent inhibition should accrue to the more

salient element than to the less salient element. Furthermore, as the number of preexposures of the stimulus compound is increased, there should be a greater gain in the amount of latent inhibition to the more salient element than to the less salient one. In addition, if compound preexposure effects and compound acquisition effects are governed by parallel processes, then with an increasing number of preexposures, both elements should lose their effectiveness. These predictions were confirmed in a study by Lubow, Wagner, and Weiner (1982).

9. Latent inhibition will be subject to extinction and spontaneous recovery. When the preexposed stimulus is repeatedly paired with a US, conditioning to that previously exposed stimulus is obtained; that is, the CS–US pairing results in extinction of the decremental effects of the preexposed CS. This, indeed, is the usual finding in latent inhibition experiments in which there are numerous CS–US pairings in the acquisition–test stage. Typically, as the number of trials increases, the differences between the stimulus-preexposed and nonpreexposed groups decrease. In a three-stage design in which the number of CS–US pairings varies *between* groups, the same effect is found (Carlton & Vogel, 1967).

Evidence for spontaneous recovery of the latent inhibition effect was provided by Wickens et al. (1983). After a 10-week delay between the CS–US conditioning phase (in which latent inhibition was abolished) and the test, latent inhibition reappeared. Those authors noted that no reference could be made to the recovery of a *specific* response, as no such identifiable response was acquired during nonreinforced stimulus preexposure. Therefore, what was reinstated was "an attitude or response of indifference or inaction toward the stimuli given during the original noncorrelated experience" (Wickens et al., 1983, p. 68). This reappearance of latent inhibition was specific to the stimulus and/or the response used in the noncontingent situation. In terms of CAT, the inattentional response conditioned to the preexposed stimulus, after being extinguished, "spontaneously" recovers with the passage of time.

10. Latent inhibition should be enhanced as a result of prior apparatus preexposure. Because apparatus preexposure facilitates learning, then preexposure to the apparatus in which stimulus preexposure subsequently will take place should facilitate the acquisition of conditioned inattention and thus potentiate the latent inhibition effect.

In light of a recent failure to show that apparatus preexposure facilitates normal, non-stimulus-preexposed learning (Hall & Channell, 1985a, Exp. 3) it is important to document the premise that is the basis of the prediction of enhanced latent inhibition. Facilitation of learning after preexposure of the conditioning–test apparatus has been reported for two-way avoidance conditioning (Dieter, 1977; McAlister et al., 1979), one-way avoidance conditioning (Dieter, 1978; Grant & Grant, 1973; Grant & Young, 1971), taste aversion (Mitchell et al., 1980), appetitive go/no-go discrimination (Mellgren & Ost, 1971, Exp. 1 and 3), and conditioned suppression of licking (Balaz et al., 1982). The anomalous failure by Hall and Channell (1985a) to find such an effect with conditioned suppression of bar pressing may be attributed to the fact that both the apparatus-preexposed and nonpreexposed groups received 10 sessions of bar-press training in the same apparatus that was later used for the experimental manipulations and test.

There appears to be ample evidence, then, for the assumption that apparatus preexposure enhances subsequent learning. As to the specific prediction of CAT that apparatus preexposure prior to stimulus preexposure should accentuate the latent inhibition effect, there have been several studies supporting this idea. Hall and Channell (1985a) directly tested the prediction and found that preexposure to the apparatus prior to stimulus preexposure enhanced the magnitude of latent inhibition. This finding was reported for both appetitively reinforced and condition suppression tests. The latter result has also been reported by Hall and Minor (1984, Exp. 6).

Predictions related to modulation of attention to S_1 by S_1-S_2 pairings

11. As shown in Figure 12, when S_1 and S_2 are paired over trials, latent inhibition will be preceded by a phase of relative facilitation. The facilitory phase, as stated earlier, is accounted for by the conditioned transfer of R_a from S_2 to S_1. The two sources of R_a are additive, thus giving S_1 from the S_1-S_2 pair more R_a than from S_1 alone. Because S_2 also acquires conditioned inattention with repeated trials, the early facilitative effect wanes and is replaced by the conditioned inattention that is manifest in the latent inhibition effect. Specifically, then, a truncated inverted U-shaped function is predicted.

11a. With a low number of S_1-S_2 pairings, subsequent associability of S_1 will be higher than if S_1 were not preexposed.

11b. The improvement in associability will increase with a relatively low number of S_1-S_2 repetitions.

11c. With a further increase in the number of S_1-S_2 pairings, latent inhibition will be obtained, and its magnitude increased, finally reaching an asymptotic level similar to that for the S_1-alone preexposure condition. These three predictions are graphically described in Figure 12. A direct test of the prediction that S_1-S_2 pairings attenuate latent inhibition was conducted by Lubow, Schnur, and Rifkin (1976). In two separate experiments, groups of rats were given one of four preexposure conditions: (1) no preexposure to the to-be-conditioned stimulus;

(2) preexposure to the to-be-conditioned stimulus, with each preexposure immediately followed by the second stimulus; (3) preexposure to the to-be-conditioned stimulus and to the second stimulus, in an unpaired relationship; (4) preexposure only to the to-be-conditioned stimulus. Conditioned suppression of licking served as the measure of latent inhibition.

The specific predictions were that the greatest conditioned suppression to S_1 would be exhibited by the nonpreexposed group, and the least suppression (i.e., the greatest amount of latent inhibition) by the group exposed to the to-be-conditioned stimulus alone. Suppression to S_1 in the paired-preexposure group should be increased by the addition of S_2 in temporal contiguity with S_1. The addition of S_2 in the absence of temporal contiguity should not have the effect of increasing suppression, as compared with the S_1-alone condition. Both experiments yielded the same pattern of results. The addition of S_2 in a conditioning relationship to S_1 during preexposure significantly reduced the amount of latent inhibition, as compared with either the S_1-alone or the S_1, S_2 unpaired condition.

Additional evidence that S_2 paired with S_1 attenuates latent inhibition was provided by Mackintosh (1973), Szakmary (1977a,b), and Best et al. (1979). Mackintosh (1973, pp. 82–83) reported that 16 preexposures to a 1-min light produced significant latent inhibition in a conditioned suppression test, but the latent inhibition was significantly reduced when the preexposed light was compounded with a tone. Rudy et al. (1977) replicated this finding. Szakmary (1977a,b) reported similar results with conditioned suppression of bar pressing for food. Greater response suppression was found to S_1 in a group that received S_1 and S_2 paired, as compared with a group in which S_1 and S_2 were explicitly unpaired. When a zero-preexposure group was compared with an S_1–S_2 group, no significant differences were found, thus suggesting that the differences observed between S_1–S_2 paired and S_1, S_2 unpaired were due to the unpairing of S_1 and S_2 and that this effect was attenuated by S_1–S_2 pairings. Finally, Best et al. (1979) showed that successive presentations of two flavors, vinegar followed by vanilla, reduced the latent inhibition effect, as compared with a vinegar-alone group. A similar design by Westbrook et al. (1982) did not produce the S_1–S_2 effect. However, this discrepancy may be attributable to the relatively long S_1–S_2 interval in the latter study. Also supportive of the S_1–S_2 analysis are the various flavor compounding studies indicating that preexposure of a compound protects the elements from acquiring latent inhibition (see chapter 3).

Both Lubow, Schnur, and Rifkin (1976) and Kaye et al. (1988) noted the possibility that the effects of S_2 may be produced as a result of omission of S_2 during acquisition. Such an interpretation, of course, would not be supportive of CAT. In a test of this hypothesis, Szakmary (1977a) demonstrated that paired preexposure of S_1 and S_2 attenuated latent inhibition of S_1 *whether or not* S_2 followed S_1 in the test. Thus, the latent inhibition decrement produced by S_1–S_2 pairing in preexposure was not a product of stimulus change between preexposure and acquisition.

In keeping with the idea that an effective latent-inhibition-attenuating S_2 may be any event that is in a conditioning relationship to S_1, there have been studies showing that positive or negative intracranial stimulation (Frey et al., 1976) and overt instrumental responding controlling S_1 onset (Lubow, Schnur, & Rifkin, 1976) or S_1 offset (Weiss & Friedman, 1975) can serve to reduce latent inhibition to the same degree as the more usual sensory S_2.

Because the S_1–S_2 predictions are critical for CAT, this evidence will be reviewed in some detail. Frey et al. (1976, Exp. 2) found that preexposure to a to-be-conditioned tone retarded subsequent eyelid conditioning in the rabbit. However, when, during preexposure, the tone was paired with a light flash or positive intracranial stimulation or negative intracranial stimulation, the latent inhibition effect was attenuated. As Frey et al. (1976) noted, the transfer effects were observed even though none of the stimuli following the CS was itself an effective reinforcer. Thus, once again, it was shown that the presence of an event contingent on the preexposed stimulus attenuated the development of latent inhibition to that stimulus whether or not the event by itself supported conditioning.

Weiss and Friedman (1975) compared three groups of rats. Group RT received 30 S_1 presentations that could be terminated by a crossing *response*. Subjects in group YT were yoked to group RT and received the same number and pattern of S_1 presentations, but without control of stimulus offset. Group NP received no S_1 presentations. Subsequent acquisition of avoidance responding to S_1 was found to be significantly poorer in the yoked group than in the RT and NP groups, which did not differ from each other.

Two experiments reported by Lubow, Schnur, and Rifkin (1976, Exp. 3 and 4) produced similar results when the subjects controlled stimulus onset. Each study, it will be recalled, had three groups. During the preexposure period, the active group could produce S_1 by pressing on a bar. Each member of the passive group was yoked to one of the active group and received S_1 every time its partner pressed the bar. Subjects of a third group were placed in the apparatus for the same period of time as were those in the active and passive groups, but they did not receive S_1. In both studies, there were no differences in number of bar presses among the three groups during preexposure. Nevertheless, when S_1 was paired with shock and then tested for conditioned lick suppression, the most latent inhibition was shown by the passive group, with a significant reduction in latent inhibition in the active group.

Recent studies of the Hall-Pearce (1979) effect also have supported the idea that S_1–S_2 pairings interfere with the development of latent inhibition (e.g., Ayres et al., 1984; Hall & Pearce, 1982; Kasprow et al., 1985; Schachtman et al., 1987). As will be recalled from chapter 3, those experiments used a weak electric shock as S_2. Such an S_2 probably is more salient than the more typical tone or light S_2. In general, those studies demonstrated that latent inhibition could occur when preexposure involved many pairings between a CS and a weak shock, but the size of the effect was less than with S_1-alone preexposures. This is exactly what is predicted by CAT.

All the foregoing studies, using either a relatively moderate number or a large number of S_1–S_2 pairings during preexposure, demonstrated, as predicted, that S_2 attenuated latent inhibition and suggested by comparison across experiments, that such attenuation might be diminished by increasing the number of such S_1–S_2 pairings. Only two studies varied the number of S_1–S_2 preexposures; and, of course, it is specifically this procedure that would allow the full prediction of an inverted U-shaped function to be tested.

The latent-inhibition-attenuating effect of S_2 as a function of number of S_1–S_2 pairings was investigated by Szakmary (1977a) using 2, 4, and 16 pairings. S_2 attained its maximum latent-inhibition-weakening effect after 4 S_1–S_2 pairings and waned with larger numbers of pairings. Similarly, two recent conditioned suppression experiments by Ben-Shahar, Weiner, Feldon, and Lubow (1986) suggested that although 20 S_1–S_2 pairings reduced the amount of latent inhibition, compared with a 20-S_1-alone group, 40 such pairings produced the same amount of latent inhibition as did 40 S_1-alone preexposures.[2]

12. The temporal relationship between S_1 and S_2 will affect the amount of latent inhibition reduction. As in normal conditioning, there should be an ISI function. However, because this function varies for different systems (e.g., eyelid conditioning, GSR conditioning, conditioned taste aversion), the exact shape of the function and its interaction with number of preexposures cannot be predicted. Although there has been one possibly relevant study with children (Lubow et al., 1975), in which a 0.5-sec ISI between S_1 and S_2 appeared more effective in attenuating the reaction-time decrement than did a 5.0-sec ISI, there have been no studies of this variable in the animal literature.

13. Attenuation of latent inhibition by addition of S_2 will be reduced by prior preexposure of S_2 alone. This procedure should produce conditioned inattention to S_2. Thus, subsequently, when paired with S_1, S_2 will have a diminished ability to maintain R_a to S_1. Indeed, Best et al. (1979) found that prior exposure of S_2 reduced its effectiveness to attenuate latent inhibition, as measured with conditioned taste aversion: Exposure of the S_2 flavor prior to S_1–S_2 pairing interfered with the ability of S_2 to disrupt latent inhibition.

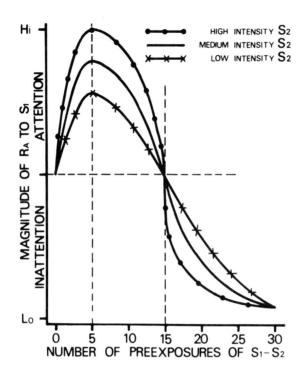

Figure 13. Magnitude of the theoretical attentional response (R_a) to S_1, when S_1 is paired with S_2, as a function of S_2 intensity. (From Lubow, Weiner, & Schnur, 1981.)

14. The efficacy of conditioning of attention to S_1 will be a positive function of S_1 intensity. Just as the conditioning of inattention should be enhanced with a higher-intensity S_1 when it is *not* followed by S_2 (prediction 5), so conditioning of attention should be enhanced with a high-intensity S_1 when it *is* followed by S_2. Thus, it is postulated that the effectiveness of S_2 in maintaining attention to S_1, and thus attenuating latent inhibition to it, will be a positive function of S_1 intensity. Such enhanced effectiveness will be reflected in either a longer-lasting or higher-magnitude latent inhibition attenuation phase. Prediction 14 has yet to be tested.

15. The predicted effects of S_2 intensity, during S_1–S_2 preexposure, on the shape of the R_a curve for S_1 are somewhat complex, but rather interesting. On the one hand, as S_2 intensity increases, there should be an increase in the conditioned attentional response to S_1. This might be reflected in a conditioned attentional response of larger magnitude, as compared with the response to a lower-intensity S_2, or the peaking of the response might occur earlier. For instance, the strongest response,

instead of occurring at the fifth trial, as in Figure 12, might occur at the fourth trial. On the other hand, because S_2 is also subject to conditioned inattention, it should suffer a more rapid acquisition of such inattention as S_2 intensity increases. Assuming that the effect of increases in S_2 intensity is to raise the magnitude of the conditioned attentional response to S_1, which is then accompanied by an increased rate of acquisition of inattention, a family of predicted curves, as shown in Figure 13, will be generated. As can be seen from Figure 13, with a low number of preexposures, an increase of S_2 intensity has a relatively facilitory effect. With a greater number of trials, this effect is predicted to reverse. This prediction has not yet been tested.

16. It follows from CAT that the facilitation resulting from the conditioning of attention and the latent inhibition resulting from conditioned inattention may be separable. If, during S_1–S_2 preexposure, S_2 could be qualitatively changed from trial to trial (e.g., on trial 1, S_2 would be a tone; on trial 2, S_2 would be a light, etc.), then conditioning of attention should be promoted over a relatively large number of trials. This follows because there should be a strong attenuation of conditioned inattention to S_2, because on each trial S_2 would be different. Thus, one would predict only the rising portion of the R_a-to-S_1 curve, which would reach some asymptotic value and remain there as long as the quality of S_2 was changed from trial to trial. This prediction also remains untested.

Conditioned attention theory and conventional conditioning

If conditioning of inattention can be attenuated by using a relatively neutral S_2, such as a tone or a light in a conditioning relationship to S_1, or a more salient weak shock as in the Hall-Pearce effect, then the pairing of S_1 with a biologically significant event, such as food or strong shock, should have a much more profound effect on the attenuation of inattention to S_1. Such pairing, of course, constitutes the conventional classical conditioning paradigm, the typical final outcome of which is an elaboration of a CR to the CS.

In addition to the specific formulations of Sokolov (1963) in regard to the role of the US in the development of the orienting response, a number of other authors have commented on an early phase during repeated CS–US pairings that is preparatory to, and independent of, the elaboration of the CR to the CS. For instance, according to Kamin (1969), the reinforcer has a dual role that is governed by two independent processes: the formation of the CS–US association, and the establishment of the CR to the CS. Likewise, Prokasy (1972) proposed two phases to describe conditioning, with different sets of rules governing response probability in each phase. According to Prokasy, it is highly unlikely that the first modification observed in the CR across trials will provide the experimenter with much information about CS–US contingency learning. In a similar vein, Bolles (1972) suggested that during conditioning, animals form response-independent representations of their environment that *then* play an important role in subsequent response selection. Schneiderman (1973) stated that "while reinforcement facilitates response emission, organisms learn more than they show us," and that the US in conditioning is involved in attention. "Conditioning apparently does not occur unless the US focuses the organism's attention upon the relationship between the CS and the US" (Schneiderman, 1973, p. 27).

Wickens, Tuber, Nield, and Wickens (1977), in their analysis of CR acquisition, maintained that a variety of learning activities occur during the pre-CR period and concluded that "generalized information about contingency relations is one of the factors that is being acquired during [this] period in the usual conditioning situation" (p. 67).

Finally, Maltzman and his colleagues, in discussing classical conditioning of the GSR, argued that such conditioning involves only discrimination, not response differentiation. "It therefore is only the first phase of what may be a more extended learning sequence when a skeletal motor R or a response other than an index of the OR is conditioned" (Maltzman et al., 1978, p. 331). Elsewhere, Maltzman (1977) wrote about the conditioning of the GSR to words as being "a problem solving discovery process. . . . It is conditioning of a form of attention" (p. 114). The notion that a period of attention and/or information processing *precedes* the elaboration of the conditioned response has been voiced most recently by Ohman (1979, 1983), in his analyses of GSR conditioning in humans, and by Campbell and Ampuero (1985), who found that in conditioning of young rats, *conditioned* cardiac decelerative responses (ORs) appeared before the final conditioning of accelerative responses.

The proposition that both response-independent and response-eliciting processes take place during conditioning is directly derivable from the postulates of CAT. According to CAT, if effective conditioning is to take place when a stimulus is followed by another stimulus, the attentional response to that stimulus, elicited on its first presentation, must be maintained. It follows that the conventional US in conditioning has a dual function: First, it maintains the attentional response to the stimulus. Second, it establishes the CR to that stimulus. These two functions of the US define two distinct sequential phases: establishment of conditioned attention to the CS, and elaboration of a CR to the CS. It is reasonable to assume that during the first phase stimulus properties are encoded, and only afterward is the associative bond between CS and US strengthened.

Conditioning of attention to the CS, though supported by the US, is not US-specific; that is, it is independent of the nature of the US (e.g., food or shock). The US serves the function of S_2 in the S_1–S_2 paradigm (i.e., maintains attention to the stimulus preceding it), and conditioning of attention to a stimulus takes place whenever it is followed by an otherwise unexpected event, whether or not the event by itself is able to support an overt conditioned response.

On the other hand, CR elaboration is US-specific; that is, it depends on the nature of the US employed. In this stage, the animal learns the appropriate response to the CS, which has been established as an event of significance by the prior conditioning of attention. The nature of the CS determines what class of responses will be strengthened.

Thus, conditioning of attention is viewed as a prerequisite for CR evocation; with conventional conditioning procedures, both maintenance of attention and CR elaboration are supported by the same US. However, following specifically from CAT, conditioning of attention may take place *without* giving rise to CR elaboration, as demonstrated in S_1–S_2 studies of latent inhibition. Whether or not a CR develops will depend on US efficacy.

As noted, conditioning of attention may be established with any type of stimulus in a conditioning relationship to the CS, including "neutral" stimuli such as light and tones, which are apparently ineffective for establishing CRs (Badia & Defran, 1970; Frey et al., 1976; Mackintosh, 1974; Schneiderman, 1973). In his discussion of sensory preconditioning, Mackintosh (1974) similarly concluded that "animals [may] learn that one stimulus follows another, but . . . formation of such an association is not sufficient to produce a CR to the first stimulus. This would imply that a motivationally significant UCS is not required for the formation of an association between CS and UCS, but only for the elaboration of a CR to the CS" (p. 96).

The function of S_2 (i.e., maintenance of attention to the CS) may also be seen in conventional conditioning when the S_2 is present intermittently or concurrently with a conventional US. In a study by Bruner (1965), rabbits that received an air puff US on 50% of the trials and light USs (by themselves insufficient to promote conditioning) on the remaining trials reached a significantly higher overall level of conditioning than did rabbits that did not receive the light presentations. Thus, light served as an effective stimulus to maintain attention to the CS and facilitated CR acquisition with a conventional US. W. Kimmel (1967), using GSR conditioning, found that subjects required to judge US intensity by moving a lever showed significantly larger conditioned GSR than did subjects conditioned without the judgment task. Apparently the motor response functioned as S_2, which increased attention to the CS.

The rules and/or the mechanisms governing the transition from attention conditioning to CR elaboration are not, at present, clear. One approach to this problem is that taken by H. D. Kimmel and Burns (1975), who, like others mentioned earlier, have proposed a two-stage model of conditioning. The first stage is described as a "preadaptive" or preconditioning stage, in which chemical couplings of sequentially repeated events take place. The second stage is described as an adaptive utilization of these couplings; this latter stage occurs if some environmental adjustment becomes possible by using the previously formed connections. In terms of Kimmel and Burns's proposition, conditioning of attention will give rise to performance if it enables the organism to produce adaptive responses on the basis of the contingencies existing during this stage.

Some modifications and elaborations

Although several strong predictions are derivable from CAT, it is quite clear that most of these simply follow from the proposition that stimulus preexposure produces some learning that is incompatible with future learning. As a result, all of the parameters that have been shown to be effective in conditioning should indeed produce more learning during the preexposure phase and thus less learning (i.e., more latent inhibition) in the test phase. As such, these predictions are somewhat uninteresting and, indeed, are either implicit or explicit in many other explanations of latent inhibition. The unique aspect of CAT comes from its formal treatment of the conditioning of inattention to S_1, with the US defined as the absence of a consequence to S_1. It is from this postulate that the various S_1-S_2 predictions are deduced, many of which are derivable only from CAT, in particular those that deal with the effect of number of S_1-S_2 pairings on the amount of latent inhibition. Unfortunately, it is exactly at this point that the amount of direct evidence supporting CAT, though positive, is far from overwhelming. The ultimate test of this central formulation of CAT must come from laborious parametric studies that systematically vary the number and type of stimulus preexposures.

Nor is this the only weak point in the theory. The theory, as originally presented (Lubow, Schnur, & Rifkin, 1976; Lubow, Weiner, & Schnur, 1981), did not take into consideration two variables that have since been shown to have strong modulating effects on latent inhibition: context and stimulus preexposure duration. In this regard, it will be recalled that in order to produce latent inhibition, one has to preexpose the stimulus in the same environment in which the conditioning–test will take place. A change in such context disrupts latent inhibition. Furthermore, it appears that latent inhibition requires that the familiarity with stimulus and environment occur conjointly. As for stimulus duration, considerable evidence has been presented indicating that the amount of latent inhibition increases with the duration of the preexposed stimulus and that there is an interaction between stimulus duration and number of stimulus preexposures. Although a few stimulus preexposures normally do not produce latent inhibition, latent inhibition can be obtained under such conditions with long stimulus durations.[3]

Stimulus duration considerations

This latter finding might appear to be exceedingly troublesome for CAT, because in the normal conditioning preparation one would expect poorer conditioning with longer CS durations. This, according to CAT, would lead to the prediction of less latent inhibition with longer S_1 preexposures. However, the fact that longer durations of preexposed stimuli produce more latent inhibition can be accounted for by continuing with the notion that the absence of a significant event provides the US for the conditioning of inattention, but noting that it is the onset of the stimulus that produces the response to be reinforced (i.e., the attentional response), perhaps even in the form of an observable OR. The long-duration stimulus, then, provides strong conditions for reinforcing inattention, because the initial attentional response occurring soon after stimulus onset is followed by a long period of time during which there is no change in stimulus conditions. This static condition should be the most potent US for conditioning inattention (being an absolute condition for *no change*, which, of course, defines the maximum imaginable condition for the "absence of a significant event").

To elaborate on the foregoing proposal, consider the possibility that stimulus preexposure engages two at least partially independent processes: attentional conditioning, as described in CAT, and a stimulus encoding process that precedes it. Such an encoding process is, in fact, demanded on logical and empirical grounds. How is it possible to condition attention, inattention, or anything to a stimulus unless that stimulus is first encoded, that is, without some qualities or properties of the stimulus being apprehended and stored? Two facts provide the evidence that some property of the preexposed stimulus is encoded during preexposure.

First, the latent inhibition effect is stimulus-specific. This, by definition, is prima facie evidence of stimulus encoding. The second source of evidence that the preexposed stimulus is encoded comes from those studies that have varied stimulus and environmental familiarity from preexposure to test (e.g., Lubow, Rifkin,, & Alek, 1976). Here it will be recalled that one of the basic findings is a facilitation effect, as in the perceptual learning studies; at the time of test, an old stimulus in a new environment results in better associative learning than does a new stimulus in a new environment. This difference ($S_oE_n > S_nE_n$) can be accounted for only by the fact that the stimulus familiarization – preexposure phase produced a representation of that stimulus that was carried over into the test phase.

Similarly, the finding that an acquisition–test conducted with a new stimulus in an old environment promotes better learning than a new stimulus in a new environment ($S_nE_o > S_nE_n$) offers proof that preexposure to an environment results in the organism learning something about that environment, and that, too, is transferred to the test phase. Finally, the finding that latent inhibition is produced only when the stimulus and environment are familiarized conjointly ($S_nE_o < S_nE_n$ or $S_nE_o > S_nE_n$), not with independent familiarization procedures (S_o, E_o), suggests that in addition to the encoding of the preexposed stimulus and the encoding of the preexposed environment, the fact that the two are preexposed *together* is also encoded.[4]

One might inquire as to which properties of the stimulus or the environment or their relationship are encoded, a refinement of the question raised in chapter 1 as to *what* is learned during nonreinforced preexposure. Although this question has not been addressed directly, at least not in the form stated above, one can identify at least some aspects of stimulus quality that are encoded during the preexposure period. Thus, in studies of the generalization of latent inhibition, Siegel (1969b) found that as the CS tone differed in frequency from the preexposed tone, the amount of latent inhibition was reduced. In other words, frequency had been encoded during the stimulus preexposure session. Similarly, in conditioned taste aversion, a number of studies have demonstrated a generalization effect for preexposed taste stimuli. However, though such generalizations were obtained (e.g., Dawley, 1979; Franchina et al., 1985), flavor quality was not dimensionalized, and thus the data merely provide additional evidence for the stimulus specificity of latent inhibition. Nevertheless, this again affirms that some aspect of stimulus quality is encoded.

One study (Klein et al., 1976) that varied the concentration of a preexposed sucrose solution and found no generalization of latent inhibition would seem to suggest that intensity characteristics of the stimulus are not encoded during stimulus preexposure. However, common sense suggests that this is not the case, and the large differences between the preexposed and conditioned sucrose concentrations (3% versus 34%) may have induced qualitative differences between the flavors. A program of research aimed at identifying those stimulus characteristics, environment characteristics, and stimulus–environment relationships that are encoded, as well as determining encoding priorities for these characteristics, would be most useful.

Given, then, the logical requirement and the empirical evidence for some type of stimulus encoding during preexposure, we may now consider the question of the relationship between the stimulus encoding process and conditioned inattention. First, it would seem reasonable to assume, as already noted, that stimulus encoding *precedes in time* the conditioning of inattention. Logic requires that in order to learn not to be attentive to a particular stimulus, the characteristics of the stimulus (i.e., the things that make it particular) must first be encoded and stored. Indeed, a simple, straightforward assumption might allow that the attentional response that is emitted by the organism when it encounters a novel stimulus is the reflection of (or by-product of) the work required for encoding. It is formally proposed, then, that the engagement of the process that leads to the conditioning of inattention occurs only *after* initiation of the encoding process.

Thus, two proposals have been offered. First, the US for the conditioning of inattention is initiated, as in the original version of CAT, when a stimulus terminates, but is followed by nothing. However, the US for the conditioning of inattention, the condition of no change, is strongest when the stimulus is of a relatively long duration. Second, there is a stimulus encoding process that precedes, or at least leads to, the conditioning of inattention. These two propositions are related in the following manner: When a novel, *long-duration* stimulus is presented, stimulus encoding is completed before the stimulus is terminated. This is the condition that provides a strong US for conditioning inattention. With a shorter stimulus duration, the stimulus is terminated *before* encoding is completed. This is the condition that provides a weak US for the conditioning of inattention. It therefore follows that in the latter case, with a short-duration stimulus, more preexposure trials are required to demonstrate latent inhibition than with long-duration stimuli.

The two systems for production of an effective US for the conditioning of inattention, one related to stimulus duration and the other related to number of trials, are shown in Figure 14. There are several features in this diagram that are common to almost all information processing models of this type. There is a short-term memory (STM) and a long-term memory (LTM), which have the usually prescribed characteristics. STM is a limited-capacity storage device whose contents rapidly decay. LTM is, by comparison, a device of unlimited storage capacity whose contents decay, if at all, at a very slow rate. When a stimulus impinges on the sensorium of the organism, a comparator process is initiated that compares the input to STM with the contents of LTM. The output of the comparator is a discrepancy signal whose magnitude is proportional to the size of the difference between the STM representation of the input and the stored LTM representation. When the value of this output is above a certain threshold level (+), two things happen: (1) There is an attentional response, which may be indexed externally by an OR.

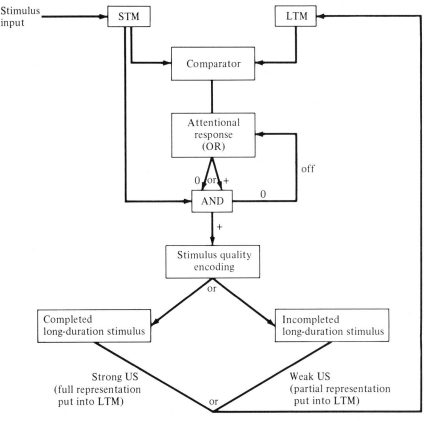

Figure 14. Schematic representation of conditioned attention theory as part of an information processing system, with emphasis on two sources of the US for inattention. With a long-duration preexposed stimulus, the full representation of the stimulus is put into LTM with few preexposures. With the assumption that quality encoding *precedes* the conditioning of inattention, then for a given number of trials there are more pairings of the long-duration preexposed stimulus with inattention than there are for the short-duration stimulus preexposures.

(2) Processing of the contents of STM is initiated (i.e., encoding occurs). If the stimulus is available for encoding for a sufficiently long time, as with a long-duration preexposed stimulus, then encoding is completed in one trial, and LTM stores a complete representation of the input stimulus. Thus, when the same stimulus is once again presented to STM, the comparator, matching the new input with its representation in LTM, produces a zero discrepancy signal. It is this zero discrepancy signal in conjunction with the presence of the external stimulus that produces one set of requirements for the conditioning of inattention. A repetition of this sequence (external stimulus input paired with zero output from the comparator, i.e., no attention) also comes to be

encoded. Note that here we are referring to the encoding of the *relationship* of events, in this case, external stimulus – nothing. The final result is that LTM now contains representations of the qualities of the stimulus *and* a representation of its relationship with other events.

It would appear, then, that there may be two different sources for latent inhibition: (1) the simple *reduction* in the discrepancy signal or loss of attention due to completed encoding of stimulus qualities and (2) an effect that occurs from the conjunction of the preexposed stimulus with the *absence* of attention. We shall return to this point after examining the foregoing processes when a short-duration-preexposure stimulus is presented to the organism.

In brief, the two situations, one for the long-duration stimulus and one for the short-duration stimulus, are quite similar. The major exception is that for the short-duration stimulus, where the stimulus is terminated in the early trials before stimulus quality encoding is completed, there is a gradual approximation (as a function of repeated trials) toward the completed representation of the stimulus in LTM. Thus, with successive trials there is an increasingly better match between STM and LTM, which results in increasingly smaller discrepancy outputs from the comparator, and smaller attentional responses.

Here, of course, the relationship between *incomplete* quality encoding (because of short stimulus duration) and the learning of the association between the two events (stimulus and nothing) is not entirely clear. But several interesting possibilities present themselves. Suppose, for example, in contrast to what was proposed earlier, that the conditioning of inattention (i.e., stimulus–nothing) were to proceed at a faster rate than stimulus quality encoding. In essence, the processing would be turned off before it had fully extracted all of the stimulus properties, and therefore LTM would contain only a partial representation of the external stimulus. To study the relationship between the two processes – conditioning of inattention and stimulus quality encoding – would first require the identification of post-stimulus preexposure tests that would independently index these processes, followed by systematic factorial studies of the effects of number and duration of stimulus preexposures.

Context effects considerations

Any theory of latent inhibition must take into account the full range of context effects, as described earlier. To briefly summarize, the following empirical generalizations seem reasonable.

1. Latent inhibition is context-specific; that is, the environment in which the preexposed stimulus is presented must be the same as the environment in which acquisition takes place. Thus, the necessary condition

for demonstrating latent inhibition is, at the time of acquisition, that of an old stimulus in an old environment, S_oE_o.
2. More specifically, the preexposures of the two elements, stimulus and context, must occur conjointly. S_oE_o will promote latent inhibition; S_o, E_o will not.
3. Preexposure of the context *prior* to context–stimulus preexposure serves to enhance the latent inhibition effect.
4. Preexposure of the context alone *following* context–stimulus preexposure has little or no effect on the amount of latent inhibition.

One way of integrating these data is to treat the context (E_o) of the prototypical latent inhibition procedure (S_oE_o) as a *conditional cue* for the presentation of the S_1–no-consequence relationship. Archer et al. (1986) made a similar suggestion, noting that context might act as a retrieval cue, and thus a change in context would result in a failure to retrieve the stimulus–no consequence association. Similarly, Hall and Channell (1986) proposed that the context can serve as a conditional cue that provides information as to whether or not the CS will be followed by a given consequence.

Although it was not developed further by those authors, there is ample reason to believe that this treatment of context and latent inhibition is viable. To begin with, one must relate the general notion of a conditional cue to the recent empirical work on *facilitation* (Rescorla, 1985; Rescorla, Durlach, & Grau, 1985) and *occasion setting* (Holland, 1985; Ross, 1983; Ross & Holland, 1981), both of which refer to the special attributes of cue that *precede* CS–US events.

As an example, in the studies of Holland and Ross, subjects were presented with tones that were followed by food *only* when the tone was preceded by a light. With this procedure, the light came to facilitate the responses normally conditioned to the tone. Note that although those studies dealt with relatively discrete conditional stimuli, the same precedence relationship holds for context and S_1. What, then, are the special qualities of an occasion-setting cue? For our purposes, there are two characteristics that are important: (1) The occasion-setting stimulus gains stimulus control of the behavioral *expression* of the CS–US relationship. (2) Unlike a normal conditioned excitatory stimulus, which is subject to extinction when it is repeatedly presented without being followed by a US, an occasion setter does not lose its power of facilitation if it is afterward repeatedly presented by itself.

If context is treated as the occasion setter for the S_1–no-consequence relationship, then these two empirical statements are sufficient to integrate the context–latent inhibition data. Thus, it follows that latent inhibition is context-specific, because when the acquisition context is different from the stimulus preexposure context, the absence, in acquisition, of the occasion setter for the S_1–no-consequence relationship does not provide the conditions for interference with the acquisition of the new

S_1–US relationship. Similarly, in an S_o, E_o preexposure procedure, where S and E are not familiarized conjointly, the subject enters the acquisition session without the presence of the appropriate occasion setter for eliciting the S_1–no-consequence relationship.

Finally, the asymmetrical effects of the order of context preexposure on latent inhibition can be treated in a most elegant fashion. Recall that context preexposure prior to stimulus–context preexposure facilitates latent inhibition, whereas context exposure after context–stimulus exposure has little or no effect. The latter result may be derived from the empirical finding that an occasion setter does not extinguish its ability as a facilitator (i.e., does not lose its power of stimulus control). On the other hand, the enhanced latent inhibition that occurs when context preexposure precedes context–stimulus preexposure appears to remain unaccounted for. To continue the analysis along these same lines, one would have to suggest that preexposure of a to-be-occasion-setting stimulus not only does not interfere with its capacity to become an occasion setter (as would be expected with "normal" latent inhibition) but also actually potentiates this ability. Although there have been no studies that have used preexposure of discrete, to-be-occasion-setting stimuli (such data would be most important in developing the current argument that context serves as an occasion setter), it is important to recognize that there is reason to expect that occasion-setting stimuli do not obey the same associative laws as do typical CSs, as noted earlier in the case of extinction. In summary, then, it is proposed that the four context–latent inhibition phenomena described earlier can be integrated into conditioned attention theory by treating the typical S_oE_o latent inhibition procedure as an instance in which E_o, the context, gains the role of a conditional stimulus, setting the occasion for expression of the conditioned S_1–no-consequence relationship and the accompanying response of conditioned inattention[5]

Parallel processes in stimulus preexposure stage and acquisition–test stage (normal conditioning)

The relationship between the stimulus preexposure phase and the acquisition–test phase should now be clear. Continuing with the general schema of CAT, which proposes a parallel in the stimulus preexposure and classical conditioning acquisition phases of the latent inhibition paradigm, it can be seen that this parallel also extends to the identification of two separate processes within each stage, two processes that should be the same for each of these phases. Thus, during stimulus preexposure there is a stimulus quality encoding process that precedes the encoding of the stimulus–no-consequence association. Similarly, during acquisition there is a stimulus quality encoding process that precedes the stimulus–consequence (reinforcement) association. Now, because there is no reason to suppose that the stimulus quality encoding processes are different in preexposure and acquisition, it would seem to follow that the stimulus preexposure group should exhibit *positive* transfer in the acquisition–test, having during the stimulus preexposure period already completed part of the stimulus quality encoding that is required to precede the acquisition of a conditioned response to that stimulus. Indeed, at least in the standard S_1–S_2 preparation and in the Hall-Pearce variation of it, this facilitation is found at least relative to the S_1-alone preexposure group. The required systematic studies that would vary the numbers of stimulus preexposures, particularly at the low end of the continuum (i.e., 1 to 10), have not been done, and thus direct evidence for the facilitation effect relative to zero preexposure is unavailable. However, a review of such evidence for preexposure to shock provides support for this contention. Furthermore, evidence for stimulus quality encoding during the preexposure period is provided by the context change studies. The fact that the S_oE_n group acquires the to-be-learned response in acquisition faster than the S_nE_n (e.g., Lubow, Rifkin, & Alek, 1976) can be interpreted only in terms of prior stimulus quality encoding.

Given this analysis, then which requires a facilitation effect with small numbers of stimulus preexposures for either the S_1-alone or the S_1–S_2 preparation, albeit a greater effect for the latter, it follows that the typical decrement in acquisition–test performance that characterizes latent inhibition is a profound effect that is traceable to negative transfer from the S_1–no-consequence association and is superimposed on what otherwise would be a facilitory outcome. In addition, it follows that in the typical latent inhibition experiment, because quality encoding is completed by the end of the preexposure period, the stimulus–no-consequence association (which produces, in terms of CAT, inattention) does indeed interfere with the *associability* of the previously preexposed stimulus in the acquisition–test phase. Again, to repeat what was proposed earlier, in the stimulus preexposure period it is the stimulus–no-consequence association that comes under control of the context in which it was experienced. It is the context that becomes the occasion setter for the expression of the stimulus–no-consequence association that is reflected in the negative transfer that describes the latent inhibition effect.

There are several important implications of this analysis. First, the various studies reviewed in chapter 6 on the neural basis of latent inhibition, most of which were concerned primarily with indexing attention, are problematic. None of those studies differentiated between (1) the effects of a drug or lesion on the stimulus quality encoding process and (2) the stimulus–no-consequence association, while assuming, implicitly, that it was the latter, reflecting inattention that was being affected by the experimental manipulation.

Thus, to take a simple example, when Ackil et al. (1969), Kaye and Pearce (1987a,b), McFarland et al. (1978), and Solomon and Moore (1975) reported that hippocampal lesions attenuated or abolished latent inhibition, those effects may have arisen for several quite different reasons. Disruption of latent inhibition, if attributable to a process occurring during preexposure, may reflect alterations in (1) the stimulus encoding process (in this case an improvement), (2) the attentional process mediated by the stimulus–no-consequence association, or (3) the process by which the context develops into the occasion setter for the expression of the stimulus–no-consequence association. Although generally it is assumed that hippocampal lesions affect latent inhibition because of disruption in attention to the preexposd stimulus, one might argue, for example, that because the hippocampus is involved in spatial mapping (O'Keefe & Nadel, 1978; cf. Foreman & Stevens, 1987), loss of hippocampal function prevents the stimulus–context association that is necessary for the context to serve as an occasion setter, at the time of test, for the stimulus–no-consequence association. Thus, in other words, the hippocampectomized animal that is stimulus-preexposed and tested in the same environment (operationally, S_oE_o) may be, functionally, in an S_oE_n situation, one that, as we know from the various context manipulation studies, does not allow the latent inhibition effect to appear.

Given these alternative possibilities, or some combination, and the failure of neurophysiological studies to differentiate between them, it may be premature to look to such studies for help in understanding the latent inhibition phenomenon. One suggestion for improving the current situation would be to use experimental designs that, in addition to the neurophysiological intervention, would cross this factor with context change and number of stimulus preexposures.

The ideas presented here concerning independent encoding of stimulus properties and the S_1–no-consequence relationship will be elaborated in chapter 9, which attempts to integrate the latent inhibition data from children and adults with data from lower organisms within the general framework of CAT.

9 Conditioned attention theory as applied to latent inhibition in humans

In the preceding chapter we developed and applied CAT to the data from latent inhibition studies with animals, but we ignored the data from experiments with human subjects. It will be recalled that these human studies present a pattern of results not entirely consistent with that obtained from lower organisms. To review, the major findings, presented in chapter 5, are as follows:

1. There are standard stimulus preexposure procedures that will produce latent inhibition in young children, but not in older children or adults.
2. If these same procedures are coupled with a masking task, then latent inhibition can be produced in older children and adults.
3. Nevertheless, there are some procedures (i.e., electrodermal conditioning, conditioned tasted aversion, and Ivanov-Smolensky conditioning) that may not require masking for the production of latent inhibition.

The last point will be dismissed, perhaps somewhat cavalierly, because of the difficulty in separating the unconditioned response to the CS (orienting or otherwise) from the conditioned response to the CS in the test phase. As a result, one cannot determine whether stimulus preexposure produces an artifactual reduction in the conditioned response or a reduction in the associability of the CS. This, it will be recalled, is similar to the problem encountered in a number of animal conditioning paradigms. In addition, these studies with humans may have included unplanned masking procedures – either by the very nature of the procedure or by instruction.

Let us now consider the effects of age and masking on latent inhibition in humans. Briefly, again, as age increases, latent inhibition decreases; however, latent inhibition can be restored in older children and adults by the use of a masking task during stimulus preexposure. The fact that there is an interaction between age and masking may provide a clue to the operations of both variables.

An analysis of the interaction between age and masking must begin with an examination of the several different masking tasks that have successfully elicited latent inhibition. Table 20 lists the appropriate studies and describes the masking task procedures. From an inspection of Table 20 it is readily apparent that all of the masking procedures serve to draw the subject's attention away from the to-be-conditioned preexposed stimulus. With the exception of Siebert's instructions (1967) to ignore the preexposed stimulus, the other manipulations appear, on

Table 20. *Different types of masking tasks successfully used to produce latent inhibition*

Experiment	Stimulus	Masking conditions	Test
Baruch et al. (1988a)	Noise	Determining number of repetitions of list of nonsense syllable pairs	Learning relationship between noise and counter
Baruch et al. (1988b)	Noise		
Ginton et al. (1975)	Noise		
Lubow et al. (1987)	Noise		
Har-Evan (1977)	Tone		Intensity discrimination
Hulstijn (1978)	Tone	Predicting sequence of light onsets	Eyelid conditioning
Schnur & Ksir (1969)	Tone		
Siebert (1967)	Tone	Instructions to ignore[a]	Ivanov-Smolensky conditioning
Lubow, Caspy, & Schnur (1982)	Visual patterns	Anagram task	Visual discrimination

[a] The Siebert (1967) study was not intentionally designed with a masking procedure.

the face of it, to be quite potent for fully occupying the subject's attentional resources. Indeed, one would expect that there would be a reciprocal relationship between attention to the preexposed stimulus and success on the masking task. To the degree that the subject can block out the nominally interfering preexposed stimulus, performance can be improved on the masking task.

There was, however, one study that appeared to be an exception to the rule that masking promotes latent inhibition. It will be remembered that Kaniel and Lubow (1986) failed to find latent inhibition in older children even though a masking task was employed. This occurred in spite of the fact that Lubow, Caspy, and Schnur (1982), with similarly aged children, did obtain the latent inhibition effect with masking.

Inspection of the procedures used in these two experiments indicates a potentially significant difference regarding the spatial relationship between the masking and the to-be-targeted stimuli. In the Lubow, Caspy, & Schnur (1982) study, the masking task, which required anagram solution, was centrally located in the visual field, and the to-be-targeted stimuli were located peripherally. In the Kaniel and Lubow (1986) study, which used a simple discrimination task for masking, the relative positions were reversed, with the masking task stimuli located in the periphery of the visual field and the to-be-targeted stimuli located centrally. The spatial relationships, as well as other characteristics of the two different preexposure conditions in which the to-be-targeted stimuli (T) and masking stimuli (M) were presented, are illustrated in Figure 15. It would seem, then, that the failure by Kaniel and Lubow (1986) to find latent inhibition in older children may have been due to the topography of the masking task stimuli and the to-be-targeted stimuli. This

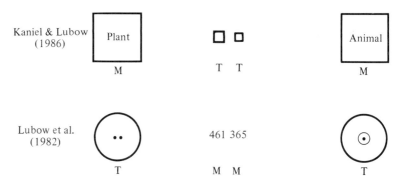

Figure 15. Relationship between masking (M) and target (T) stimuli in the Kaniel and Lubow (1986) and Lubow, Caspy, and Schnur (1982) studies.

hypothesis was tested by Meiri (1984), who found that the spatial relationship between stimuli and mask determined the effectiveness of the masking task in producing latent inhibition. When masking task items were peripheral to the target items, latent inhibition was not obtained. When these conditions were reversed, target items peripheral to masking task items, latent inhibition was obtained. Unfortunately, we have not been able to consistently replicate these findings, and therefore we shall refrain from commenting on their potential theoretical significance. However, it remains an interesting and plausible possibility that the Kaniel and Lubow (1986) procedure, because of the central position of the target stimulus relative to the masking materials, did elicit attentional responses, which thereby interfered with the accrual of latent inhibition.

If masking works because it prevents the subject from paying attention to the target stimulus, it follows, conversely, that inattention to the target stimulus is a critical condition for producing latent inhibition. Indeed, conditioned attention theory, developed to explain and integrate the animal data on latent inhibition, is based on the proposition that latent inhibition is a function of conditioned inattention. Thus, in regard to the age differences in human latent inhibition, one could conclude that young children do not attend to the preexposed stimulus, whereas adults do. There are several lines of support for this contention: (1) the effects of masking in adults, as discussed earlier, (2) the absence of an S_1–S_2 effect in adults, and the presence of an S_1–S_2-like effect in children, at least in a reaction-time study (Lubow et al., 1975), (3) the reasonable notion that, without being instructed otherwise, the demand characteristics of the experimental situation are such that adults will continue to pay attention to a preexposed stimulus, with the expectation that something has to happen soon, and that, contrary to adults, young children will quickly lose interest in such a stimulus. In fact, in the Kaniel and Lubow (1986) study, the primary function of the masking task for the young children was to ensure that they were at least facing the target stimulus. A study of eye fixation for young children and adults during repetitive presentations of a single visual stimulus should reveal gross differences between the two populations. Children should display considerably fewer fixations on the target than adults. If this were found, it would support the proposal that the natural pattern of inattention to the target stimulus in children is functionally equivalent to that produced by the masking task in adults.

Indeed, any measure that reflects attention directly (as with eye movements) or indirectly (as with autonomic indices of the orienting response) should yield the same results; children without masking and adults with masking should show *reduced* levels of the attentional index to the target stimulus, as compared with adults without masking. As indicated earlier, for the normal adult serving as a subject in an experimental psychology laboratory, every stimulus is potentially a significant signal. As a result, attention to that stimulus is maintained throughout the course of preexposure.

Iacono and Lykken (1983) made a similar suggestion in regard to habituation: "Under such circumstances, where the instructions are ambiguous and the purpose of the of the experiment inscrutable, it must be expected that some subjects (e.g., those most cooperative or skeptical about the intentions of the psychologists) will assume that the task calls for *careful attention* to the tones" (1983, p.72; emphasis added). In two experiments, Iacono and Lykken (1983, 1984) succeeded in showing that subjects who received instructions to actively ignore a series of tones, either by watching a videotape of an old silent film (Iacono & Lykken, 1984) or by listening to a radio-play (Iacono & Lykken, 1983), produced fewer electrodermal responses, habituated those responses more quickly, and rated the tones as less loud than did subjects in several different control groups. For our purposes, it would have been useful to continue from the stimulus-preexposure–habituation phase to an associative learning task, preferably not with GSR as the dependent measure. Such a study, if it were to vary the number of stimulus preexposures, would go a long way in providing critical evidence regarding many of the theoretical conjectures that have heretofore been proposed. To further elaborate on the predictions, and, thereby, also on the theory, not only would the stimulus-preexposure–masking interaction be expected, for which ample evidence has already been presented, but also one would expect to find two dissimilarly shaped curves. For the nonmasked group there should be an actual improvement in performance after a few stimulus preexposures, followed by either maintenance of the peak level or a gradual decline. These two hypothetical curves, describing two possible courses of attention during stimulus preexposure in a nonmasked group, are shown in Figure 16, panel A. Figure 16

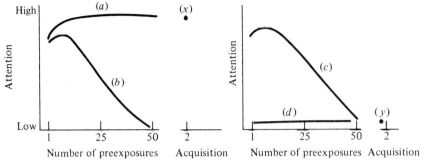

Figure 16. Hypothetical curves showing the possible courses of attention during preexposure, and the level of attention in acquisition, as a function of the absence of a masking task (panel A) or the presence of a making task (panel B).

also indicates the presumed level of attention to the preexposed stimulus after the first acquisition trial (x). The high level of the attentional response (x) simply accounts for the superior performance of this nonmasked group as compared with the masked group (y), as shown in Figure 16, panel B. The question of theoretical interest is the source of the elevated attention. We accept that the adult human, in a nonmasked condition, can override the attentional decline that normally accompanies stimulus preexposure in animals and young children. However, as indicated in Figure 16, there are two ways that this override can be accomplished: by maintaining attention during stimulus preexposure (curve a) or by quickly reinstating attention after its decline in preexposure [i.e., at the initiation of the acquisition session (curve b)].

A similar analysis of the attentional changes during *masked* stimulus preexposure is instructive and is shown in Figure 16, panel B. Here it can be seen that the attentional level at the beginning of acquisition (y) is relatively low. Once again, this low level is introduced simply to account for the relatively poor learning of the masked-stimulus-preexposed group, as compared with the nonmasked group. As with the nonmasked group, the question remains how this level is reached. Is it through a gradual decline of attention (as in curve c), or is the attentional level low throughout the masked preexposure period (as in curve d)?

Although one may conclude that masking prevents the adult human from reinstating attention, the source of that effect still must be ascertained. Masking may operate to prevent rearousal of attention, as in c–y compared with b–x, or masking may operate by not allowing the attentional response to be elicited in the first place, as in d–y.

The choice of function depends, at least in part, on one's ability to identify the locus of operation of the masking procedure. Does it serve to obscure the transition between preexposure and acquisition, as Spence (1966) would have suggested, or does it operate primarily during the preexposure phase itself? Whereas some studies that have obtained latent inhibition in older children and adults have used masking only during the preexposure period (Har-Evan, 1977; Lubow, Caspy, & Schnur, 1982), and others during both preexposure and acquisition (Baruch et al., 1988a, b; Ginton et al., 1975; Hulstijn, 1978; Lubow et al., 1987; Schnur & Ksir, 1969), there has been no experiment that has systematically varied the location of the masking task. If the successful masking condition were to prevent the detection of a *change* from preexposure to acquisition (i.e., the transition), then the (*b*) function would receive support. If, on the other hand, masking were to operate only during stimulus preexposure, then the (*a*) function would be implicated. The fact that some successful latent inhibition studies used masking only in the preexposure phase suggests that, indeed, it is the (*a*) function that is responsible for the latent inhibition effect.

Attentional decrement overrides

How can we relate these considerations to the conditioned attention theory that was proposed to account for latent inhibition in animals? An inclusive theory of such stimulus preexposure effects might start with the basic proposals of conditioned attention theory and merely incorporate the notion of an attentional override that becomes operational at a particular developmental stage in the human. In this manner, once can account for the fact that latent inhibition normally appears only in young children, not in adults, but can also be elicited in adults by the use of a masking task. The masking task, of course, serves to cancel the operation of the attentional override process in adults.

What more might be said about attentional override? With what other ideas and/or data is such a concept congruent? What are some other attributes of attentional override? What are the phylogenetic and ontogenetic implications of the development of an attentional override concept? These questions will be addressed in the following sections.

Which attention?

One of the difficulties in using the concept of attention is that we do not always have a clear understanding of what is meant by the term. In trying to unravel the various meanings of "attention," many discourses on the subject, including this one, are compelled to open with an apt quotation from William James. James, like the Bible before him, had some appropriate, if sometimes contradictory, things to say on just about all subjects psychological. The obligatory quotation can serve diverse purposes. In addition to embellishing the scholarly credentials of the writer, it can provide authoritative support for, or a convenient point of departure from, the thesis that inevitably follows.

As a reference point, then, there is this statement by James that, though perhaps not profound, is certainly pithy: "Everyone knows what attention is" (James, 1890/1950, p. 403). William James, of course, recognized that in the very truth of his assertion lay the dilemma of the usefulness of the concept. And though Wundt, James, and Titchener gave "attention" a prominent role in their psychologies, it was usually defined in common-sense parlance: "It is the taking possession by the mind, in clear and vivid form, of one of what may seem several simultaneously possible objects or trains of thought. Focalization, concentrating of consciousness, are of its essence. It implies withdrawal from some things in order to deal effectively with others: (James, 1890/1950, pp. 403–404). James continued and defined attention by describing its absence. It "is a condition which has a real opposite in the confused, dazed, scatterbrained state which in French is called *distraction*" (James, 1890/1950, p. 403). Or, elsewhere, "without it the consciousness of every creature would be a gray chaotic indiscriminateness, impossible for us even to conceive" (James, 1890/1950, pp. 402–403).

The rise of behaviorism in the 1920's, with its emphasis on environmental determinism, its simple associative principles of learning defined in terms of stimuli and responses external to the organism, and its general impatience with literary concepts that fit within the bounds of neither operational definitions nor intervening variables, caused the term "attention" to recede from the American psychological scene. In the late 1950's and early 1960's, with the development of cognitive psychology, "attention" was rehabilitated. A psychology that explains behavior in terms of processes *within* the organism provides a comfortable, indeed fertile, setting for miracles of resurrection.

Today, "attention" is very much with us, again. Unfortunately, however, many of the old behaviorist criticisms are still valid. The concept of attention is often used with an almost studied ambiguity, as though if they are not directly confronted, the problems of definition will disappear. Indeed, there are at least three different general usages of the term: (1) as a process underlying, causing, or accompanying a behavioral state, (2) as an explanation for a particular behavior, (3) as a behavior itself. Very often the three usages, especially that of process and explanation, remain vague and undifferentiated. As an example, in the classical dichotic listening paradigm (e.g., Moray, 1959), when a subject tracks a message in one ear, the subject is oblivious to the input to the other ear, except under special circumstances, such as when the subject's name is called. In that case, the name is heard (1) because "we can switch our attention to another channel" (Solso, 1979, p. 127) or (2) because the name, as a significant stimulus, "attracts" our attention. In the preceding explanations, the difference is that in (1), hearing the name *results from* an attentional process, and in (2), hearing the name *produces* an attentional response. In one case, attention is a cause; in the other, it is an effect. There is an almost inevitable confusion when the behavior to be explained (hearing the name) and the explanation of the behavior are indexed by the same response. Mostofsky (1970) described this problem as a "failure to select a dependent variable as an external referent for ascertaining the presence of attention, independent of that which is to be measured" (p. 10).

One of the advantages of beginning with an animal model of attention, as we have done with conditioned attention theory, is to ensure that the introspective, phenomenal aspects of attention will not interfere with an objective evaluation of the concept. In the conditioned attention model, it will be recalled, the manipulation of attention level is independent of the measured effects of the attentional changes. This is achieved by the very nature of the latent inhibition paradigm, which consists of two stages, stage 1 involving the controlled differentiation of attentional responses, and stage 2 involving the measurement of the effects of those changes. As such, the latent inhibition paradigm provides a clear, parametrically sensitive procedure that relates "attention" to established stimulus manipulations, which in many cases are in accord with the common-sense meaning of the term. Thus, we should expect, for instance, that with repeated exposures to a stimulus, the amount of attention elicited by a stimulus will decline, or if we increase the complexity or significance of the stimulus, the rate of decline of attention will be attenuated. And, as indicated, one has the opportunity (at least theoretically) to measure attentional level independent of its subsequent effects. A recent example was supplied by Kaye and Pearce (1987a,b), who showed that the level of the orienting response during preexposure foreshadowed the amount of latent inhibition in the test. Furthermore, they found that a manipulation (hippocampectomy), in addition to stimulus preexposure, that affected orienting responses in preexposure also influenced the amount of latent inhibition.

What is lacking, at least up to this point, is an attempt to make contact between the animal model of attention and related concepts in the human literature. A theory of latent inhibition that would encompass a wide range of animal data, together with the significant findings in the human literature, particularly age and masking effects, would provide a useful approach to a more comprehensive understanding of a variety of phenomena, including not only latent inhibition itself but also attention and learning.

Returning, then, to William James, who described six varieties of attention, we begin with the distinction between nonvoluntary and voluntary attention. The former is characterized as being effortless and passive, and the latter as active and effortful. These processes could be applied to sensorial or intellectual events. The passive–active categories are mirrored in a number of modern theories of attention, as, for example, in the distinction that Schneider and Shiffrin make between automatic and controlled processing (Schneider & Shiffrin, 1977; Shiffrin & Schneider, 1977; Schneider, Dumais, & Shiffrin, 1984). Whereas Schneider and Shiffrin are particularly concerned about the role of practice in causing a shift from controlled to automatic modes of processing, it is quite evident that some form of automatic processing is also available for certain types of events independent of any past history with those events. It is such processing, reflexive in nature, that characterizes much of a lower animal's interaction with its environment. In such a case, the initial presentation of a stimulus elicits an attentional response. On further repetition of this stimulus without any change or consequence, the attentional response declines. This habituation of the passive attentional response to a repeated stimulus, the ultimate effect of which is to reduce the subsequent associability of that stimulus, is the fundamental process that is built on and modulated by other attention-related override processes. In particular, it is suggested that there are two types of attentional decrement overrides, corresponding roughly to James's distinction between internal and external. However, whereas James used the dichotomy between internal (intellectual) and external (sensorial) to identify two different sources of attention, we have chosen to modify this approach by emphasizing one basic decremental process that deals with subject-interactive stimuli independent of their source, but that can be modified by the two types of overrides. These overrides serve to change the normal decremental effects such that attention is prolonged and/or intensified. In addition, it is suggested that the external override may operate at all phylogenetic and ontogenetic levels, whereas the internal override is primarily operative in the older child and adult. The two overrides and the empirical evidence that justifies their identification are presented next.

External attentional decrement override

This override is initiated when there is any change in the prevailing conditions of presentation of the repeated stimulus. It is most probably accompanied by at least one of the classical components of the OR. In agreement with Sokolov (1960, 1963), the OR can be re-evoked after repeated presentations of the target stimulus "when the signal is intensified, weakened, lengthened, shortened; when it is presented before the usual time; when it has been omitted at the usual time" (Sokolov, 1963, p. 674).[1] There have been numerous studies investigating the effects of such changes on the habituation of various orienting responses, particularly GSR, that have provided evidence for a rearousal of the orienting response; for recent summaries, see Kimmel et al. (1979) and Siddle (1983); for a recent empirical demonstration, see Siddle (1985). We identify this OR rearousal with an incrementing of attention.

However, in addition to re-evoking attention, the external override can be deployed to *maintain attention*. Indeed, this function was the primary focus of the chapter on conditioned attention theory. In this latter mode of operation, a second stimulus, S_2, is placed in a conditioning relationship to the preexposed target stimulus. If we were to compare the iteration of S_1–S_2 with that for S_1 alone, there is little doubt that appropriate OR differences would be revealed (i.e., slower OR habituation to S_1–S_2 than to S_1 alone). The behavioral consequences of such stimulus preexposure procedures have already been described. For lower animals S_1–S_2 preexposure, as compared with S_1-alone preexposure, produced faster subsequent learning to S_1. This was found to be true in conditioned suppression, conditioned taste aversion, and eyeblink and nictitating membrane response conditioning (see chapter 2). Furthermore, Lubow et al. (1975) found that young children's reaction times increased as a function of the number of preexposures to S_1, but when S_2 was placed in a conditioning relationship to S_1 (i.e., when S_1 was closely followed in time by S_2), then not only was the reaction time significantly reduced for the S_1–S_2 groups, relative to the comparable S_1 groups, but also the function was altered. Thus, for the S_1–S_2 groups, there was a *decrease* in reaction time from zero preexposures to three preexposures, whereas for the comparable S_1 groups there was no change.

In terms of conditioned attention theory, it seems that the initial attentional response to S_1 is not at the maximum possible value with zero preexposures (complete novelty), but it can be enhanced further by a contiguous temporal relationship between S_1 and S_2. The effectiveness of this enhancement relative to zero preexposures was found to be short-lived, for although 3 S_1–S_2 pairings reduced reaction time relative to zero preexposures, 10 S_1–S_2 pairings were about as effective as zero preexposures, and further presentations of the S_1–S_2 pair resulted in slower reaction times. The two groups should reach similar asymptotic response-time levels with additional increases in the number of preexposures.

In addition to the S_1–S_2 effects in young children and animals, there are other manipulations that can be categorized as belonging to external override, as, for example, the well-known facilitation of reaction time when the target stimulus is *preceded* by a warning stimulus. However, a distinction may have to be made on the basis of different mechanisms for S_2–S_1 and S_1–S_2 effects. In the latter case, we would argue that S_2 serves as a US to reinforce the attentional response that S_1 by itself elicits, thus using a mechanism similar to that described for S_1-alone effects, in which the absence of a consequence allows for a diminution of the attentional response and serves as a reinforcer of an inattentional response. Whether or not S_2–S_1 procedures are amenable to the same analysis will depend on the strength of evidence concerning the reliability of backward conditioning.[2] Irrespective of these considerations, the data for S_1–S_2 effects, in both animals and young children, provide sufficient evidence to warrant postulation of an externally driven override of the normal attentional decrement that occurs with iterative stimulation.

Internal attentional decrement override

The internal override loosely corresponds to James's voluntary attention and Schneider and Shiffrin's controlled process. It is characterized by a subjective awareness of effort and/or active engagement with the target stimulus. The internal override of the attentional decrement is engaged most easily by verbal instructions, either given by the experimenter, as in directions such as "Look at the polygon and count the corners," or in self-directed instructions, as may occur in the ambiguous experimental situation that is created when the experimenter refrains from giving specific directions. This latter situation is, of course, most prominent in the nonmasked stimulus preexposure phase of a latent inhibition experiment. It is during stimulus preexposure that the adult may initiate the internal attentional decrement override that prevents habituation of attention and thus also precludes latent inhibition.

The successful masking task fully occupies the attentional resources of the subject, as it is specifically directed to events *other than* to the to-be-targeted stimuli. The internal override is thus prevented from being deployed against the to-be-targeted stimulus. Therefore, an attention response to that stimulus is not elicited, even at the initial presentation; or, if attention is elicited, it follows the normal decremented path with repeated presentations, allowing for the appearance of latent inhibition in the test.

The pattern of data that we see in the animal literature and human literature, namely, the presence of latent inhibition in animals without using a masking task, and the presence of latent inhibition in adult humans only with a masking task, strongly suggests (albeit circularly) three points in regard to the phylogenetic distribution of attentional processes: (1) In regard to the proposition that a relatively novel stimulus will elicit an attentional response and that this response will decrement over repeated presentations of that stimulus, there is some evidence, as discussed in preceding chapters, that this fundamental attentional process does not appear in phyla other than mammals.[3] Among mammals, it is present in subhuman species as well as humans. (2) Attention may be maintained at a level higher than that supported by iterative stimulation alone by the engagement of an external override. The external override would follow the same phylogenetic distribution as the basic decremental process. (3) Attention may be maintained at a level higher than that supported by iterative stimulation alone by the engagement of an internal override. It would appear that this internal override is to be found only in humans, particularly older children and adults.

The internal override is related to the ability of older children and adults to use language to mediate between stimulation and response. This receives some support from the basic differences in latent inhibition between lower animals and humans and the reasonable assumption that the masking task serves to disengage linguistic responses to the target stimulus. In addition, the fact that young children show latent inhibition without the use of a masking task, and perhaps even delayed expression of latent inhibition in children from low socioeconomic conditions (Kaniel & Lubow, 1986), adds some indirect support to the supposition that language may serve as a mediator of these effects.

These conjectures appear to be quite congruent with the findings and theories of Kendler (Kendler, 1979a,b). Starting with the fact that the proportion of children who exhibit reversal shifts increases with age, Kendler has proposed two basic information processing levels or modes, nonselective and selective encoding, with the following characteristics: The primary level, stimulus encoding, "nonselectively encodes all information registered by the peripheral receptors" (1979a, p. 237). This level is common to lower animals and humans. The selective second level, however, which receives input from the first, and in which "immediately relevant information is abstracted and processed further" (1979a, p. 237), develops only in the human. Thus, the primary level matures first, both phylogenetically and ontogenetically, whereas the second level appears late in phylogenesis and matures relatively slowly in the developing human organism.

Within this framework, it would appear that the normal decline in associability with repeated stimulus presentations, that which characterizes latent inhibition, reflects the operation of the primary passive process; with increasing utilization of the higher, abstractive, selective encoding mode, the once-observable primary process effects become less apparent. Kendler's nonselective and selective modes would correspond to a passive–active distinction.

Although we have gone one step further by positing an internal override and external override, we would now suggest that the external override operates on the passive system in lower mammals and merely resets a parameter value for rate of decline of the attentional level. In the human, however, the external override may have one of two different effects. It may operate in the animal mode, as described earlier, so as to maintain attentional level during iterative stimulation, or it may operate when there is a change of conditions, as when going from the preexposure session to the test session, in which case the external override also serves to engage the internal override such that the human subject becomes aware of the change in contingencies. Continuing with the passive–active distinction, it is proposed that in the typical latent inhibition procedure, children normally operate in the passive mode, and adults in the active mode. However, the introduction of an S_2 (as, for example, implicit or explicit labeling) in a conditioning relationship to the preexposed stimulus will, at least temporarily, terminate the passive mode and initiate the active mode processing of S_1. In adults, the reverse can be accomplished, switching from active to passive mode, by means of a masking task that diverts attention from the preexposed to-be-targeted stimulus.

Again, this conceptualization fits well with the fact that latent inhibition is found in lower animals and in young children, but disappears in older children and adults. In animals and younger children, attention is controlled by external stimuli (stimulus-bound), and with changing conditions, such as the introduction of new reinforcement contingencies, it is difficult to redirect that attention. The absence of latent inhibition in older children and adults is related to the increased flexibility of attention, where, with the development of language, attentional deployment comes under internal control. In this stage, attention escapes the deterministic action of external stimulation, and the internally controlled attentional processes allow for the mediated expressions of dimensional control rather than the expediency of S–R domination.

The Kaniel and Lubow (1986) studies, in which latent inhibition persisted for an additional year in children from low socioeconomic groups, as compared with middle socioeconomic groups, suggested a slower development of the active selective mediational stage (or, in terms of CAT, the internal attentional decrement override) in the former than in the latter population. If one accepts the general premise of conditioned attention theory as a basis for the modulation of attentive processes, then one should compare the natural histories of reinforcement contingencies as they appear in these two groups. Such a comparison might aid in explaining test performance differences between the two socioeconomic groups, specifically the differences in the development of latent inhibition and, more generally, the differences in the delay of internal as opposed to external control of attentional deployment.

Stimulus preexposure engages two different encoding processes

This section presents an elaboration of the ideas mentioned in chapter 8 in the section on modifications of conditioned attention theory. Specifically, this section treats stimulus quality encoding (here termed "property extraction") and stimulus relationship encoding (in the case of latent inhibition, the encoding of the stimulus–no-consequence relationship).

Given that a novel stimulus elicits an attentional response and that repeated stimulus presentations produce a decline in that attentional response that can be attenuated by the external and internal overrides, what, then, is the nature of the encoding process that accompanies the various levels of attention? Here we are faced with a paradox. Starting with the same conditions of stimulus preexposure, Posner and Boies (1971) and Kraut and Smothergill (1978) have proposed that the encoding process *facilitates reaction time*, whereas Lubow and others claim that the encoding process *debilitates associability*. In other words, given the same repeated presentations of a stimulus, with one subsequent test (reaction time) an enhancement effect is predicted, and with another (associative task) a decremental effect is expected.

Before attempting to resolve the problem, a description of the Posner theory is in order: Briefly, there are two processes, alertness and encoding, that are engaged during stimulus preexposure. The reaction-time response reflects the interaction of these two processes, such that the stronger the encoding and/or the higher the alertness, the faster will be the reaction time. With an increase in the number of stimulus preexposures, the alertness factor decreases, and the encoding factor increases. Normally the former has a stronger effect than the latter and serves to mask the facilitating effects of encoding on reaction time. However, the alertness decrement is transient, and it may be vitiated by a number of manipulations. Indeed, the strongest tests of this two-factory theory have examined the effects of alertness, rather than encoding manipulations. Nevertheless, a number of studies from Kraut and Smothergill, with experimental designs that were directed at minimizing the effects of alertness, showed that reaction time was *faster* to a familiarized stimulus than to a novel stimulus (e.g., Cecil et al., 1984; Kraut, 1976, Exp. 3; Kraut & Smothergill, 1978, Exp. 2).[4] Following Posner and Boies (1971), they attributed the facilitated reaction time to the operation of an encoding process during the preexposure period.

There are, then, two theories that deal with stimulus preexposure effects. One attributes facilitation of reaction time to encoding during preexposure, and the other postulates an encoding process that produces lower associability for the preexposed stimulus. Can the apparent discrepancy between these two positions be resolved? One approach to answering the question might begin by asking two additional questions: What is it that is being encoded? What is the relationship between the encoded stimulus and the test performance requirements? First, it must be remembered that *what* is encoded in the stimulus preexposure stage, for any given stimulus, must be the same, independent of whether the test to follow is one of reaction time or associative learning. This follows simply from the fact that during the stimulus preexposure phase, the subjects do not know what task they will be called on to perform in the test phase.

In regard to the first question, it is proposed that there are at least two encoding processes, each of which operates on and encodes different aspects of the stimulus – one deals with stimulus properties, and the other with stimulus relationships. The first will be called a property extraction process, and the second a stimulus relationship process. For reasons already alluded to, but to be more fully explained later, it is assumed that the property extraction process precedes the stimulus relationship process.

Property extraction

Property extraction is that process by which the organism learns about the qualitative and quantitative aspects of the stimulus (i.e., the stimulus is red, bright, large, round, etc.). Property extraction makes those aspects of the stimulus that are encoded more readily available for assimilation into new categories. Thus, with alertness kept constant, after many exposures to a red circle, when the subject enters the choice reaction-time test and is told to press the right button when the familiar red stimulus appears and the left button when a novel blue stimulus appears, the subject will respond faster to the red than to the blue. The faster response to the familiar stimulus than to the novel stimulus will occur because the red property of the preexposed stimulus is more *accessible*[5] and therefore more easily identified with the category "familiar" than is the blue property with the category "novel." In these choice reaction-time experiments the instructions to subjects would, indeed, seem to promote this facilitated categorization. For example, subjects often are specifically instructed to press one button when "the familiar stimulus" appears, and the other when "the novel stimulus" appears (e.g., Cecil et al., 1984; Kraut et al., 1981), or to press one for the blue light and the other for the red light (e.g., Kraut, 1976). The same effect is achieved when the response buttons are colored so that one colored button represents the preexposed stimulus and one the novel stimulus. Here the more readily accessible qualities of the preexposed stimulus make it easier to "remember" that the response goes to that button colored the *same* as the preexposed stimulus; see Cantor and Cantor (1965, 1966), Cantor and Fenson (1968), and Miller (1969) for examples of this type of experiment.

The key to understanding the reaction-time data lies in recognizing that the subject must associate the familiar or novel stimulus with an event (a given button) that is *already revealed*. The subject does not have to discover the correct association, but simply exercise a preexisting association. It is this exercising of an association, with the connection between its two elements being *given*, that characterizes the choice reaction-time studies. By comparison, in the latent inhibition learning tests, the subject has to learn or *discover* the appropriate association.

The effect of stimulus preexposure on encoding in the reaction-time paradigm, then, is through the operation of property extraction, which facilitates the exercising of an already established association and therefore results in accelerated test reaction times (again, of course, in the absence of alertness decrements).

Stimulus relationship processing

The second encoding operation concerns the development of relationships between the preexposed stimulus and other events that occur in its presence. It is this encoding process, associative in nature, modulated by the laws of classical conditioning, that operates to produce latent inhibition. Several theories have attempted to define the critical relationships that are learned during the stimulus preexposure period of a latent inhibition experiment, some emphasizing context–stimulus associations, and others stimulus–consequence associations. For a better understanding of the possibilities, let us consider an exhaustive list of all the relationships that can be encoded during the preexposure period. The list items, of course, are determined by how one categorizes events of the preexposure period. It will be assumed here that there are only two event categories during this period: the nominal, to-be-conditioned stimulus (S_1) and the environmental stimuli (context). That being the case, three possible relations can be encoded: (1) stimulus–no consequence, (2) context–no consequence, and (3) context–stimulus.

Here it should be noted that just as it has been proposed that stimulus quality or property encoding must precede stimulus relationship encoding (as in learning something about the quality of a preexposed tone must come prior to the conditioning of inattention to that tone), so also learning something about the stimulus quality, and learning something about environment characteristics, must occur before the organism associates the two. There is yet one additional point. Just as the subject eventually develops conditioned inattention to the stimulus, the subject likewise develops conditioned inattention to the environmental context.[6] We can now summarize all of the events that take place during the preexposure period of a latent inhibition experiment.

The encoding of the stimulus properties of context is initiated before that of the nominal stimulus. This follows from the facts that (1) the organism is placed into the preexposure environment for some period of time before the nominal stimulus is presented, and (2) the context precedes *each* presentation of the nominal stimulus. The context is, of course, also present after each offset of the nominal stimulus. Another characteristic of the context is that it is typically more complex than the nominal stimulus. Thus, property encoding of environmental stimuli will require more time than encoding of the nominal stimulus. It should be noted that although such encoding may require more time, nevertheless it may well be completed *before* the property encoding of the nominal stimulus. This follows simply from the fact that context stimuli are continuously available for processing, whereas the nominal stimulus is available only for discrete, usually short, periods of time.

Thus far, then, a simple description of the obvious has been presented. During the preexposure period, there must inevitably be stimulus property encoding of the attributes of both the nominal stimulus and the contextual stimuli. Once this has been completed, relationship encoding is engaged. Here it is proposed that all three relationships (context–no consequence, S_1–no consequence, and context–S_1) are encoded.

Both stimulus and context have the same consequence: nothing. Two sets of stimuli presented conjointly, *and* with both having the same set of consequences, should produce an association between these stimuli. The common conditioning of inattention, then, to these stimuli is, in part the cause of the context–S_1 association. It is this context–S_1 association that provides the basis for context playing the role of an occasion setter for the expression of conditioned inattention to S_1 during the test phase.

As elaborated earlier, this type of analysis can account for the major context effects in the latent inhibition literature. This point receives some confirmation from a recent conditioned taste aversion study by Archer et al. (1986) using the noisy-bottle technique to simulate a local context. In that study, all groups were conditioned and tested with a noisy-saccharin solution. The group that was preexposed to the noisy-saccharin condition showed significantly more latent inhibition than did groups preexposed to noisy bottle without saccharin, or saccharin without noisy bottle. In apparent opposition to this, it will be recalled that Holland and Forbes (1980) found an opposite effect, with more latent inhibition to the elements than to the compound. However, this discrepancy is easily resolved when one recalls that in the Holland and Forbes (1980) study, the two stimuli were presented sequentially, S_1–S_2, a procedure that, according to conditioned attention theory, should, indeed, reduce latent inhibition. Note the critical differences between the two studies. In the Archer et al. (1986) study, the S_2 (noise context) was likely to have formed an association with S_1 because the two stimuli completely overlapped and resulted in the *same* consequences. In the study by Holland and Forbes (1980), S_1 and S_2 did not overlap, and they did not have the same consequences. We would propose, then, that normal stimulus preexposure produces a stimulus–context association and that the conditioning of inattention to both of these events is, in part, the cause of the compounding, as well as the final product of the association. It would seem to follow from the foregoing that the context specificity of latent inhibition would also be some function of number and duration of stimulus–context preexposures. A small number of stimulus preexposures in a particular context should produce a latent inhibition effect that is less context-specific than that produced by a larger number of stimulus preexposures. In fact, one might expect that the context specificity of latent inhibition would be differentially modulated by number, as compared with duration, of stimulus preexposures.

More about context

The effects of context on latent inhibition are very clear and can be easily summarized: (1) A change in context from preexposure to acquisition–test will disrupt the usual stimulus preexposure effect (i.e., it will reduce or abolish latent inhibition). (2) For context and stimulus familiarity to be effective in promoting latent inhibition, the two have to be preexposed together. (3) The available data suggest that the effects of context extinction following stimulus preexposure in that same context are, at best, weak. (4) However, preexposure of the context, by itself, prior to acquisition, does interfere with subsequent learning, thus demonstrating latent inhibition of context–CS and –US associations. (5) Preexposure of the context prior to stimulus preexposure in the same context increases the magnitude of the latent inhibition effect.

In spite of the considerable data on these context effects, the various theories of latent inhibition have ascribed to context considerably less than the importance it deserves. Indeed, some theories have virtually ignored the role of context (Hall and Pearce, and Mackintosh). Others have given it a general status (i.e., context should affect latent inhibition in the same manner that context affects normal conditioning) (Lubow, Weiner, & Schnur, 1981). Only in Wagner's priming theory is context accorded a major role in the development of latent inhibition. As a result of preexposure of stimulus A in context Y, A–Y become associated, so that at the time of acquisition of the new A–US association, context Y primes A into STM, and such priming interferes with subsequent processing of the new A–US association. The evidence supporting this theory, particularly in regard to the effects of context preexposure prior to stimulus preexposure in that context, and extinction of the presumed stimulus–context association, has been discussed elsewhere (see chapter 3). Such evidence has been considerably less than overwhelming.

Although agreeing with Wagner as to the importance of context for an understanding of latent inhibition, here we shall suggest a completely different theoretical mechanism for context effects, indeed, one that operates exactly the opposite from that proposed by Wagner. As a result of stimulus–context preexposure, the context develops the properties of an occasion setter for the S_1–no consequence association. As an occasion setter during the test phase, the context serves to *prevent* S_1 from entering working memory, rather than, as Wagner postulates, to prime S_1 into STM. There are several advantages to this new approach.

First, it solves some logical problems that are raised by Wagner's formulation. As discussed earlier, there is concern as to how, in Wagner's model, a context–stimulus association is learned during preexposure if priming in STM interferes with rehearsal. The context, always present, should be continuously represented in STM via self-generated priming, and thus it should interfere with the context–stimulus rehearsal. The new formulation accepts the idea that there is a joint rehearsal of context and stimulus representations in STM and that such rehearsal promotes the development of a context–stimulus association. However, a critical addition is proposed. Once this association is established, the context *prevents* the priming of the stimulus into STM.

Second, the prevention hypothesis is in accord with observations with human subjects. Subjects who have been preexposed to the stimulus under conditions of masking are not aware of the target stimulus in the test session (Ginton et al., 1975). Because it is agreed by many cognitive theorists that one of the characteristics of a stimulus in working memory is that it is consciously perceived or in a state of awareness, these data suggest that the preexposed stimuli *do not reside* in STM, at least not at the time of test.

Third, and related to the foregoing points, common sense would seem to demand that unimportant stimuli (stimuli that have been followed by no consequence) be kept away from STM, which by the nature of its limited capacity can ill afford to be constantly aggravated by ecologically trivial events, events that have been previously experienced as unimportant. Indeed, because these make up the vast majority of the stimuli impinging on our sensors, every moment of our existence would be characterized by the Jamesian booming, buzzing confusion if contexts were continuously priming into the conscious awareness of STM the very stimuli that characterized the context's insignificance.

The problem, then, is how to keep these stimuli out of STM, and it is proposed that the role of context in latent inhibition can be formulated around the general notion that context serves to minimize the amount of information deployed in STM. More specifically, it is proposed that during the stimulus preexposure stage, the organism (1) establishes an S_1–no consequence association according to the rules of conditioned attention theory; (2) in addition, the organism develops an associative link between the stimulus and the context, such that (3) the context becomes an occasion setter for the expression of the S_1–no consequence relation; (4) as a result, in the acquisition–test phase the S_1 is *kept from* STM by the presence of the context in which the S_1–no consequence association was developed. The net effect of this latter process is to interfere with the acquisition of any *new* association involving S_1 as displayed in the latent inhibition phenomenon. It may very well be that the role of context as an occasion setter for the expression of associations is either special to or stronger for S_1–no consequence associations than for S_1–reinforcement associations. Although such a conclusion would destroy the CAT-proposed symmetry between stimulus preexposure and normal classical conditioning, let the data lead where they may.

Final comments on conditioned attention theory

A theory to account for one particular phenomenon, latent inhibition, has been presented. Although in this sense the theory is narrowly conceived, nevertheless it can be used to gain more complete understanding of other phenomena, including habituation, sensory preconditioning, blocking, and the feature positive effect (Lubow, Weiner, & Schnur, 1981), as well as learned helplessness and perhaps even schizophrenia (see chapter 10). As such, a very specific theory may have rather general implications. To summarize, conditioned attention theory is based on the application of conditioning principles during the stimulus preexposure phase of a transfer experiment. By considering the absence of a significant event following the preexposed stimulus as an unconditioned stimulus for conditioning inattention, we derive a relatively large number of specific predictions. Empirical support for these predictions was drawn from a series of experiments specifically designed to evaluate the theory, as well as from experiments conducted outside the theory's framework.

Although the apparent generality of the theory is complemented by the formality with which it is stated and its ensuing testability, the use of the word "attention" to describe the locus of the effect may, to some, be suspect. In order to preempt the accusation that the concept of attention, as employed in conditioned attention theory, is devoid of utility and/or is circular, it is necessary to explicitly address that problem. To begin with, it is recognized that "attention" is indeed a much abused term. We appreciate Thompson's claim (1972) that "to assert that an animal makes a differential response to one of two stimuli because he selectively attends to that stimulus represents no significant advance over the assertion that God . . . makes it happen" (p. 121).

However, the term "attention" is used in this book as an intervening variable, in the same sense that Hull used such concepts as "drive" or "habit strength." To this end, the antecedent conditions by which attention is varied, and the subsequent conditions within which that variation is manifest, have been prescribed; that is, the rules for manipulating and measuring attention have been independently identified. In addition, however, the attention construct is used as a shorthand notation for some underlying process that must necessarily coordinate the functional relationship between input and output. As such, it derives from several different sources. On the other hand, as indicated in the review of theories of latent inhibition, several different explanations of latent inhibition have been refuted, such as competing response, conditioned inhibition, and so forth. Almost by elimination, then, one is left with an effect that appears to reflect reduced stimulus salience that is translated into a loss of stimulus associability. This, in turn, is treated as an attentional effect, for several reasons: (1) It is not contradicted by any of the data. (2) It reflects the renewed interest in cognitive aspects of learning, where the organism actively processes information. (3) It complements the *general* terminology already in use in modern learning theory, particularly the α of the Rescorla-Wagner (1972) model, and the various theories derived either directly or indirectly from it (e.g., Mackintosh, 1975; Pearce & Hall, 1980). (4) It corresponds with the effects of a variety of physiological manipulations. Thus, for example, hippocampectomy (Ackil et al., 1969; Kaye & Pearce, 1987a,b; Solomon, 1977; Solomon & Moore, 1975; Solomon et al., 1980; Weiss et al., 1974) and electrical stimulation of the hippocampus (Salafia & Allan, 1980b) disrupt latent inhibition. The effects of these lesions are widely regarded as resulting in a failure to produce attention decrements where such degradations should be expected. Because the same lesions also prevent the development of latent inhibition, it is only reasonable to assume that for the normal subject, during nonreinforced preexposure, there is a centrally mediated decrement of attention to the preexposed stimulus, and that decrement is responsible for the subsequent latent inhibition effect. (5) Finally, identification of the underlying process as attentional allows one to make contact with other areas of research. Some of these points of contacts are described in the next chapter, particularly in regard to learned helplessness and schizophrenia.

10 Some applications of conditioned attention theory: learned helplessness and schizophrenia

Although conditioned attention theory was developed specifically to account for latent inhibition effects, it is also relevant to other phenomena in the area of learning, particularly to those situations where the target stimulus is presented under conditions in which attention is diverted from it by competing stimuli, such as in blocking and overshadowing, or in which stimuli are presented repeatedly, as S_1 in the preexposure phase of the learned helplessness paradigm, or S_1–S_2 in the first phase of sensory preconditioning. In regard to the first of these areas, Lubow, Weiner, and Schnur (1981) have already commented on the implications of conditioned attention theory for understanding blocking and overshadowing, as well as the feature positive effect. Similarly, in the second area, the relationships among CAT, learned helplessness, habituation, and sensory preconditioning have been explored. With the exception of learned helplessness, there is little to be added to the already published analysis, and the interested reader is referred to the original source (Lubow, Weiner, & Schnur, 1981). However, in regard to learned helplessness, new and important materials are available, enough to warrant a reconsideration of the relationship between latent inhibition and learned helplessness. In addition, a literature has recently developed on the relationship between latent inhibition, as it reflects normal attentional processes, and schizophrenia, often regarded as characterized by attentional dysfunction. These two topics, learned helplessness and schizophrenia, will be explored in the following two sections.

Learned helplessness

Animals preexposed to a series of unavoidable and inescapable shocks, as compared with animals receiving no shock preexposure, exhibit marked interference in subsequent tests, particularly tasks involving learning. Animals receiving the same amount of escapable shock show no such interference. This phenomenon has been labeled "learned helplessness" (Maier & Seligman, 1976; Maier et al., 1969; Seligman, Maier & Solomon, 1971).

Learned helplessness and latent inhibition paradigms both involve noncontingent preexposure to stimuli later to be used in a learning task, and, in both, subsequent learning decrements to the preexposed stimuli are obtained. The apparent similarity between the two phenomena has been acknowledged by a number of authors (Baker, 1976; Lubow, Caspy, & Schnur, 1982; Mackintosh, 1974; Wickens et al., 1977). "Just as a stimulus which is uncorrelated with changes in reinforcement is only with great difficulty established as a CS for reinforcement during subsequent conditioning, so, when changes in behavior are uncorrelated with changes in reinforcement, any subsequent association between behavior and reinforcement is hard to establish" (Mackintosh, 1974, p. 218).

One example of a within-experiment demonstration of the similarities between latent inhibition and learned helplessness was provided by Wickens et al. (1977), who conducted a series of experiments to investigate the effects of prior noncontingent experience with eventual CSs and USs on subsequent acquisition of conditioned responses to these stimuli. They concluded that if either the to-be-CS or the to-be-US occurs during a preexposure session, it subsequently becomes somewhat ineffective in the establishment of a CR, and especially ineffective in the maintenance of a CR over time.

Seligman et al. (1971) and Maier and Seligman (1976) provided an early theoretical interpretation of the learned helplessness phenomenon. According to these authors, during shock preexposure, animals learn that responding is independent of shock termination. This learning interferes with the subsequent response–outcome acquisition. The theory emphasizes lack of response–outcome contingency during the preexposure session and the resulting development of a cognitive set of uncontrollability, which in turn affects subsequent behavior through emotional, motivational, and associational interference.

Although Seligman and his colleagues defined the uncontrollable shock preexposure as a lack of contingency between a repeatedly presented stimulus and other events in the environment, the learned helplessness paradigm can also be considered as an instance of simple stimulus preexposure. Under such circumstances, according to CAT, there should be conditioning of inattention to the stimulus, thereby reducing its associability in a subsequent test. On the other hand, when animals are given the opportunity to respond to the shock during preexposure, no subsequent decremental effects are found. In terms of CAT, the controlling response constitutes an event of consequence in the environment that is correlated with the preexposed stimulus (i.e., an S_2). As such, it should maintain attention to the shock and serve to attenuate the future learning decrement. It will be recalled that when animals are given an opportunity to respond to the to-be-CS during preexposure, latent inhibition to that CS is also attenuated (Lubow, Schnur, & Rifkin, 1976).

Like the theory proposed by Maier and Seligman (1976), CAT emphasizes a learning process during the stage of preexposure and a transfer of this learning to the response acquisition stage. However, CAT and learned helplessness theory differ in the identification of what is learned and in the specification of which theoretical construct best describes the acquisition of such learning. Whereas learned helplessness theory emphasizes the lack of contingency between response and outcome, and consequently the development of a cognitive set of uncontrollability during shock preexposure, CAT addresses itself to lack of contingency between the repeatedly presented simulus (shock) and *any other event* in the environment, and to the development of an associability deficit to the preexposed stimulus due to the conditioning of inattention to that stimulus. What is suggested by CAT is that response–outcome contingency is but one instance of a broader category of contingency that can be applied to all events and that can be analyzed in terms of general properties of such contingencies, rather than in terms of controllability–uncontrollability.

An additional difference between the two theories is that helplessness theory emphasizes, in addition to the associative interference, general motivational and emotional deficits due to the experience with uncontrollable USs. CAT, however, focuses on primary learning processes during the stage of preexposure to shock, and it focuses on associative deficits only as the effects of such preexposure during subsequent learning about the stimulus. This is not to say that emotional and motivational disturbances due to shock preexposure do not exist. Exposure to shock is an aversive situation that clearly elicits strong emotional responses. However, if lack of contingency among events is emphasized, rather than uncontrollability of events, the emotional and motivational effects due to the latter become of secondary importance in producing the observed learning decrement. In fact, without appealing to possible negative emotional and motivational aftereffects of shock preexposure, CAT would still predict more profound decremental effects due to such preexposure, as compared with those obtained with CS preexposure. It will be recalled that one of CAT's postulates states that conditioning of inattention to the preexposed stimulus is a positive function of stimulus intensity. More latent inhibition has been obtained to preexposed stimuli of high intensity than to those of low intensity (Crowell & Anderson, 1972; Schnur & Lubow, 1976). Shock, the preexposed stimulus in the typical learned helplessness procedure, is an intense stimulus. Consequently, CAT would predict considerable conditioning of inattention to the shock, and thus a marked learning decrement in a subsequent transfer task.

The test of the applicability of CAT to learned helplessness is straightforward. If, as the theory maintains, the decremental effects of shock preexposure are due to conditioning of inattention to the shock, then *any* event in the environment correlated with the shock should attenuate the learned helplessness effect, *even though it does not terminate shock presentation* (i.e., does not establish a response–outcome contingency and does not increase controllability).

This critical prediction, derived from CAT, was subjected to several tests in our laboratory. In one such study, five groups of rats were run. In the preexposure phase, three of the groups constituted the traditional "triadic" design. The animals in the fourth group received a short-duration light as an S_2. The light appeared immediately on each shock offset. Termination of the shock was independent of any response that the animal might make. In the fifth group, the same numbers of lights and shocks were presented, but the time interval between them was randomized. Both of these groups were yoked to the shock presentations of the other three groups. All animals subsequently were tested on an FR-2 escape task. The results of the experiment confirmed the CAT predictions: Animals that received light immediately following each shock presentation did not show a decrement in escape learning, as compared with the nonshock group or with the group for which shock and light were unpaired. In fact, they performed as well as those of the active-escape control group (i.e., the animals that, during preexposure, could terminate the shock by responding). Thus, presentation of an event, S_2, in a contingent relationship to the preexposed stimulus, shock, prevented "learned helplessness," even though the animals in this group did not have the opportunity to "control" shock termination (Lubow, Weiner, Rosenblatt, Lindenbaum, & Margolit, 1979).

Although we have had some difficulty in replicating these results in our laboratory, primarily because of occasional failures to obtain learned helplessness itself, several recently published studies support our conclusions (Mineka, Cook, & Miller, 1984; Starr & Mineka, 1977; Volpicelli, Ulm & Altenor, 1984). Those studies are described next.

Starr and Mineka (1977, Exp. 2) found that a feedback signal provided to yoked rats at the termination of the passively received shocks (given at the same time that the animals with control made the shock-terminating response) reduced the amount of conditioned fear, as measured in an avoidance test. In fact, this S_1-S_2 group was not different from the group that controlled shock offset.

Mineka et al. (1984) pursued the suggestion that control of shock is important in reducing fear, at least in part because of the stimulus feedback generated by the controlling response, rather than the control per se. Using the multivariate fear assessment techniques developed by Corriveau and Smith (1978) (e.g., readiness to leave a safe ledge for a previously shocked floor; amount of time spent on the previously shocked floor), Mineka et al. (1984, Exp. 1 and 2) found that a light-off feedback stimulus immediately following shock termination produced the same amount of reduced fear (compared with the yoked group) as did active control.[1] Similar results were found when fear was measured by amount of freezing behavior in a signaled shock test (Mineka et al., 1984, Exp. 3 and 4).

Whereas in the foregoing experiments the associability of the shock was indexed by various classically conditioned fear responses, Volpicelli et al. (1984) used an instrumentally conditioned response as the dependent variable. In two experiments, four groups were differentiated on the basis of preexposure conditions. Three groups represented the traditional triadic design: a group that received shock and that could make an escape response (control); a yoked group that did not have the opportunity to escape (no control); a non–shock-preexposed group. A fourth group was treated like the no-control group, but received a light-offset feedback signal at the end of the inescapable shock. The two experiments differed in the type of test that followed the preexposure session.

In both experiments the usual learned helplessness effect within the triadic design was obtained, with the groups receiving preexposure to uncontrollable shock performing more poorly than the groups with escapable shock or no shock. The groups with the feedback signal (S_2) did not exhibit any decrement in performance compared with the latter two groups (i.e., there was no learned helplessness effect). This was the case when either an FR-3 response or an FR-1 with a 3-sec delay between the response and the shock offset was required to escape shock in the test.

The same pattern of results has been obtained with human subjects (Lubow, Rosenblatt, & Weiner, 1981, Exp. 1 and 2; Barber & Winefield, 1986, Exp. 1), but not on all occasions (Barber & Winefield, 1986, Exp. 2, 3, and 4). The successful demonstrations of learned helplessness reduction in humans with an S_1–S_2 procedure employed passive presentations of a noxious stimulus during the preexposure phase (high-intensity tones to which the subjects were instructed just to listen). The unsuccessful S_1–S_2 procedures used innocuous stimuli and/or frustration-producing conditions (moderate-intensity tones and false instructions to the subject that there was some action that could be taken that would terminate the tone). Table 21 describes the four possible combinations of stimulus intensity and instructions, as well as the use of these combinations in the Lubow, Rosenblatt, and Weiner (1981) and Barber and Winefield (1986) studies.

It should be noted that only the treatment shown in the upper right-hand cell of Table 21 (high-intensity tone and passive instructions) produced attenuation of learned helplessness in the S_1–S_2 condition, and this occurred in all three experiments listed in that cell. This, of course, is the treatment cell that also best describes the successful S_1–S_2 manipulation in those animal studies that examined the effects of the S_1–S_2

Table 21. *Experimental conditions during S_1–S_2 preexposure in the Lubow, Rosenblatt, and Weiner (1981) and Barber and Winefield (1986) studies (L and B, respectively)*

Instructions	Intensity of tone	
	Low	High
Passive	B (Exp. 3)	L (Exp. 1)
		L (Exp. 2)
		B (Exp. 1)
Frustrating	B (Exp. 2)	B (Exp. 4)

manipulation on fear (e.g., Mineka et al., 1984; Starr & Mineka, 1977) and learned helplessness (shock escape) (Volpicelli et al., 1984). On the other hand, S_1–S_2 manipulations within the latent inhibition paradigm would appear to be situated in the adjacent cell: passive preexposure of a low-intensity stimulus. However, the successful S_1–S_2 latent inhibition procedures used S_1s of fairly high intensity or salience (e.g., Lubow, Schnur, & Rifkin, 1976; Szakmary, 1977a, b). Indeed, Mercier and Baker's failure (1985, Exp. 1) to find an S_1–S_2 effect in rats *may* have been due to the use of a 79-dB SPL clicker against a background noise of 60 dB SPL.

It will be recalled that CAT predicts, and there is evidence for, more latent inhibition with an intense preexposed stimulus than with a less intense stimulus. It would appear to follow that if S_2 in the S_1–S_2 preparation is effective in reducing latent inhibition, it should be more effective with a high level of latent inhibition. In other words, S_1–S_2 effects should be more visible with a high-intensity S_1 than with a low-intensity S_1.

The tenor of the arguments presented earlier suggests that learned helplessness and latent inhibition are, at least in part, the same (i.e., governed by the same underlying process). To support this contention, we have attempted in the preceding paragraphs to show that similar manipulations have similar effects within both procedures – in particular, reduction of the test performance decrement as a result of S_1–S_2 preexposure. The failures to find such an effect, as in the studies of Barber and Winefield (1986, Exp. 2, 3, and 4), are attributable to the particular stimulus preexposure induction procedure, which may not have produced latent inhibition. This argument, of course, is quite circular if one maintains that there is complete identity between the latent inhibition and learned helplessness processes. In fact, there is ample evidence of important differences between the two phenomena. One such difference concerns breadth of transfer or generalization. There are considerable data, from animal studies that we have reviewed at length, indicating that latent inhibition is stimulus-specific and context-specific, whereas learned helplessness effects appear across stimuli and contexts that are quite different from the preex-

Table 22. *Amount of test decrement (D)*[a]

Preexposure procedure	Test stimulus	
	Same	Different
LH	4	2
LI	2	0

[a] Computed from equation (9) for latent inhibition (LI) and learned helplessness (LH) procedures in which the test stimulus is either the same as or different from the preexposed stimulus (for $A = 1$ and $B = 2$).

posure conditions; see Maier and Seligman (1976) for a review.[2] Indeed, we confirmed this conclusion in a series of studies with 10-year-old children that directly compared latent inhibition and learned helplessness effects (Lubow, Caspy, & Schnur, 1982).

To handle these differences in generality, we have argued (Lubow, Caspy, & Schnur, 1982) that CS preexposure produces a stimulus-specific associative deficit, whereas US preexposure produces *both* a stimulus-specific associability deficit and a generalizable performance deficit, perhaps stemming from emotional-motivational factors. Thus, learned helplessness would have two sources of performance decrement, but latent inhibition only one. This formulation, after the fact, accounts for the generalization of learned helplessness and the absence of generalization of latent inhibition. A refinement of the formulation, as shown in equation (9), would also seem to integrate a number of the findings discussed in the preceding paragraphs:

$$D = f(A \times B) + B \tag{9}$$

D represents the amount of performance decrement exhibited in the postpreexposure test; A represents the stimulus-specific associative deficit; B represents the nonspecific generalizable effect. As can be seen, the parenthetical value $(A \times B)$ suggests that in addition to the presence of an independent B function, the associative deficit is modulated by the B factor. A further assumption of the model is that the A process can be elicited only by the presence of the preexposed stimulus in the test situation, whereas the B process is not under such stimulus control. An additional assumption, though not critical to the model, is that B is a function of the intensity of the preexposed stimulus.

Let us examine several situations in order to evaluate the model. The first set of examples compares the amount of latent inhibition and learned helplessness exhibited in a test that uses either the same stimulus as that employed during preexposure or a different stimulus. If we assign arbitrary values to A and B such that $A = 1$ and $B = 2$ and solve equation (9) for the four possibilities (latent inhibition or learned helplessness, same stimulus or different stimulus at test), we obtain the figures shown in Table 22. The pattern of simulated results fits nicely with

Table 23. *Predicted order of results from the model for LI and LH procedures, with added S_2, on two tests*

Preexposure procedure	Test stimulus Same	Different
LH + S_2	2	2
LI + S_2	0	0

that obtained empirically: (1) Learned helplessness procedures are generally assumed to produce a greater deficit than do latent inhibition procedures. (2) Learned helplessness is found to generalize to new situations. (3) Latent inhibition is found not to generalize to new situations.

Now let us look at the effects of adding a second stimulus, S_2, during preexposure, so that each S_1 is followed by S_2. The effects of that S_2 have been proposed to be stimulus-specific, to interfere with latent inhibition. Assuming that S_2 is complete in its antagonistic effect, and retaining the same values for A and B as in the first set of examples, we obtain the simulated results shown in Table 23.

These results suggest that the S_1–S_2 manipulation should *reduce*, but not obliterate, the amount of learned helplessness, as compared with the S_1-alone manipulation as described in Table 22, but only in the specific test, not in the generalization test; and the gradient of generalization should be eliminated. Again, this general pattern of results has been found in our laboratory (Lubow, Rosenblatt, & Weiner, 1981; Lubow et al., 1979).

Additional tests of the model are required, especially tests that will employ comparable procedures for producing latent inhibition and learned helplessness, as well as stimulus-specific and generalization tests that will allow for a direct comparison of the relative amounts of performance decrement. Furthermore, it is of critical importance to vary the number of preexposures. As described earlier, all of the effects of stimulus preexposure are strongly modulated by the number of preexposures, so that, for example, excessive numbers of S_1-alone and S_1–S_2 preexposures may produce the same amounts of decremented test performance.

In summary, then, it is proposed that latent inhibition and learned helplessness are related in that the latter typically incorporates a stimulus-specific latent inhibition effect that is added to a nonspecific emotional-motivational component. These separate processes can be differentiated by using "same" and "different" tests. Such tests should result in identifying a family of variables that are differentially effective in modifying latent inhibition and learned helplessness.

Schizophrenia, latent inhibition, and attention

Once attention is allowed to participate as an explanatory aid to an understanding of latent inhibition, it is inevitable that researchers should try to discover or develop links between latent inhibition and schizophrenia. This simply follows from the fact that both scientists and clinicians have given a prominent role to attentional processes in schizophrenia. If latent inhibition reflects the effects of normal attentional functioning, and schizophrenia is characterized by attentional dysfunction, then (as long as one is describing the same attentional process, and this indeed is the core of the problem) schizophrenics should exhibit a different pattern of latent inhibition than do normals. A continuation of this optimistic logic would suggest that if the foregoing is true, then to the extent that one has a good theory of latent inhibition, one also has a better understanding of the schizophrenic process.

Because the arguments for the utility of the conditioned attention theory have already been presented, and a considerable effort has been expended in detailing the meaning of "attention" in this theory, a brief summary will suffice here.

Latent inhibition in the nonhuman organism is assumed to be under control of an automatic mode of operation, so that with repeated presentations of a stimulus, two or more processes are engaged: stimulus property extraction and stimulus relationship encoding, with property encoding preceding relationship encoding. The latent inhibition effect is a product of the operation of the relationship encoding process. Specifically, the associability deficit that results from stimulus preexposure is a function of the normal coding of the stimulus–no-consequence relationship, the final state of which is the conditioning of inattention to the nominal stimulus. Furthermore, as has been shown, the expression of this stimulus-specific conditioned inattention is under the control of the context in which it was developed. This description holds for subhuman organisms and young children. However, the situation is more complicated with human adults, in whom latent inhibition does not occur unless the preexposed stimulus is presented in conjunction with a masking task, the purpose of which is to divert attention from the normal stimulus. To explain these discrepancies between animal and human studies, two additional attentional constructs were introduced: the external attentional override and the internal attentional override. The former is a process common to both humans and lower animals, and it is concerned with the relative maintenance or restoration of the attentional response by external stimulation, characterized by, for example, stimulus change, complexity, intensity, and so forth. In contrast, the internal attentional override is viewed as a uniquely human process. It serves the same purpose as the external override, but the source of information for maintenance or restoration of attention is within the organism. The internal attentional override is conceived as being basically verbal and under voluntary control. In the adult human, latent inhibition fails to develop without the use of a masking task, because the normal conditioning of inattention is prevented by the engagement of the internal attentional override. This, by definition, maintains the attentional response to the preexposed stimulus (or, conversely, precludes the conditioning of inattention to that stimulus). The function of the masking task, then, is to disengage the internal attentional override. When this is accomplished, normal (i.e., as in lower animals) conditioning of inattention ensues, and therefore latent inhibition will be manifest in the test phase.

The foregoing description succinctly describes the role of attention in latent inhibition. What can be said about attention in schizophrenia? More specifically, is there any contact between the usages of the word "attention" in latent inhibition and in schizophrenia? As we shall see, the answer is, at least tentatively, affirmative.

It has long been recognized that the behaviors of schizophrenics exhibit some patterns that can be described in terms of an attentional dysfunction. Thus, Kraepelin wrote

It is quite common for them [patients] to lose both inclination and ability on their initiative to keep their attention fixed for any length of time.

The patients digress, do not stick to the point, let their thoughts wander *without voluntary control* in the most varied directions. On the other hand the *attention is often rigidly fixed* for a long time, so that the patients stare at the same point, or the same object, continue the same line of thought, or do not let themselves be interrupted in some definite piece of work. Further it happens that they deliberately turn away their attention from those things to which it is desired to attract it, turn their backs when spoken to, and turn away their eyes if anything is shown to them. But in the end there is occasionally noticed a kind of irresistible attraction of the attention to casual external impressions. [Kraepelin, 1919/1971, p. 5; emphasis added.]

Although Bleuler's descriptions (1911/1966) of schizophrenics' behavior emphasized disorders of thought, what he called "the loosening of associated threads," it is not difficult to portray such behaviors in terms of a failure to attend to appropriate context information. Indeed, Chapman and Chapman (1973) placed such an interpretation on a number of studies that demonstrated schizophrenics' inability to use contextual restraints to organize verbal material that was subsequently to be recalled (e.g., Honigfeld, 1963; Lawson, McGhie, & Chapman, 1964; Rutter, Wishner, & Callaghan, 1975).

The appeal to attentional processes in describing and explaining schizophrenic behavior was, at least in part, a response to the renewed general interest in cognitive psychology. Broadbent's early model (1958)

of information processing stimulated a number of searches for attentional malfunctions in schizophrenics. Thus, a phenomenon of schizophrenia, such as overinclusive thinking, has been described in terms of a general breakdown of the filter process (e.g., Payne, Matussek, & George, 1959). Shortly after Broadbent's seminal book, a monograph by McGhie and Chapman (1961), which was to become very influential, summarized a number of clinical manifestations and case reports of schizophrenia. They concluded that perception and thinking disturbances in acute schizophrenics were incidental to the primary disturbances in the control of the direction of attention.

The elaboration of a variety of other information processing models of cognitive behavior during the next 25 years, for almost all of which the concept of attention was central, led to a further increase in the number of attempts to specify an attentional deficit in the schizophrenic; see Gjerde (1983) for a recent review. Although this is not the place to review the extensive research literature on this topic, it is quite evident that today, "many writers regard a deficit in selective attention as one of the central psychological lesions in schizophrenia" (Anscombe, 1987), and "many theorists . . . view faulty attention as the fundamental cognitive deficit" of schizophrenia (Mirsky & Duncan, 1986).

With this as background, it is but one step to make the association between latent inhibition and schizophrenia: Both involve some type of attentional process – latent inhibition reflects the normal operation of an attentional process, whereas schizophrenia reflects a dysfunction of an attentional process. However, until now, there has been no assurance that it is the *same* attentional process that is engaged in both latent inhibition and schizophrenia. That, indeed, was the point of departure for the series of animal studies in our laboratory involving the effects of amphetamine and haloperidol on latent inhibition.

The logic of the argument is derived from the premise that amphetamine, a dopaminergic drug, can produce a schizophrenia-like state in humans. The evidence for this statement, which had led other investigators to create an animal–amphetamine model of schizophrenia (e.g., Kokkinidis & Anisman, 1980; Segal & Janowsky, 1978), is fourfold: (1) Amphetamine intoxication in normal humans elicits schizophrenic symptoms that often are clinically indistinguishable from paranoid schizophrenia (Angrist & Gershon, 1969, 1970; Bell, 1973; Connell, 1958; Griffith, Cavanaugh, Held, & Oates, 1972; Griffith, Oates, & Cavanaugh, 1968). (2) Amphetamine administration exacerbates existing schizophrenic symptoms[3] and reinduces psychoses in schizophrenics who are in remission (Alpert & Friedhoff, 1982; Janowsky, Huey, Storms, & Judd, 1977; Segal & Janowsky, 1978). (3) Amphetamine administered to rats produces a peculiar series of behavioral disturbances said to resemble the picture of human amphetamine psychosis and clinical schizophrenia (e.g., Ellison & Eison, 1983; Kokkinidis & Anisman, 1980; Robinson & Becker, 1986).[4] (4) Antipsychotic drugs (neuroleptics) used in the treatment of schizophrenia (e.g., haloperidol) antagonize amphetamine-induced psychoses in humans (e.g., Angrist, Lee, & Gershon, 1974) and in animals Segal & Schuckit, 1983).

The evidence presented earlier argues for the viability of an animal–amphetamine model of schizophrenia. At the same time, by adding that the effects of neuroleptic drugs are achieved by blocking dopamine receptors, this becomes the basis of the more general dopamine hypothesis of schizophrenia (e.g., Snyder, 1973; Swerdlow & Koob, 1987).

To briefly summarize, in addition to the data offered in favor of latent inhibition representing a normal attentional function, and schizophrenia representing an abnormal function, there is also reason to believe that some schizophrenic symptoms can be induced by amphetamine. It follows from these points that (1) amphetamine should disrupt latent inhibition in animals, and (2) schizophrenia in humans should disrupt latent inhibition. Evidence for both of these effects should provide support for the idea that the attentional dysfunction in schizophrenia is related to disruption of the normal attentional process that subserves latent inhibition. Indeed, the data presented next argue persuasively for just that interpretation.

Effects of amphetamine and haloperidol on latent inhibition in rats

Our first studies began with the use of the conditioned suppression preparation. There were two reasons for that choice. First, conditioned suppression had been used extensively to demonstrate latent inhibition. Second, the index of conditioning is the degree of not responding. As such, the frequently reported activity-enhancing effects of low doses of amphetamine (e.g., Robbins & Iversen, 1973; Scheele-Kruger, 1971) would not be likely to bias the learning score in favor of the hypotheses. Nevertheless, to ensure that the final interpretation would be free from confounding with the drug's effects on activity, a conditioned avoidance procedure also was employed. It would be expected that increased activity, if it occurred at all, would have an influence on learning that was in the opposite direction from that to be found with conditioned suppression.

The results of several different experiments with conditioned suppression (Weiner et al., 1984, 1988, Exp. 1 and 3; Weiner, Israeli-Telerant, & Feldon, 1987) and conditioned avoidance (Weiner et al., 1988, Exp. 2) led to the conclusion that amphetamine does disrupt latent inhibition. More specifically, as described in chapter 6, amphetamine administered in both the stimulus preexposure stage and the acquisition stage affected stimulus-preexposed animals so that they behaved as though they had not been exposed to the stimulus. They learned as well as the nonpreexposed, nonamphetaminized control groups. It is important to note that this effect was observed only for groups that received amphetamine in both the stimulus preexposure and acquisition stages. Stimulus-preexposed groups that were treated with amphetamine in either the preexposure stage or the acquisition stage exhibited normal latent inhibition. Similar results were obtained by Crider, Solomon, and McMahon (1982) and Solomon et al. (1981). That they obtained such an effect when drug administration was limited to the preexposure period can be accounted for by the relatively high dose (4.0 mg/kg, compared with 1.5 mg/kg in our studies) and the fact that the acquisition stage immediately followed the stimulus preexposure stage. Interestingly, Hellman et al. (1983) showed that stress produced by pinching of rats' tails (which also enhanced dopaminergic activity) could substitute for amphetamine administration to attenuate latent inhibition.

If a dopaminergic drug such as amphetamine disrupts latent inhibition in the direction of removing the stimulus preexposure effect, then a dopaminergic antagonist such as haloperidol should increase the effectiveness of such stimulus preexposure. That, indeed, was what happened in conditioned suppression. With a "normal" amount of stimulus preexposure (40 tone presentations), haloperidol in both stages produced more latent inhibition than did the absence of haloperidol (Weiner & Feldon, 1987, Exp. 1). Furthermore, with a typically less than effective number of stimulus preexposures (10 tone presentations), haloperidol did produce the latent inhibition effect (Weiner & Feldon, 1987, Exp. 2). The same pattern was found for avoidance learning, where it was also shown that the effect was dependent on the presence of haloperidol in both the stimulus preexposure and acquisition stages (Weiner, Feldon, & Katz, 1987).

Taken together, the data indicating that latent inhibition is disrupted by the dopamine agonist amphetamine, and enhanced by the dopamine antagonist haloperidol, present a convincing argument for the proposal that latent inhibition is modulated by the same attentional process that is dysfunctional in schizophrenia. To complete the circle, one need only demonstrate that schizophrenic patients do not display latent inhibition.

Schizophrenia and latent inhibition

Although the first study in this area found no differences in the amounts of latent inhibition among normal, paranoid schizophrenic, and nonparanoid schizophrenic populations, it is worth describing the procedures in full because it provides the basis for two experiments that did report the expected results. The general method, therefore, would appear to be a powerful one for further elaboration of the relationship involving latent inhibition, attention, and schizophrenia.

The Lubow et al. (1987) study. This study investigated whether or not there was a disturbance in attentional function, as indexed by disruption of latent inhibition, in paranoid and nonparanoid schizophrenics, as compared with normals. To produce latent inhibition in adults, the procedure developed by Ginton et al. (1975) was employed. They found that a stimulus-preexposed group learned a new association much more slowly than did a group not preexposed to the stimulus (i.e., the former group demonstrated the latent inhibition effect).

The schizophrenic subjects were 39 inpatients of a psychiatric institution. Inclusion of subjects in the study was limited to patients with an unequivocal diagnosis of schizophrenia, as judged independently by two psychiatrists. All patients were on medication. Patients were excluded from the study if they had auditory difficulties or organic brain syndrome or if they had been treated with electroconvulsive shock during the month preceding the experiment.

Assignment of patients to the paranoid and nonparanoid groups was carried out in two ways: (1) by formal psychiatric diagnosis according to the patient's hospital chart and (2) by the Maine scale (Magaro, Abrams, & Cantrell, 1981).[5]

Matched subjects from the psychiatric diagnoses groups were randomly assigned to stimulus-preexposed and nonpreexposed groups. At the completion of the study, the data were reanalyzed using the Maine scale diagnostic criteria.

Two lists, consisting of the same 40 pairs of nonsense syllables, were recorded separately, each on one track of a stereophonic tape recorder. The lists were repeated five times, with no indication as to the termination or restart of each list. The interval between the syllables ranged from 1 to 2 sec. Both tracks were rerecorded, with the target stimulus, a white noise, randomly superimposed 30 times on one of the recordings. Noise duration, which varied from 0.5 to 2.0 sec, averaged 1.25 sec. The volume of the syllables was at the level of normal speech, whereas that of the white noise was approximately half the subjective loudness of the verbal material. Thus, two recordings were created, one with the nonsense syllable list alone (L), and one with the nonsense syllable list plus target stimulus (L + S). The recordings were presented to the subject through stereo earphones.

Data collection from the normal subjects took place at the university laboratories, and collection from the clinical groups took place at the hospital. The basic procedure was the same for both populations, with one exception. For the clinical groups, the latent inhibition session was preceded several days earlier by a session in which the experimenter introduced himself and explained, in general terms, the nature and purpose of the experiment. In the second half of this session, the Maine scale was administered via a semistructured interview.

During the preexposure phase, subjects of all groups were instructed to listen carefully to the recording and to count the number of syllables. They were told that they would have to report those numbers at the end of the recording. Subjects were required to monitor the syllables to ensure that they directed their attention to the masking material, not to the target stimulus, a necessary condition for producing latent inhibition in adults (Ginton et al., 1975; Lubow, Caspy, & Schnur, 1982). The preexposed groups were exposed to recording L + S, which included the to-be-conditioned stimulus together with the nonsense syllables, whereas the nonpreexposed groups were exposed to recording L, which included only the nonsense syllables. In each of the preexposed groups, half of subjects had the to-be-conditioned stimulus presented to the right ear, and half to the left ear. Both groups were later tested with the to-be-conditioned stimulus presented to the ear in which it had been preexposed. For the nonpreexposed groups, half the subjects were tested with the to-be-conditioned stimulus presented to one ear, and half to the other ear.

The test phase began at the termination of the preexposure phase. Each subject was instructed as follows:

We are now starting a new task. During presentation of the recording you will see me raising points on this scoreboard. The adding of the points is dependent on what you will be hearing on the recording. Listen to the recording and watch the score rise. The moment you think you have caught the rule according to which I am raising the points, raise your hand. Repeat this whenever you expect me to raise a point.

In this phase, all groups were exposed to recording L + S. A scoreboard was placed in front of the subject, and a point was added to the board at the termination of each noise stimulus.

The session ended either after the subject had raised his/her hand on five consecutive presentations of the noise stimulus or after 30 presentations of the stimulus. The final score was the number of noise stimuli remaining to the termination of the recording after the subject had raised his/her hand correctly five consecutive times. Subjects who did not reach criterion received a score of zero.

Test performance was dichotomous. With the exception of two subjects from the normal nonpreexposed group who received intermediate scores, all of the remaining subjects either reached the learning criterion within several trials or did not reach it at all. Figure 17 displays the mean number of trials to reach the learning criterion for each of the stimulus-preexposed and nonpreexposed groups (including the two

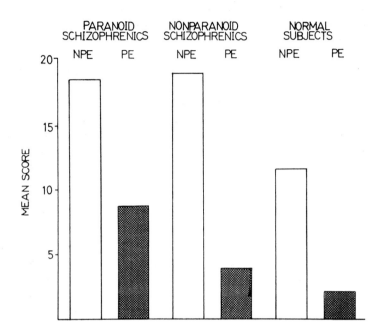

Figure 17. Mean number of trials to reach learning criterion for stimulus-preexposed groups (PE) and nonpreexposed groups (NPE) for three subject populations: paranoid schizophrenics, nonparanoid schizophrenics, and normals. (From Lubow et al., 1987.)

intermediate subjects). As can be readily seen, for each of the subject populations, the preexposed group performed more poorly than the nonpreexposed group.

A comparison of numbers of normal subjects who learned or did not learn indicated that performance was significantly poorer for the preexposed group than for the nonpreexposed group (i.e., the expected latent inhibition effect was obtained). The combined schizophrenic groups performed in a manner similar to that for normals. Comparisons within the nonparanoid and paranoid schizophrenic groups indicated that for both groups the difference between preexposure and nonpreexposure was reliable. Thus, the latent inhibition occurred not only in the normal group but also in schizophrenic subjects, both paranoid and nonparanoid. When the data were reclassified according to the Maine scale and reanalyzed, the same overall pattern as that described earlier was obtained. Normal, paranoid, and nonparanoid subjects all showed significant latent inhibition, with some suggestion that there might be a potentiation of the latent inhibition effect in the schizophrenic subjects. However, this is difficult to evaluate, because in both the preexposed and nonpreexposed conditions, the normal group performed more poorly than did the combined schizophrenic groups.

As in the Ginton et al. (1975) study, normal subjects receiving a series of stimulus preexposures failed to learn a new association to that stimulus, as compared with a nonpreexposed group. Contrary to our expectations, which were based on results obtained with amphetamine-treated animals, schizophrenic subjects, both paranoid and nonparanoid, also learned to ignore or not to attend to the noise stimulus, as reflected in their subsequent failure to learn the association between the stimulus and the change in the value of the counter.

There are two possible explanations for these results. First, the latent inhibition phenomenon, although successful in reflecting attentional deficits in animals, may not be suitable for tapping such deficits in humans. This disassociation may be related to the fact that the procedures used for obtaining latent inhibition in animals differ from those used with humans; the procedures with humans require masking (Lubow, Capsy, & Schnur, 1982). Indeed, the latent inhibition procedure, as employed with human subjects, resembles the dichotic listening task in that it includes two types of information, only to one of which are the subjects explicitly directed to attend. In general, these experiments (e.g., Korboot & Damiani, 1976; Payne, Hochberg, & Hawks, 1970; Pogue-Geile & Oltmanns, 1980; Schneider, 1976) indicate that schizophrenics can attend successfully to a message when there is competing information from another channel. Those deficits that are found in shadowing performance when task requirements are made more difficult (e.g., fast presentation rate, explicit instructions to attend to the irrelevant message, delayed recall of the shadowed material) do not appear to be due to simple selective attention or filtering deficits (e.g., Payne et al., 1970; Pogue-Geile & Oltmanns, 1980; Schneider, 1976). In view of the foregoing, it may not be surprising that schizophrenics in the study by Lubow et al. (1987), with very low competing task demands, exhibited the latent inhibition effect.

The second possible explanation derives from the fact that the patients in the Lubow et al. (1987) study were under medication. Many studies have shown that the amelioration of schizophrenic symptoms treated with antipsychotic drugs is related to normalization of the attentional deficits (e.g., Braff & Saccuzzo, 1982; Kornetsky, 1972; Maloney, Sloane, Whipple, Razani, & Eaton, 1976; Oltmanns, Ohayon, & Neale, 1978; Rappaport, Silverman, Hopkins, & Hall, 1971; Spohn, Lacoursiere, Thompson, & Coyne, 1977). In line with this evidence, and directly relevant to the results of Lubow et al. (1987), are the findings that amphetamine-treated animals that receive chlorpromazine (Solomon et al., 1981) or haloperidol (Weiner & Feldon, 1987) show a normal latent inhibition effect. Moreover, we have recently shown that

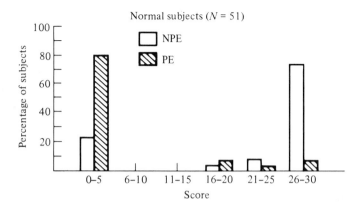
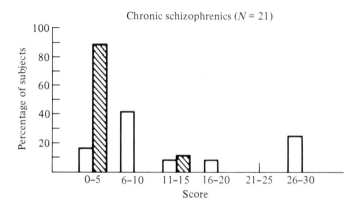
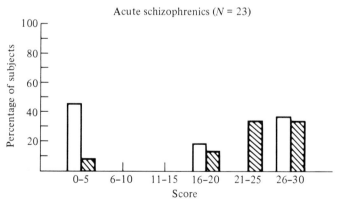

Figure 18. Percentages of normal subjects and chronic and acute schizophrenics who learned to associate white noise with counter increments as a function of preexposure (PE) to the white noise or no preexposure (NPE). (From Baruch et al., 1988a.)

haloperidol, administered on its own, dramatically facilitates latent inhibition in animals (Weiner & Feldon, 1987; Weiner, Feldon, & Katz, 1987). These results suggest that neuroleptics may enhance attentional processes, and they support the possibility that the presence of latent inhibition in schizophrenics may be due to antipsychotic medication. If this supposition is correct, it will strengthen even further the analogy between the animal-amphetamine model and the clinical syndrome.

The Maudsley studies. Considerable support for the foregoing contention has been provided by two recent studies from the Maudsley Institute of Psychiatry (Baruch, et al., 1988a, b), both of which used the procedures described in the previous section. The first study directly examined the hypothesis that the absence of differential latent inhibition between normal and schizophrenic subjects in the Lubow et al. (1987) study was a result of the patient population being on neuroleptic drugs, the effects of which would normalize the defective attentional process and thus produce a normal latent inhibition effect.

To test this hypothesis, Baruch et al. (1988a) compared the performances of three groups: normals, chronic schizophrenics, and acute schizophrenics. The acute patients either were in an acute phase of a chronic disorder or were having a first psychotic breakdown. They all were tested within 2 weeks of the start of the psychotic episode. The chronic group consisted either of remitted schizophrenics attending a rehabilitation center or patients regularly followed up in an outpatient clinic. At the time of testing, chronic patients were free of all major psychiatric symptoms.

Because it is well known that many days are required for the ameliorating action of neuroleptic drugs to take effect (e.g., Klein & Davis, 1969; Lipton & Nemeroff, 1978), it is reasonable to assume that the attentional responses of the acute schizophrenic subjects had not yet been normalized. On the other hand, the chronic schizophrenics were tested at a time when their symptoms were in remission, and thus it can be assumed that their attentional processes were normalized.

We would expect, of course, under these conditions, that the acute schizophrenic group would exhibit a disruption of latent inhibition, whereas the chronic group would not, or at least would be more similar to the normal group than to the acute group. In addition, if the disruption of attention in the acute schizophrenic group were governed by the same process as the amphetamine-induced disruption of latent inhibition in animals, then that should be made manifest by improved test performance in the stimulus-preexposed groups. Indeed, that was exactly what Baruch et al. (1988a) found (Figure 18). Furthermore, a retest of the acute schizophrenic group conducted 6 to 7 weeks after admission, and therefore after the neuroleptics already had had an opportunity to become effective, indicated that in the drug state, latent inhibition was again present, as it should have been if the attentional processes had been normalized. Indeed, there was even some suggestion of an enhancement of latent inhibition in this group. It will be recalled that just such potentiation was present in the stimulus-preexposed groups of rats that were treated and tested with haloperidol (Weiner & Feldon, 1987; Weiner, Feldon, & Katz, 1987). That the neuroleptics were indeed effective was independently indicated by the fact that the acute group also showed a significant decline in psychopathology, as measured by difference scores on the Brief Psychiatric Rating Scale (Overall & Gorham, 1962).

It is important to remember that the normal latent inhibition effect reflects poorer learning in the stimulus-preexposed group than in the nonpreexposed group. Thus, acute, "nonmedicated" schizophrenic subjects, when they display no latent inhibition, are actually showing better test performance, faster learning. Similarly, but in the opposite direction, the normal latent inhibition in the neuroleptic-responding patients results from poorer learning in the stimulus-preexposed group. From these data it is clear that the disruption in latent inhibition in schizophrenics is not a result of a nonspecific effect, a problem that has plagued much of the research on schizophrenia (Spring & Zubin, 1978). Rather, these effects are directly attributable to a more precise dysfunction, in particular a disorder of the same attentional process that mediates latent inhibition in normal humans and lower organisms.

Additional evidence for this proposition can be developed from the view of psychopathology that holds that normal and abnormal behaviors lie on a single underlying dimension, with psychotic states, such as schizophrenia, occupying an extreme position on the continuum (Claridge & Broks, 1984; Eysenck & Eysenck, 1975). It would follow from this position that if it is correct that latent inhibition reflects the normal attentional process that is disrupted in schizophrenia, then normal subjects with different underlying tendencies toward psychotic breakdown should exhibit different amounts of latent inhibition. More specifically, on the basis of the previously reported data, one would expect that subjects who score high on a test that predicts psychotic behavior should exhibit less latent inhibition than a group that scores low on such a test. Baruch et al. (1988b) tested just such a hypothesis. Subjects who did not have a history of mental illness or drug or alcohol abuse (the same subjects who constituted the normal group in the previous study) were required to respond to three questionnaires: (1) the P sale of the Eysenck Personality Questionnaire (Eysenck & Eysenck, 1975), which measures "psychoticism"; (2) a scale from the Schizotypal Personality Questionnaire (Claridge & Broks, 1984), which measures "schizotypal personality"; (3) The Launey-Slade Hallucination Scale

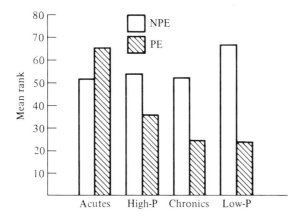

Figure 19. Mean rank score on the auditory task (higher rank = better learning) as a function of preexposure (PE) or no preexposure (NPE) to the white noise, as well as grouping (acute schizophrenics; high-P subjects; chrnoic schizophrenics; low-P subjects). The data for the schizophrenic groups are taken from Baruch et al. (1988a). The groups are ordered from left to right on the abscissa according to the size of the obtained latent inhibition effect. (From Baruch et al., 1988b.)

(Launey & Slade, 1981), which measures the tendency to hallucination. All subjects were then assigned to either the stimulus-preexposed group or the nonpreexposed group, and the usual procedures for demonstrating latent inhibition were performed. After data collection, subjects were divided on the basis of a median split for each of the three questionnaires. Thus, there were three separate 2 × 2 analyses, with one factor being stimulus preexposure versus nonpreexposure, and the other factor being high psychotic proneness versus low psychotic proneness. For both psychoticism score and the schizotypal personality score, highly psychotic-prone (high-P) subjects showed less latent inhibition than did subjects low on the scale of psychosis proneness (low-P subjects).

The pattern of the data from the Maudsley studies indicates that high-P subjects are similar to acute nondrugged schizophrenic subjects, whereas low-P subjects are similar to normal subjects and to neuroleptic-treated chronic schizophrenic patients. These relationships are shown in Figure 19.

One of the important consequences of the data from high-P and low-P normal subjects is that they help clarify the categorization of the schizophrenic subjects in the previous study (Baruch et al., 1988a), in which drug treatment versus nontreatment was confounded with acute versus chronic. The fact that nondrugged high-P normal subjects performed on the latent inhibition test more like the nondrugged acute schizophrenic group than like the drugged chronic group suggests that the effects in the latter group were indeed drug-induced.

Some speculations

The data presented in this section strongly support the idea that the attentional process that governs latent inhibition in normal organisms is the same as the process that is malfunctioning in schizophrenics. This apparently simple statement, together with its supporting data, may provide one of the keys to understanding schizophrenia and its wide variety of symptoms, many of which appear to be contradictory. This is particularly noteworthy in the descriptions of attentional difficulties in schizophrenics. On the one hand, attention may be extremely labile, being captured momentarily first by this event and then by that; on the other hand, attention may be sluggish and even inert, resulting in excessive dwelling on unimportant incidental stimuli. One might speculate that the attenuated latent inhibition produced by amphetamine/dopaminergic activity may be related to the "labile" condition, whereas the super-latent-inhibition state produced by haloperidol, a dopamine antagonist, may be related to the "inert" condition of attention. In the human, both attentional deficits will be characterized by a feeling of loss of voluntary control, a condition in which the patient "increasingly becomes the audience of his mental life, rather than its initiator, so that he becomes passive and displaced" (Anscombe, 1987), a state that may well account for the schizophrenic's loss of identity, passivity, feelings of being an object and being manipulated, and all the emotional responsivity, or lack thereof, that necessarily accompany disengagement. Anscombe (1987) has recently presented an extremely lucid account of such a state. The point here is simply that whereas schizophrenia may be characterized by attentional deficits that are either overly labile or overly inert, both may result from a process that subverts the voluntary attentional override, and though they are different types of attentional deficits, they well may share all of the symptoms that accompany the self-perception of loss of such voluntary attentional control.

On another level, this type of analysis suggests that the dopamine hypothesis of schizophrenia may present only one side of the picture. Excess dopamine activity may account for overly labile attention, whereas curtailed dopamine activity may account for overly inert attention; for a related view, see Matthysse (1978). This contention has received some support from several studies that have successfully treated some of the symptoms of schizophrenia by using a dopamine agonist (Angrist, Peselow, Rubinstein, Corwin, & Rotrosen, 1982; Kay & Opler, 1985). Crow (1980) offered another refinement of the original dopamine hypothesis that may be related to the notions presented earlier. He proposed that positive symptoms of schizophrenia (e.g., agitation, delusions, hallucinations, and various thought disorders) are related to dopaminergic overactivity, which may be reversed by a neuroleptic dopaminergic antagonist, but that negative symptoms (e.g., emotional apathy, absence of volition) are not reversible and possibly are due to structural, as opposed to functional, brain disease. A biphasic dopamine account of psychopathology has recently been presented by Swerdlow and Koob (1987).

To continue the direction of these speculations, let us return to the relationship involving schizophrenia, latent inhibition, and conditioned attention theory. First, on a substantive level, the case has been made for the major premise of this section: The attentional processes that account for latent inhibition in normals are the same processes that are dysfunctional in schizophrenia. This identification of a locus of disruption should prove to be of major importance in understanding schizophrenia. Furthermore, the data presented in the preceding paragraphs not only provide the basis for another animal model of schizophrenia but also offer a model with significantly richer representation, one that provides instructions for inducing and/or ameliorating the pathological behaviors (as, for example, administering dopamine agonists and antagonists). More important, the formulation suggested here offers a clear set of directions for identifying the critical attentional processes. The model suggests that to the extent that one explains latent inhibition, one also can understand schizophrenia, and that the pathologies of schizophrenia are bidirectional deviations from the normal attentional functions as exhibited in latent inhibition. To the degree that conditioned attention theory provides an adequate explanatory model for latent inhibition, it should also contribute to an explication of schizophrenic behavior.

How do we proceed? Several possibilities present themselves. On the empirical level, one might start by listing all of those manipulations that disrupt latent inhibition, with the assumption that there is a reasonable prima facie case that one or more of the variables involved in altering normal latent inhibition may also be related to the disrupted behavior that we call schizophrenia. Indeed, but with considerably more logical support, that was the reasoning behind our work with amphetamine-induced attenuation of latent inhibition and haloperidol-induced exaggeration of latent inhibition. Along the same lines, the following variables, for example, have been shown to modulate latent inhibition: preexposure stimulus intensity and stimulus duration, both of which increase latent inhibition; preexposure of complex stimuli and S_1–S_2 pairings, both of which decrease latent inhibition. One type of speculation that follows from these findings would suggest that the amphetaminized hyperdopaminergic pathological state that corresponds to reduced latent inhibition might be characterized by any one or more of the following phenomenally analogous states: subjective reports of subnormal stimulus intensities, stimulus durations, and stimulus complexities. Conversely, the haloperidol-hypodopaminergic pathological state that corresponds to a super-latent-inhibition effect might be characterized by one or more of the antithetical phenomenal states: subjective reports of abnormally high stimulus intensities, stimulus durations, and stimulus complexities. Clearly, it would be of considerable value to have data from studies that had factorially manipulated amphetamine and haloperidol, or other dopamine agonists and antagonists, with extrinsic stimulus preexposure variables such as the ones noted. Ideally, because of the nonmonotonic relationship between number of stimulus preexposures and latent inhibition, and because of the assumption that stimulus preexposure initiates both property extraction processing and stimulus relationship processing, with the former preceding the latter, such studies should also be crossed with the factor of number of preexposures. As can be appreciated, though the experimental designs are not necessarily complicated, they are certainly heroic. A typical study might consist, for example, of three factors (drug, stimulus intensity, and number of stimulus preexposures), with two levels for the first factor (drug, no drug), three levels for the second factor (low, medium, and high stimulus intensity), and five levels (because we still know so little about its effect) of the third factor (1, 5, 10, 15, and 30 stimulus preexposures). This, of course, makes 30 groups. It is no wonder that such parametric studies are hardly to be found anywhere in the literature; had they been there, this book would have been much shorter, and my colleagues much wiser.

Further evidence for the relationship involving latent inhibition, conditioned attention theory, and schizophrenia comes from studies that have examined the neural basis of latent inhibition (see chapter 6). The dopamine hypothesis of schizophrenia requires that damage to dopaminergic pathways produce schizophrenic symptoms. Such pathways are found in the limbic system, and indeed Weinberger, Wagner, and Wyatt (1983), among others, have concluded that schizophrenia is associated with abnormalities of this system. More specifically, several researchers have suggested that the hippocampus and the nucleus accumbens may be involved in schizophrenic symptoms; for a recent review, see Schmajuk (1987). The thesis for a relationship between latent inhibition and schizophrenia has received considerable support from studies showing that damage to those systems affects latent inhibition. Hippocampal lesions have abolished or reduced latent inhibition in a variety of tests (Ackil et al., 1969; Kaye & Pearce, 1987a, b; McFarland et al., 1978; Solomon & Moore, 1975), and hippocampal stimulation may either attenuate or facilitate latent inhibition (Salafia, 1987; Salafia & Allan, 1980a, b, 1982). Similarly, septal lesions, total (Weiss et al., 1974) or lateral (Burton & Toga, 1982; Toga & Burton, 1979), have disrupted latent inhibition, as has microinjection of amphetamine into the nucleus accumbens (Solomon & Staton, 1982). Furthermore, attentional processes have been directly implicated by the finding that hippocampal lesions not only attenuated latent inhibition but also retarded the decline of the orienting response (Kaye & Pearce, 1987a).

The other road to relating latent inhibition, conditioned attention theory, and schizophrenia also begins with data, but connects these topics primarily at the level of theory. One such speculation centers on the role of context. In the chapter on conditioned attention theory and latent inhibition in humans, four basic points regarding the role of context were noted: (1) The organism establishes an S_1-no-consequence association according to the rules of conditioned attention theory. (2) In addition, the organism develops an associative link between the stimulus and the context, such that (3) the context becomes an occasion setter for the expression of the S_1-no-consequence relation. As a result, (4) in the acquisition–test phase the S_1 is kept from STM by the presence of the context in which the S_1-no-consequence association was developed. The net effect of this latter process is to interfere with the acquisition of any new association involving S_1. This effect is displayed as the latent inhibition phenomenon.

Such a set of processes has direct implications for schizophrenia. The hyperdopaminergic state will be characterized by a breakdown in the relationship between the preexposed stimulus and the context. The context no longer serves as the occasion setter for the expression of the stimulus–no-consequence relationship, the result of which is that STM is inundated with experimentally familiar but phenomenally novel stimuli, each competing for its share of the limited attentional resources. The net effect of this situation is what I have earlier called overly labile attention. "The patients digress, do not stick to the point, let their thoughts wander without voluntary control in the most varied direction" (Kraepelin, 1919/1971, p. 5).

On the other hand, the hypodopaminergic state will be characterized by a strengthening of the relationship between the preexposed stimulus and the context. The context continues to serve as an occasion setter for the expression of the stimulus–no-consequence relationship, and that relationship becomes almost absolute. Within such a context, the attentional state becomes inert, the antithesis of the overly labile state: "The attention is often rigidly fixed for a long time, so that the patients stare at the same point, or the same object, continue the same line of thought, or do not let themselves be interrupted in some definite piece of work" (Kraepelin, 1919/1971, p. 5).

According to this formulation, the super-latent-inhibition effect should be even more context-dependent than is normal latent inhibition. That is, a change in context from the stimulus preexposure session to the acquisition session should disrupt the haloperidol-induced super latent inhibition more than a non-drug-induced latent inhibition effect.

The type of symptoms that fit a hypodopaminergic state, characterized by the strengthened occasion-setting function of context and super latent inhibition, might include the following: concretized thoughts, stimulus-bound behaviors such as stereotypies (which should be subject to modification by context change, internal or external), and context-inappropriate behaviors. Such inappropriate responses will appear when a transfer of behavior from one context to another would be expected and normal, but where, because of overly strict context control, such transfer does not occur. Stated somewhat differently, under conditions of strict context control, the organism cannot generalize from one context to another. After one learns that a particular stimulus is of no significance in one environment, then when that same stimulus subsequently reappears in a different environment, there should be little or no savings in acquiring the stimulus–no-consequence relationship.

Interestingly, there have been no experiments addressing this question directly. All of the studies of the effects of context change on latent inhibition utilized only a single such change, and they focused on the finding that latent inhibition was reduced under such conditions; they did not attribute any theoretical import to the fact that although context change reduces latent inhibition, often there still remains a residual latent inhibition effect (e.g., Lovibond et al., 1984).

At this point it should be quite apparent that although these proposals are indeed speculations, they are presented not as idle ones but rather as empirically testable hypotheses. These hypotheses, as well as the more speculative suggestions, derive directly from the data on latent inhibition and the conditioned attention theory that was developed to explain them. As such, they offer some support for the utility of both enterprises and perhaps they lend credence to a more general hypothesis about the workings of science: To have knowledge of any area in depth, no matter how narrowly specified, provides tools, insights, and directions for exploring and understanding other subjects, no matter how broadly defined. Thus stands latent inhibition to schizophrenia: "inexorable habits of inattention" protecting us against "the contagious madness of this cosmic tarantella" (Berenson, 1952/1980, p. 104).

Notes

Chapter 1

1. The prototypical latent learning experiment involves two groups, one of which is given unreinforced preexposure to a multiple T-maze, and the other not. Subsequently, both groups, hungry or thirsty, are given repeated trials in the maze, with the goal box containing the appropriate reinforcer. When the maze-preexposed group learns the route to the goal box faster than the nonpreexposed group, latent learning is said to have occurred – "latent" because the superior performance of the maze-preexposed group reflects the positive transfer of some unobserved learning that accrued during the preexposure period.

2. A typical sensory preconditioning experiment uses two groups of subjects in a three-stage procedure. In the first stage, for the group from which sensory preconditioning (SPC) is supposed to be elicited, two neutral stimuli are paired for a number of trials (e.g., tone–light). The control group receives the same stimuli for the same number of trials, but unpaired. In the second and third stages, both groups are treated identically. In the second stage, the second stimulus (in this case, the light) is paired with an unconditioned stimulus (US) in a traditional classical conditioning procedure, usually until a stable conditioned response is obtained. In the third stage, the first of the stage 1 paired stimuli (in this case, the tone) is paired with the US. SPC is demonstrated when the group that received the paired neutral stimuli shows better conditioning in phase 3 than the control group. When this occurs, the superior phase 3 performance is due to positive transfer from a presumed learned association between the two neutral stimuli, sensory preconditioning.

3. It is true that Hull also had theoretical concepts such as reactive inhibition and conditioned inhibition that were designed to account for a *relative* reduction in performance. However, this reduction was relative, and the net effect was to reduce the total amount of *excitatory* strength. Likewise, Spence (1936), employing these concepts most elegantly for his theory of discrimination learning, clearly used inhibitory strength only in the context of other, excitatory effects. Indeed, this provides the basis for the definition and empirical demonstration of conditioned inhibition – a concept quite different from latent inhibition, the latter of which has as a reference point not an excitatory state but a neutral state.

4. The quotation comes from a Berenson essay written about the turn of the century, entitled "The Central Italian Painters." It was recently reissued together with three other essays in the cited book (Berenson, 1980).

Chapter 2

1. Although the vast majority of these studies used rats as subjects, it is worthwhile to note that there is reasonable evidence for latent inhibition of conditioned taste aversion in humans. Retrospective questionnaire studies with adult humans have provided data that taste familiarity interferes with the acquisition of a conditioned taste aversion response (Garb & Stunkard, 1974; Logue, 1985; Logue, Logue, & Strauss, 1983; Logue, Ophir, & Strauss, 1981). These findings are supported by laboratory experiments (Arwas, Rolnick, & Lubow, 1989; Cannon, Best, Batson, & Feldman, 1983).

2. Archer and his associates have demonstrated similar effects for pre-exposed "noisy" saccharin solutions (Archer, Sjoden, & Nilsson, 1985). For additional references, see the section "Role of Context" in the next chapter.

3. Theoretically, the preexposure stage can be given either before stage 1 or before stage 2 of conditioned suppression. In practice it is usually presented prior to the beginning of stage 2. The preexposure to the CS also may be accomplished simultaneously with the development of a stable operant (i.e. on baseline). See Crowell and Anderson (1972) for an example of this procedure. Similarly, the conditioning stage may be presented on baseline, in which case it is the *acquisition* of conditioned suppression that becomes the critical measure. See Domjan and Siegel (1971) for an example.

4. In the listings that follow, only the most recent studies are cited for bar pressing for food and water-tube licking. The list for bar pressing for tap water and sweetened water is complete.

5. The Solomon, Lohr, and Moore (1974) study is omitted from this list because the acquisition test followed a test for summation.

6. Lubow and Moore (1959) failed to find reliable latent inhibition to a preexposed moving rotor. In this same study, goats and sheep did show the latent inhibition effect to a flashing light.

7. Following the experimental treatment, the stimulus that is being assessed is simply paired with a US. The rate at which the stimulus–US association is formed is noted. If the rate of acquisition (or some other measure of conditioning strength) is lower than that for the comparison group, the experimental treatment is said to produce *retardation* of the acquisition of the conditioned response.

8. In the test for summation, after the experimental treatment, the stimulus that is being assessed is superimposed on another stimulus that has been previously treated in order to endow it with either known excitatory or inhibitory properties. If the experimental treatment produces conditioned inhibition, then it should summate with the known excitor or inhibitor, such that for the former excitation should be reduced, and for the latter inhibition increased, again, of course, relative to an appropriate comparison group. These states will be reflected in measures of the strengths of the conditioned responses to the compounded stimuli.

9. Cohen and Sparber (1985) and Cohen, Messing, and Sparber (1987) have reported latent inhibition of autoshaping in rats. However, their data are difficult to interpret because of a confounding of apparatus preexposure and presentations of the retractable lever (the to-be-conditioned stimulus).

10. The OR is considered by many to differ from the adaptation reflex (AR) and the defensive reflex (DR). For a full discussion of these differences, see Lynn (1966), Sokolov (1963), and, more recently, Graham (1979) and Siddle, Kuiack, and Kroese (1984). For a general overview of the OR and its interrelationship with other phenomena such as conditioning, attention, etc., see the recently edited volumes by Kimmel, Van Olst, and Orlebeke (1979) and Siddle (1983).

11. Experiment 1 of Baker and Mackintosh (1977) was not truly a discrimination study, because the S⁻ was the absence of the tone rather than the presence of a formal stimulus. The study is mentioned here because it is a companion to the more traditional discrimination procedure reported by Baker and Mackintosh (1977, Exp. 2), as discussed in the previous section.

Chapter 3

1. The stimulus and situational generality of the learned helplessness effect is kept within the confines of a learning phenomenon by postulating that the animal acquires the general cognition of response-reinforcer independence (Maier & Seligman, 1976).
2. One set of studies has questioned, although indirectly, the proposition that the latent inhibition effect is stimulus-specific (Rudy, Krauter, & Gaffuri, 1976). In these experiments, prior exposure to a light, a stimulus different from the tone to be used in the standard preexposure and conditioning phases, interfered with the acquisition of latent inhibition. However, these very unusual findings have not been reported elsewhere in the literature and have failed to be replicated in a number of unpublished experiments from several different laboratories.
3. Similarly, in studies by DeVietti and his associates, in which latent inhibition was obtained with as few as six preexposures, the duration of the preexposed stimulus was quite long, 15 sec (DeVietti, Emmerson, & Wittman, 1982; DeVietti, Wittman, & Comfort, 1980; DeVietti, Wittman, Emmerson, & Thatcher, 1981).
4. This same general design rule would, in fact, be applicable to other preexposure stimulus parameters, as, for example, stimulus duration and interval between stimulus presentations.
5. In this regard it should be noted that Swartzentruber and Bouton (1986) have suggested that the failure of Ayres, Moore, and Vigorito (1984) to obtain Hall-Pearce negative transfer, a latent-inhibition-like effect, using conditioned nictitating membrane response in rabbits may have been due to the relatively short interstimulus interval during preexposure.
6. See the section on discrimination learning in chapter 2.
7. A recent study by Wright, Skala, and Peuser (1986) compared stimulus preexposure and *acquisition* context similarity with stimulus preexposure and *test* similarity. While purporting to find latent inhibition only in the latter condition, the absence of a number of control groups made this conclusion questionable, as the authors themselves recognized in a subsequent paper (Wright & Gustavson, 1986).
8. Sullivan (1984) made a similar distinction between stimulus attributes associated with food objects and those associated with the conditioning chamber. Long-delay taste conditioning procedures produced aversions to the former, but not the latter.
9. The S_1–S_2 and S_2–S_1 preexposure procedures may be conceptualized in terms of the effects on S_1 of S_1 *predicting* S_2, and the effects on S_1 of S_1 *being predicted by* S_2, respectively. In the first case, S_1 is a predictor of a set of consequences; in the second case, S_1 is itself the consequence of a prediction. Although important theoretical distinctions have been made in terms of these very different conceptualizations (e.g., compare Mackintosh, 1975, and Pearce and Hall, 1980), I prefer to avoid that terminology in this section for two reasons: (1) to maintain the distinction between the empirical and theoretical; (2) to avoid the problem of having to decide if a single S_1–S_2 or S_2–S_1 presentation is sufficient for the organism to have established the relationship in question. Certainly there are effects of S_1–S_2 or S_2–S_1 presentations that are independent of the attainment of any prediction concept. The use of simple operational definitions, as employed in this section, avoids any prejudgment on these issues.
10. Particularly in the light of a recent comment that the S_1–S_2 effect reported by Lubow, Schnur, and Rifkin (1976) "is surprising (and to my knowl-

edge has not always been replicated by others)" (Mackintosh, 1983, p. 231), it is important to document the various replications of the effect.

11. With conditioned suppression in rats, but not with the conditioned nictitating membrane response preparation in rabbits.

12. Although all of the foregoing procedures employed perfect contingency between S_1 and S_2, Bouton and King (1986) compared the effects of preexposure to a *partially* reinforced stimulus (12 trials; trials 1, 4, 7, and 10 reinforced) and the effects in a situation in which the first 8 trials were nonreinforced and the last four were reinforced. The subsequent acquisition of conditioned suppression was considerably retarded in the latter group, compared with the former, a finding that perhaps points to some type of primacy effect.

13. The authors reported the Hall-Pearce effect only with conditioned suppression preparation with the rat, not with classical conditioning of the nictitating membrane response in the rabbit.

14. As an aside, it should be mentioned that the US preexposure effect, which, like CS preexposure, leads to subsequently poorer learning with that same US, is disrupted when the US is predicted: Baker and Mackintosh (1979), but not Randich and LoLordo (1979) with conditioned suppression; Tomie (1976a, b) with autoshaping. These studies avoided the design problem encountered by Szakmary by employing a predictor stimulus that was not used in conditioning.

15. Related studies that employed the noisy bottle technique as an S_2 were treated by their authors (e.g., Archer, Sjöden, & Nilsson, 1985) as context manipulations, and as such will be reviewed in the section on context.

16. Table 15 includes experiments with compound stimulus preexposure. Studies that preexposed only elements and then conditioned to either the element or a compound are omitted (e.g., Carr, 1974; Schnur, 1971, 1975). Experiments that preexposed a stimulus element and employed a compound stimulus during acquisition for purposes of assessing *conditioned* inhibition in a summation test (e.g., Reiss & Wagner, 1972; Rescorla, 1971) likewise are omitted.

Chapter 4

1. A recent study by Kraemer, Hoffman, and Spear (1988) compared 6- and 12-day-old rats. Latent inhibition to a chocolate milk solution was found in the older group, but not the younger group.

2. Estimated from Braud's graph (1971, Fig. 1).

3. An additional study (Shishimi, 1985, Exp. 2), using an instrumental appetitive conditioning procedure, and testing a preexposed tone as a conditioned inhibitor *or* as a conditioned excitor, also found, in both cases, no effect of preexposure.

Chapter 5

1. A full description of the procedure, without the preexposure variable, can be found in a study by Siebert, Nicholson, Carr-Harris, and Lubow (1969).

2. Similar results have been reported by Bjorkstrand (1987) for fear-relevant and fear-irrelevant CSs.

3. For ethical reasons, the commands were delivered in French.

4. For ideas on how the latent inhibition effect might be used prophylactically (to precent development of fears, phobias, etc.), see Lubow (1973b), Poser (1970), Poser, Baum, and Skinner (1970), Poser and King (1975), Surwit (1972), and Surwit and Poser (1974).

Chapter 6

1. Identical experiments were reported by Mason and Fibiger (1979).
2. This suggestion is based exclusively on the similarity of behavioral outcomes (disruption of latent inhibition following hippocampal lesions and NAcc activation, e.g., by amphetamine microinjection). Although the hippocampal input to NAcc is believed to be excitatory (glutaminergic) (e.g., Totterdell & Smith, 1987; Groenewegen et al., 1987), the effects of interrupting this input are at present impossible to deduce, because little is known about the interaction of this input with the accumbal mechanisms (i.e., the local, e.g., GABAergic, connections within the NAcc or possibly different DA receptor populations) (e.g., Cools & van Rossum, 1980).

Chapter 7

1. For general reviews of habituation phenomena, see Harris (1943), Peeke and Herz (1973), Horn and Hinde (1970), Thompson and Spencer (1966), Siddle (1983), and Tighe and Leaton (1976).
2. Reported in Russo, Reiter, and Ison (1975).
3. These conclusions in regard to stimulus intensity appear inconsistent with the findings of Lubow, Markman, and Allen (1968) on the habituation of the pinna reflex in the rabbit. Except for differences in species and response systems, I see no obvious reason for this discrepancy.
4. Seligman, Maier, and Solomon (1971) provided an exception to this statement.
5. For reviews, see Kimmel, Van Olst, and Orlebeke (1979), Lynn (1966), and Siddle (1983).
6. The Maltzman, Weissbluth, and Wolff (1978) data were reported in brief form by Wolff and Maltzman (1968).
7. For a discussion of the relationships among OR and defensive and startle reflexes, see Graham (1979).
8. The galvanic skin response is elicited by a variety of stimulus conditions, as described in the text. The response consists of a relatively slow change in the electrical conductivity of the skin that is governed by automatic nervous system activity. The response has a latency of a few seconds and an even longer duration. It may be subdivided into a variety of components, including a differentiation between responses occurring early after stimulus onset (first-interval responses) and those that occur later (second-interval responses).
9. P300 is a relatively large amplitude, positive, evoked brain potential that is recorded from electrodes placed on the scalp. The amplitude of the response appears to be correlated with a state of subjective uncertainty, with lower response amplitudes reflecting an increased match between present and expected stimulation, and higher amplitudes indicating an increased discrepancy between present and expected stimulation (e.g., Squires, Wickens, Squires, & Donchin, 1976; Sutton, Braren, Zubin, & John, 1963).
10. For differential effects of septal lesions on avoidance conditioning

and OR, at least as indexed by unconditioned lick suppression, see Weiss, Friedman, and McGregor (1974).

11. This section is adapted from an unpublished manuscript by Schnur and Lubow (1983a).

12. This section is adapted from an unpublished manuscript by Weiner, Schnur, and Lubow (1983).

13. For a related approach, see Moore and Stickney (1980, 1985).

14. $\bar{V}$ is the sum of the associative strengths of all stimuli that are present, in this case the nominal CS and the context stimuli.

15. The feature positive effect refers to the fact that discrimination learning proceeds faster when the particular feature distinguishing positive from negative stimuli appears on reinforced as compared with nonreinforced trials. For example, a feature positive discrimination might involve reinforced trials to AX and nonreinforced trials to X alone, the distinguishing feature, A, appearing on reinforced trials. A comparable feature negative discrimination would involve reinforced trials to X and nonreinforced trials to AX, the distinguishing feature, A, appearing on negative trials. The feature positive discrimination develops rapidly, with responding on positive trials and a rapid cessation of responding on negative trials. The feature negative discrimination develops slowly, if at all. The feature positive effect has considerable generality across species, situations, and experimental procedures; for a review, see Hearst (1978).

16. The catastrophe rule provides a response mapping rule for conditioning that is separate from the associative process. Unlike in the Rescorla-Wagner model, where associative strength and therefore response probability accrue as a negatively accelerated function, the catastrophe rule allows for an abrupt change in response probability.

17. Ohman (1979) considers the OR as a by-product of that processing, representing an index of the degree to which the input stimulus is expected or is surprising.

18. It should be recognized that this formulation of latent inhibition makes it a special case of blocking.

19. An explicit statement regarding the applications of empirical laws of conditioning to the conditioning of attention to S_1 in preexposure is made in conditioned attention theory (Lubow, Weiner, & Schnur, 1981).

20. A recent conditioned suppression study by Wright, Skala, and Peuser (1986) suggested that context change from preexposure to acquisition-test does not affect the acquisition of the conditioned response (i.e., associability), but rather the retrievability of the S_1-no-consequence association. However, the design appears to be flawed.

21. A similar finding of absence of a context extinction effect was reported by Marlin and Miller (1981) for long-term habituation, which according to Wagner (1976, 1978, 1979) should obey the same rules as latent inhibition.

Chapter 8

1. The first section of this chapter is adapted, in part, from a chapter by Lubow, Weiner, and Schnur (1981).

2. Related findings with children were reported by Lubow, Alek, and Arzy (1975). The young subjects were preexposed to 0, 3, 10, 20, or 40 S_1–S_2 pairings. Three such pairings facilitated subsequent reaction time to S_1, as compared with reaction time in the nonpreexposed group. On the other hand, there were no differences between the 0- and 10-preexposure groups, and the 20- and

40-preexposure groups which showed significant reaction-time decrements.

3. For direct evidence of this interaction, see a recent paper by DeVietti, Bauste, Nutt, Barrett, Daly, and Petree (1987).

4. This, of course, is a major assumption of Wagner's explanation of latent inhibition (1976, 1978). The evidence supporting the assumption seems quite strong. However, it does not follow from this that such encoding is used in the manner suggested by Wagner's priming theory.

5. This analysis receives some support from a recent study by Bouton and Swartzentruber (1986), which presented evidence that context, indeed, has occasion-setting properties.

Chapter 9

1. The latter condition, frequently referred to as producing the "missing-stimulus effect," has not stood the test of empirical verification. Whereas, according to Sokolov (1963), the symmetrical procedures of stimulus addition and stimulus omission should elicit equal comparator-generated differences, and thus equal ORs, recent evidence fails to support such a prediction (Barry, 1984; Foreman & Thinus-Blanc, 1987). Indeed, the failure to find a "missing-stimulus effect" after repeated stimulus presentations provides unexpected support for the basic premise of CAT, namely, that stimulus preexposure produces *inattention* to the stimulus. If the preexposed stimulus is no longer attended to, then its absence should *not* be fully processed.

2. S_1–S_2 and S_2–S_1 effects do not exhaust the list of stimulus conditions that can change the typical decremental course of attention with repeated stimulation. There are all of those variables that Berlyne (1960) called collative and that include movement, color, stimulus intensity, and change, as well as stimulus complexity and novelty. It should also be recalled from the animal data that the effects of preexposure stimulus compounding and complexity can be treated as examples of S_1–S_2 effects.

3. This is not to imply that habituation does not occur in submammalian species. There is, of course, ample evidence that such a process is virtually ubiquitous. Rather, what is being suggested here is that (1) the stimulus-specific habituation processes are not accomplished by an attentional decrement, perhaps because attentional processes do not occur in these lower organisms, or (2) if such attentional processes and their decrementing do occur, these lower organisms do not have the capacity to associate the repeated stimulus with the inattentional response.

4. Schnur and Lubow (1987) have failed to replicate these results.

5. Mandler, Nakamura, and Van Zandt (1987) made a similar point by demonstrating that repeated exposures of irregular shapes not only generated judgments of familiarity and preferences (Zajonc, 1980) but also generated nonspecific effects, such as judgments of brightness or darkness, independent, of course, of stimulus intensity.

6. Support for this point comes from two types of evidence: direct and indirect. The former includes studies demonstrating that preexposure of the context interferes with subsequent context–US associations, as when context exposure precedes shock presentations in that context, and it has been found that escape responding is slower than in a group that has not been preexposed to that context (e.g., Nakaya, 1982). Indirect support for latent inhibition to the context has been provided by various tests of the context-blocking hypothesis of US preexposure effects (e.g., Hinson, 1982; Mowrer, 1987).

Chapter 10

1. These results have been replicated by Rosellini, DeCola, and Warren (1986), who, in addition, found that the intertrial interval modulated the effectiveness of the S_2 feedback signal.

2. Other differences between latent inhibition and learned helplessness include the fact that whereas latent inhibition is easily demonstrated in animals and young children, but requires a masking task in adults, learned helplessness is induced without special masking procedures in all subjects. Indeed, the effects of masking on US preexposure have not been studied, perhaps because of the conceptual difficulty in devising an appropriate masking task – one that also can be used for a subject who has "control" over a stimulus!

3. Recent evidence suggests that these effects are quite selective, with amphetamine increasing the positive symptoms of schizophrenia, but decreasing the negative symptoms (e.g., Angrist, Peselow, Rubinstein, Corwin, & Rotrosen, 1982; Kay & Opler, 1985); also see Van Kammen et al. (1982) for bidirectional effects of amphetamine on schizophrenic patients.

4. It is important to note here that the animal models of schizophrenia have almost exclusively been concerned with stereotyped behavior produced by relatively high doses of amphetamine (Segal & Janowsky, 1978). Our own approach has been to look for attentional deficits (as reflected in, for example, latent inhibition) produced by low doses of amphetamine. There are data suggesting that different dopaminergic systems mediate these effects, with the nigrostriatal system being involved in the former, and the mesolimbic-cortical in the latter. Matthysse (1974) and Iversen (1977), as well as others, have proposed that nigrostriatal dopamine pathways control motor action patterns (e.g., stereotypy), whereas mesolimbic-cortical dopamine pathways control phenomena related to cognitive processing (e.g., attention).

5. Because classification of the patients according to cutoff scores recommended by Magaro, Abrams, and Cantrell (1981) yielded a significant inconsistency, a different scoring procedure was employed. Patients were classified according to their relative scores on the two subscales. Patients receiving higher scores on the paranoid subscale were classified as paranoids, and patients receiving higher scores on the nonparanoid subscale were classified as nonparanoids. Patients who received the same scores on the two subscales were excluded from those analyses that used the Maine scale to differentiate between the two clinical populations. This procedure, used by Brennan and Hemsley (1984), provided greater consistency with the psychiatrists' diagnoses that did that of Magaro, Abrams, and Cantrell (1981).

References

Abramson, C. I., & Bitterman, M. E. 1986. Latent inhibition in honey bees. *Animal Learning & Behavior, 14*, 184-189.

Ackil, J. E., & Mellgren, R. L. 1968. Stimulus preexposure and instrumental learning. *Psychonomic Science, 11*, 339.

Ackil, J. E., Mellgren, R. L., Halgren, C., & Frommer, G. P. 1969. Effects of CS preexposure on avoidance learning in rats with hippocampal lesions. *Journal of Comparative and Physiological Psychology, 69*, 739-747.

Ahlers, R. H., & Best, P. J. 1971. Novelty vs. temporal contiguity in learned taste aversions. *Psychonomic Sciences, 25*, 34-36.

Allen, J. A. 1967. Latent inhibition: Generalization during eyelid conditioning. Unpublished master's thesis, North Carolina State University.

Allen, J. A. 1971. Relationships between orienting response level, conditioners and voluntary responders. In *Proceedings of the 79th Annual Convention of the APA*, (pp. 29-30).

Alleva, E., De Acetis, L., Amorico, L., & Bignami, G. 1983. Amphetamine, conditioned stimulus, and non-debilitating preshock effects on activity and avoidance: Further evidence for interactions between associative and non-associative changes. *Behavioral and Neural Biology, 39*, 78-104.

Alpert, M., & Friedhoff, A. J. 1982. An un-dopamine hypothesis of schizophrenia. *Schizophrenia Bulletin, 6*, 387-390.

Andelman, L., & Sutherland, N. S. 1970. The effects of nondifferential reinforcement on two-cue discrimination learning. *Psychonomic Science, 18*, 37-38.

Anderson, D. C., Cole, J., & McVaugh, W. 1968. Variations in unsignaled inescapable preshock as determinants of responses to punishment. *Journal of Comparative and Physiological Psychology, Monograph, 65* (3, Pt. 2).

Anderson, D. C., O'Farrell, T., Formica, R., & Caponigri, V. 1969. Preconditioning CS exposure: Variation in place of conditioning and presentation. *Psychonomic Science, 15*, 54-55.

Anderson, D. C. & Paden, P. 1966. Passive avoidance response learning as a function of prior tumbling-trauma. *Psychonomic Science, 4*, 129-130.

Anderson, D. C., Tyson, H. W., & Williams, F. 1966. Acquisition of a passive avoidance response as determined by variations in prior aversive stimulation. *Psychonomic Science, 4*, 131-132.

Anderson, D. C., Wolf, D., & Sullivan, P. 1969. Preconditioning exposure to the CS: Variation in place of testing. *Psychonomic Science, 14*, 233-235.

Angrist, B., & Gershon, S. 1969. Amphetamine induced schizophreniform psychosis. In D. V. Sira Sanka (Ed.), *Schizophrenia: Current Concepts and Research* (pp. 508-524). Hicksville, N.Y.: P.J.D. Publications.

Angrist, B., & Gershon, S. 1970. The phenomenology of experimentally induced amphetamine psychosis. Preliminary observations. *Biological Psychiatry, 2*, 95-107.

Angrist, B., Lee, H. K., & Gershon, S. 1974. The antagonism of amphetamine-induced symptomatology by a neuroleptic. *American Journal of Psychiatry, 131*, 817-819.

Angrist, B., Peselow, E., Rubinstein, M., Corwin, J., & Rotrosen, J. 1982. Partial improvement in negative schizophrenic symptoms after amphetamine. *Psychopharmacology, 78*, 128-130.

Anisman, H., de Catanzaro, D., & Remington, G. 1978. Escape performance following exposure to inescapable shock: Deficits in motor response maintenance. *Journal of Experimental Psychology: Animal Behavior Processes, 4,* 197–218.

Anisman, H., & Waller, T. G. 1971a. Effects of inescapable shock upon subsequent one-way avoidance learning in two strains of rats. *Psychonomic Science, 24,* 101–102.

Anisman, H., & Waller, T. G. 1971b. Effects of methamphetamine and shock duration during inescapable shock exposure on subsequent active and passive avoidance. *Journal of Comparative and Physiological Psychology, 77,* 143–151.

Anscombe, R. 1987. The disorder of consciousness in schizophrenia. *Schizophrenia Bulletin, 13,* 241–260.

Archer, T. 1982a. DSP4 – (N-2-chloroethyl-N-2-ethyl-bromobenzylamine) a new noradrenaline neurotoxin, and the stimulus conditions affecting acquisition of two way active avoidance. *Journal of Comparative and Physiological Psychology, 96,* 476–490.

Archer, T. 1982b. The role of noradrenaline in learned behaviors: Studies using DSP4. *Scandinavian Journal of Psychology, Suppl., 1,* 61–71.

Archer, T., Järbe, T. U. C., Mohammed, A. K., & Priedite, G. 1985. The effects of stimulus preexposure upon the context effect in taste aversion learning in noradrenaline depleted rats. *Scandinavian Journal of Psychology, 26,* 158–169.

Archer, T., Mohammed, A. K., & Järbe, T. U. C. 1983. Latent inhibition following systemic DSP: Effects due to presence and absence of contextual cues in taste aversion learning. *Behavioral and Neural Biology, 38,* 287–306.

Archer, T., Mohammed, A. K., & Järbe, T. U. C. 1986. Context-dependent latent inhibition in taste aversion learning. *Scandinavian Journal of Psychology, 27,* 277–284.

Archer, T., Sjödén, P. O., & Nilsson, L. G. 1985. Contextual control of taste aversion conditioning and extinction. In P. D. Balsam & A. Tomie (Eds.), *Context and Learning* (pp. 225–271). Hillsdale, N.J.: Lawrence Erlbaum.

Arwas, S., Rolnick, A., & Lubow, R. E. 1989. Conditioned taste aversion in humans using motion-induced sickness as the US: The effects of CS familiarity. *Behavior Research and Therapy.*

Asin, K. E., Wirtshafter, D., & Kent, E. W. 1980. The effect of electrolytic median raphe lesions on two measures of latent inhibition. *Behavioral and Neural Biology, 28,* 408–417.

Ayers, J. J. B., Axelrod, H., Mercker, E., Muchnik, F., & Vigorito, M. 1985. Concurrent observations of barpress suppression and freezing: Effects of CS modality and on-line vs. off-line training upon posttrial behavior. *Animal Learning & Behavior, 13,* 44–50.

Ayres, J. J. B., Benedict, J. O., & Witcher, E. S. 1975. Systematic manipulation of individual events in a truly random control in rats. *Journal of Comparative and Physiological Psychology, 88,* 97–103.

Ayres, J. J. B., Moore, J. W., & Vigorito, M. 1984. Hall and Pearce negative transfer: Assessment in conditioned suppression and nictitating membrane conditioning experiments. *Animal Learning & Behavior, 12,* 428–438.

Ayres, J. J. B., & Vigorito, M. 1984. Posttrial effects of presenting vs. omitting expected shock US in the conditioned suppression procedures: Concurrent measurement of barpress suppression and freezing. *Animal Learning & Behavior, 12,* 73–78.

Azmitia, E. C. 1978. The serotonin-producing neurons of the midbrain median and dorsal raphe nuclei. In L. L. Iverson, S. D. Iverson & S. H. Snyder (Eds.), *Handbook of Psychopharmacology*, (Vol. 9, pp. 233–314). New York: Plenum Press.

Azmitia, E. C., & Segal, M. 1978. An autoradiographic analysis of the differential ascending projections of the dorsal and median raphe nuclei in the rat. *Journal of Comparative Neurology, 179,* 641–667.

Babb, H. 1956. Proportional reinforcement of irrelevant stimuli and transfer value. *Journal of Comparative and Physiological Psychology, 45,* 586–589.

Badia, P., & Defran, R. H. 1970. Orienting response and GSR conditioning. *Psychological Review, 77,* 171–181.
Baker, A. G. 1976. Learned irrelevance and learned helplessness: Rats learn that stimuli, reinforcers, and responses are uncorrelated. *Journal of Experimental Psychology: Animal Behavior Processes, 2,* 130–141.
Baker, A. G., & Mackintosh, N. J. 1977. Excitatory and inhibitory conditioning following uncorrelated presentation of CS and UCS. *Animal Learning & Behavior, 5,* 315–319.
Baker, A. G., & Mackintosh, N. J. 1979. Preexposure to the CS alone, US alone, or CS and US uncorrelated: Latent inhibition, blocking by context or learned irrelevance. *Learning and Motivation, 10,* 278–294.
Baker, A. G., & Mercier, P. 1982. Extinction of the context and latent inhibition. *Learning and Motivation, 13,* 391–416.
Baker, A. G., & Mercier, P. 1984. Prior experience with the conditioning events: Evidence for a rich cognitive representation. In M. L. Commons, R. J. Herrnstein, & A. R. Wagner (Eds.), *Quantitative Analyses of Behavior. Vol. 3: Acquisition Processes* (pp. 117–143). Cambridge, Mass.: Ballinger.
Balaz, A. M., Capra, S., Kasprow, W. I., & Miller, R. R. 1982. Latent inhibition of the conditioning context: Further evidence of contextual potentiation of retrieval in the absence of appreciable context–US associations. *Animal Learning & Behavior, 10,* 242–248.
Balaz, M. A., Gustin, P., Cacheiro, H., & Miller, R. R. 1982. Blocking as a retrieval failure: Reactivation of associations to a blocked stimulus. *Quarterly Journal of Experimental Psychology, 34B,* 99–113.
Balsam, P. D., & Tomie, A. (Eds.) 1981. *Autoshaping and Conditioning Theory.* New York: Academic Press.
Barber, J. G., & Winefield, A. H. 1986. Learned helplessness as conditioned inattention to the target stimulus. *Journal of Experimental Psychology: General, 115,* 236–246.
Barker, L. M., Best, M. R., & Domjan, M. (Eds.) 1977. *Learning Mechanisms in Food Selection.* Waco, Tex.: Baylor University Press.
Barnett, S. 1963. *The Rat: A Study in Behavior.* Chicago: Aldine.
Barrett, R. J., Leith, N. J., & Ray, O. S. 1972. Permanent facilitation of avoidance behavior by *d*-amphetamine and scopolamine. *Psychopharmacology, 25,* 321–331.
Barry, R. J. 1984. Preliminary processes in OR elicitation. *Acta Psychologica, 55,* 109–142.
Baruch, I., Hemsley, D. R., & Gray, J. A. 1988a. Differential performance of acute and chronic schizophrenics in a latent inhibition task. *Journal of Nervous and Mental Disease, 176,* 598–606.
Baruch, I., Hemsley, D. R., & Gray, J. A. 1988b. Latent inhibition and "psychotic proneness" in normal subjects. *Personality and Individual Differences, 9,* 777–784.
Bateson, P. P. G., 1973. Internal influences on early learning in birds. In R. A. Hinde & J. S. Stevenson-Hinde (Eds.) *Constraints on Learning* (pp. 101–116). London: Academic Press.
Bateson, P. P. G., & Chantrey, D. F. 1972. Retardation of discrimination learning in monkeys and chicks previously exposed to both stimuli. *Nature, 237,* 173–174.
Batson, J. D., & Best, M. R. 1982. Lithium-mediated disruptions of latent inhibition: Overshadowing by the unconditioned stimulus in flavor conditioning. *Learning and Motivation, 13,* 167–184.
Baum, M. 1969. Dissociation of respondent and operant processes in avoidance learning. *Journal of Comparative and Physiological Psychology, 67,* 83–88.
Becker, D. E., & Shapiro, D. 1980. Directing attention toward stimuli affects the P300 but not the orienting response. *Psychophysiology, 17,* 385–389.
Bell, D. S. 1973. The experimental reproduction of amphetamine psychosis. *Archives of*

General Psychiatry, 29, 35–40.
Bell, J. A., & Livesey, P. J. 1977. The effects of prior experience with visual shapes under differing conditions of reinforcement on subsequent discrimination learning in the rat. *Psychological Record, 27,* 683–691.
Ben-Shahar, O., Weiner, I., Feldon, J., & Lubow, R. E. 1986. Latent inhibition attenuation as a function of number of S1-S2 preexposures. Unpublished manuscript.
Bennett, T. L., & Ellis, H. C. 1968. Tactual-kinesthetic feedback from manipulation of visual forms and nondifferential reinforcement in transfer of perceptual learning. *Journal of Experimental Psychology, 77,* 495–500.
Bennett, T. L., Rickert, E. J., & McAllister, L. E. 1970. Role of tactual-kinesthetic feedback in transfer of perceptual learning for rats with pigmented irises. *Perceptual and Motor Skills, 30,* 916–918.
Berenson, B. 1980. *The Italian Painters of the Renaissance.* Ithaca, N.Y.: Cornell University Press. (Original work published 1952).
Berlyne, D. E. 1960. *Conflict, Arousal, and Curiosity.* New York: McGraw-Hill.
Bernstein, I. L., & Borson, S. 1986. Learned food aversions: A component of anorexia syndromes. *Psychological Review, 93,* 462–472.
Bernstein, I. L., & Goehler, L. E. 1983. Vagotomy produces learned food aversions in the rat. *Behavioral Neurosciences, 97,* 585–594.
Best, M. R. 1975. Conditioned and latent inhibition in taste aversion learning: Clarifying the role of learned safety. *Journal of Experimental Psychology: Animal Behavior Processes, 104,* 97–113.
Best, M. R. 1982. Nonassociative and associative sources of interference with the acquisition of a flavor aversion. In M. L. Commons, R. J. Herrnstein, & A. R. Wagner (Eds.) *Quantitative Analyses of Behavior. Vol. 3: Acquisition Processes* (pp. 67–83). Cambridge, Mass.: Ballinger.
Best, M. R., & Barker, L. M. 1977. The nature of "learned safety" and its role in the delay of reinforcement gradient in taste aversion learning. In L. M. Barker, M. R. Best, & M. Domjan (Eds.), *Learning Mechanisms in Food Selection* (pp. 295–317). Waco, Tex.: Baylor University Press.
Best, M. R., & Domjan, M. 1979. Characteristics of the lithium-mediated proximal US-preexposure effect in flavor-aversion conditioning. *Animal Learning & Behavior, 7,* 433–440.
Best, M. R., & Gemberling, G. A. 1977. Role of short term processes in the conditioned stimulus preexposure effect and the delay of reinforcement gradient in long-delay taste aversion learning. *Journal of Experimental Psychology: Animal Behavior Processes, 3,* 253–263.
Best, M. R., Gemberling, G. A., & Johnson, P. E. 1979. Disrupting the conditioned stimulus preexposure effect in flavor aversion learning: Effects of interoceptive distractor manipulations. *Journal of Experimental Psychology: Animal Behavior Processes, 5,* 321–334.
Best, M. R., & Meachum, C. L. 1986. The effects of stimulus preexposure on taste-mediated environmental conditioning: Potentiation and overshadowing. *Animal Learning & Behavior, 14,* 1–5.
Bignami, G., Giardini, V., & Alleva, E. 1985. Treatment-behavior interactions in activity/habituation and avoidance: State changes and preexposure to CS, US and drug. In C. F. Lowe, M. Richelle, D. E. Blackman, & C. M. Bradshaw (Eds.), *Behavior Analysis and Contemporary Psychology* (pp. 187–204). Hillsdale: N.J.: Lawrence Erlbaum.
Bitterman, M. E., Menzel, R., Fietz, A., & Schäfer, S. 1983. Classical conditioning of proboscis extension in honeybees. (*Apis mellifera*). *Journal of Comparative Psychology, 97,* 107–119.
Bjorkstrand, P. A. 1987. Conditioned stimulus preexposure and latent inhibition in

human electrodermal conditioning: Effects of fear-relevant and fear-irrelevant stimuli. Presented at the 27th annual meeting of the Society of Psychophysiological Research, October.

Bleuler, E. 1966. *Dementia Praecox or the Group of Schizophrenias* (translated by J. Zinkin). New York: International Universities Press (originally published in 1911).

Blodgett, H. C. 1929. The effect of the introduction of reward upon the maze performance of rats. *University of California Publications in Psychology, 4*, 113–134.

Blowers, G. H., Spinks, J. A., & Shek, D. T. L. 1986. P300 and the anticipation of information within an orienting response paradigm. *Acta Psychologia, 61*, 91–103.

Bolles, R. C. 1970. Species-specific defense reactions and avoidance learning. *Psychological Review, 77*, 41–52.

Bolles, R. C. 1972. Reinforcement, expectancy and learning. *Psychological Review, 79*, 394–409.

Bolles, R. C., Riley, A. L., & Laskowski, B. A. 1973. A further demonstration of the learned-safety effect in food-aversion learning. *Bulletin of the Psychonomic Society, 1*, 190–192.

Bond, N., & DiGiusto, E. 1975. Amount of solution drunk is a factor in the establishment of taste aversion. *Animal Learning & Behavior, 3*, 81–84.

Bond, N. W., & Westbrook, R. F., 1982. The role of the amount consumed in flavor preexposure effects and neophobia. *Animal Learning & Behavior, 10*, 511–515.

Bostock, E., & Gallagher, M. 1982. Naloxone-induced facilitation of latent inhibition in rabbits. *Society for Neuroscience Abstract.*

Bouton, M. E. 1987. Context, encoding, and retrieval in latent inhibition. Unpublished manuscript.

Bouton, M. E., & Bolles, R. C. 1979. Contextual control of the extinction of conditioned fear. *Learning and Motivation, 10*, 445–466.

Bouton, M. E., & King, D. A. 1986. Effect of context on performance to conditioned stimuli with mixed histories of reinforcement and nonreinforcement. *Journal of Experimental Psychology: Animal Behavior Processes, 12*, 4–15.

Bouton, M. E., & Swartzentruber, D. 1986. Analysis of the associative and occasion-setting properties of contexts participating in a Pavlovian discrimination. *Journal of Experimental Psychology: Animal Behavior Processes, 12*, 333–350.

Bower, G. H. (Ed.) 1981. *Progress in Learning and Motivation.* New York: Academic Press.

Braff, D. L., & Saccuzzo, D. P. 1982. Effect of antipsychotic medication on speed of information processing in schizophrenic patients. *American Journal of Psychiatry, 139*, 1127–1130.

Brandeis, R. 1974. The influence of number of preexposures to one stimulus and to two stimuli on the latent inhibition effect in human GSR conditioning. Unpublished master's thesis, Bar-Ilan University.

Braud, W. G. 1971. Effectiveness of "neutral," habituated, shock-related, and food-related stimuli and CSs for avoidance learning in goldfish. *Conditioned Reflex, 6*, 153–156.

Braun, J. J., Slick, T. B., & Lorden, J. F. 1972. Involvement of gustatory neocortex in the learning of taste aversions. *Physiology and Behavior, 9*, 637–641.

Braveman, N. S., & Jarvis, P. J. 1978. Independence of neophobia and taste aversion learning. *Animal Learning & Behavior, 6*, 406–412.

Brennan, J. F., & Barone, R. J. 1976. Effects of differential cue availability in an active avoidance CS for young and adult rats. *Developmental Psychobiology, 9*, 237–244.

Brennan, J. H., & Hemsley, D. R. 1984. Illusory correlation in paranoid and nonparanoid schizophrenia. *British Journal of Clinical Psychology, 23*, 225–226.

Broadbent, D. E. 1958. *Perception and Communication.* London: Pergamon.

Brogden, W. J. 1939. Sensory pre-conditioning. *Journal of Experimental Psychology, 25*, 323–332.

Brown, P. L., & Jenkins, H. M. 1968. Auto-shaping of the pigeon's keypeck. *Journal of Experimental Analysis of Behavior, 11,* 1–8.

Brown-Su, A. M., Matzel, L. D., Gordon, E. L., & Miller, R. R. 1986. Malleability of conditioned associations: Path dependence. *Journal of Experimental Psychology: Animal Behavior Processes, 12,* 420–427.

Bruner, A. 1965. UCS properties in classical conditioning of the albino rabbit's nictitating membrane response. *Journal of Experimental Psychology, 69,* 186–192.

Burton, H. A., & Toga, A. W. 1982. Successive discrimination performance improves with increasing numbers of stimulus preexposures in septal rats. *Behavioral and Neural Biology, 34,* 141–151.

Campbell, B. A., & Ampuero, M. X. 1985. Conditioned orienting and defensive responses in the developing rat. *Infant Behavior and Development, 8,* 425–434.

Cannon, D. S., Best, M. R., Batson, J. D., Brown, E. R., Rubenstein, J. A., & Carrell, L. A. 1985. Interfering with taste aversion learning in rats: The role of associative interferences. *Appetite, 6,* 1–19.

Cannon, D. S., Best, M. R., Batson, J. D., & Feldman, M. 1983. Taste familiarity and apomorphine-induced taste aversion in humans. *Behavior Research and Therapy, 21,* 669–673.

Cantor, G. N. 1969a. Stimulus familiarization effect and the change effect in children's motor task behavior. *Psychological Bulletin, 71,* 144–160.

Cantor, G. N. 1969b. Effects of stimulus familiarization on child behavior. In J. P. Hill (Ed.), *Minnesota Symposia on Child Psychology* (Vol. 3, pp. 3–30). Minneapolis: University of Minnesota Press.

Cantor, G. N., & Cantor, J. H. 1965. Discriminative reaction time performance in preschool children as related to stimulus familiarization. *Journal of Experimental Child Psychology, 2,* 1–9.

Cantor, G. N., & Cantor, J. H. 1966. Discriminative reaction time in children as related to amount of stimulus familiarization. *Journal of Experimental Child Psychology, 4,* 150–157.

Cantor, G. N., & Fenson, L. 1968. New data regarding the relationship of amount of familiarization and the stimulus familiarization effect (SFE). *Journal of Experimental Child Psychology, 6,* 167–173.

Carlton, P. H., & Vogel, J. R. (1967). Habituation and conditioning. *Journal of Comparative and Physiological Psychology, 63,* 348–351.

Carlton, P. H. 1969. Brain acetylcholine and inhibition. In J. T. Tapp (Ed.), *Reinforcement and Behavior* (pp. 286–327). New York: Academic Press.

Carr, A. F. 1974. Latent inhibition and overshadowing in conditioned emotional response conditioning with rats. *Journal of Comparative and Physiological Psychology, 86,* 718–723.

Carter, C. J., & Pycock, C. J. 1978a. The effects of 5-hydroxytryptamine lesions of extrapyramidal and mesolimbic sites on spontaneous motor behavior and amphetamine-induced stereotypy. *Naunyn-Schmiedeberg's Archives of Pharmacology, 308,* 51–54.

Carter, C. J., & Pycock, C. J. 1978b. Differential effects of central serotonin manipulation on hyperactive and stereotyped behavior. *Life Science, 23,* 953–960.

Cecil, L. S., Kraut, A. G., & Smothergill, D. W. 1984. An alertness decrement hypothesis of response inhibition to repeatedly presented stimuli. *American Journal of Psychology, 97,* 397–398.

Chacto, C., & Lubow, R. E. 1967. Classical conditioning and latent inhibition in the white rat. *Psychonomic Science, 9,* 135–136.

Channell, S., & Hall, G. 1981. Facilitation and retardation of discrimination learning after exposure to the stimuli. *Journal of Experimental Psychology: Animal Behavior Processes, 7,* 437–446.

Channell, S., & Hall, G. 1983. Contextual effects in latent inhibition with an appetitive conditioning procedure. *Animal Learning & Behavior, 11*, 67–74.

Chantrey, D. F. 1972. Enhancement and retardation of discrimination learning in chicks after exposure to the discriminanda. *Journal of Comparative and Physiological Psychology, 81*, 256–261.

Chantrey, D. F. 1974. Stimulus preexposure and discrimination learning by domestic chicks: Effects of varying interstimulus time. *Journal of Comparative and Physiological Psychology, 87*, 517–525.

Chapman, L. J., & Chapman, J. P. 1973. *Disordered Thought in Schizophrenia*. New York: Appleton-Century-Crofts.

Chiodo, L. A., & Bunney, B. S. 1983. Typical and atypical neuroleptics: Differential effects of chronic administration on the activity of A9 and A10 midbrain dopaminergic neurons. *Journal of Neuroscience, 3*, 1607–1619.

Chronister, R. B., Sikes, R. W., & White, L. E., Jr. 1976. The septohippocampal system: Significance of the subiculum. In J. F. De France (Ed.), *The Septal Nuclei* (pp. 115–132). New York: Plenum Press.

Claridge, G. S., & Broks, P. 1984. Schizotypy and hemisphere function. I: Theoretical consideration and the measurement of schizotypy. *Personality and Individual Differences, 5*, 633–648.

Clarke, M. E., & Hupka, R. B. 1974. The effects of stimulus duration and frequency of daily preconditioning stimulus exposures on latent inhibition in Pavlovian conditioning of the rabbit nictitating membrane response. *Bulletin of the Psychonomic Society, 4*, 225–228.

Cohen, C. A., Messing, R. B., & Sperber, S. B. 1987. Selective learning impairment of delayed reinforcement autoshaped behavior caused by low doses of trimethyltin. *Psychopharmacology, 93*, 301–307.

Cohen, C. A., & Sparber, S. B. 1985. The utility of a latent inhibition paradigm in combination with a delayed reinforcement autoshaped behavior. Presented before the Society for Neurosciences, Dallas.

Cohen, D. H., & MacDonald, R. L. 1971. Some variables affecting orienting and conditioned heart-rate responses in the pigeon. *Journal of Comparative and Physiological Psychology, 74*, 123–133.

Cone, D. M. 1974. Age effects in preexposure of male albino rats to auditory and light CSs. Presented before the Psychonomic Society, Boston.

Connell, P. H. 1958. *Amphetamine Psychosis*. Maudsley Monograph No. 5. London: Oxford University Press.

Conti, L. H., & Musty, R. E. 1986. The effects of dorsal noradrenergic bundle lesions on the latent inhibition of a taste aversion in rats. *Appetite, 6*, 310.

Cools, A. R., & van Rossum, J. M. E. 1980. Multiple receptors for brain dopamine in behavior regulation: Concept of dopamine-E and dopamine-I receptors. *Life Sciences, 27*, 1237–1253.

Corrigan, J. G., & Carpenter, D. L. 1979. Early selective visual experience and pattern discrimination in hooded rats. *Developmental Psychobiology, 12*, 67–72.

Corriveau, D. P., & Smith, N. F. 1978. Fear reduction and "safety-test" behavior following response-prevention: A multivariate analysis. *Journal of Experimental Psychology: General, 107*, 145–158.

Costall, B., Hui, S. C. G., & Naylor, R. J. 1979. The importance of serotonergic mechanisms for the induction of hyperactivity by amphetamine and its antagonism by intra-accumbens (3, 4-dihydroxyphenylamino)-2-imidazoline (DPI). *Neuropharmacology, 18*, 605–609.

Costall, B., Naylor, R. J., Marsden, C. D., & Pycock, C. J. 1976. Serotonergic modulation of the dopamine response from the nucleus accumbens. *Journal of Pharmaceutic Pharmacology, 28*, 523–526.

Coulombe, D., & White, N. 1982. Post-training, self-stimulation and memory: A study of some parameters. *Physiological Psychology, 10,* 343–349.

Cousins, L. S., Zamble, E., Tait, R. W., & Suboski, M. D. 1971. Sensory preconditioning in curarized rats. *Journal of Comparative and Physiological Psychology, 77,* 152–154.

Creese, I., & Iversen, S. D. 1975. The pharmacological and anatomical substrates of the amphetamine response in the rat. *Brain Research, 83,* 419–436.

Crider, A., Solomon, P. R., & McMahon, M. A. 1982. Disruption of selective attention in the rat following *d*-amphetamine administration: Relationship to schizophrenic attention disorder. *Biological Psychiatry, 17,* 351–361.

Crow, T. J. 1980. Molecular pathology of schizophrenia; more than one disease process. *British Medical Journal, 280,* 66–68.

Crow, T. J., & Alkon, D. L. 1978. Retention of an associative behavioral change in *Hermissenda. Science, 201,* 1239–1241.

Crowell, C. R., & Anderson, D. C. 1972. Variations in intensity, interstimulus interval and interval between CS exposure and conditioning with rats. *Journal of Comparative and Physiological Psychology, 79,* 291–298.

Dale, P. S. 1969. Color naming, matching and recognition by preschoolers. *Child Development, 40,* 1135–1144.

Danielson, T. J., & Boulton, A. A. 1976. Distribution and occurrence of amphetamine and *p*-hydroxyamphetamine in tissues of the rat after injection of *d*-amphetamine sulfate. *European Journal of Pharmacology, 37,* 257–264.

Danielson, T. J., Petrali, E. H., & Wishart, T. B. 1976. The effect of acute and chronic injections of *d*-amphetamine sulfate and substantia nigra lesions on the distribution of amphetamine and parahydroxyamphetamine in the rat brain. *Life Science, 19,* 1265–1270.

Davis, M. 1970. Effects of interstimulus interval length and variability on startle response habituation in the rat. *Journal of Comparative and Physiological Psychology, 72,* 177–192.

Davis, M., & Wagner, A. R. 1968. Startle responsiveness after habituation to different intensities of tone. *Psychonomic Science, 12,* 337–338.

Dawley, J. M. 1979. Generalization of the CS-preexposure effect transfers to taste aversion learning. *Animal Learning & Behavior, 7,* 23–24.

Deakin, J. F. W. 1983. Roles of serotonergic systems in escape, avoidance and other behaviors. In S. J. Cooper (Ed.), *Theory in Psychopharmacology, Vol. 2* (pp. 144–193). London: Academic Press.

DeVietti, T. L., & Barrett, O. V. 1986a. Latent inhibition with one preexposure trial. Presented before the Psychonomic Society, November.

DeVietti, T. L., & Barrett, O. V. 1986b. Latent inhibition: No effect of intertrial interval of the preexposure trials. *Bulletin of the Psychonomic Society, 24,* 453–455.

DeVietti, T. L., & Bauste, R. L. 1982. Latent inhibition: No effect of stimulus duration change in rats. Presented before the Psychonomic Society, November.

DeVietti, T. L., Bauste, R. L., Nutt, G., Barrett, O. V., Daly, K., & Petree, A. D. 1987. Latent inhibition: A trace conditioning phenomenon? *Learning and Motivation, 18,* 185–201.

DeVietti, T. L., Emmerson, R. Y., & Wittman, T. K. 1982. Disruption of latent inhibition by placement of an electrode in the dorsal hippocampus. *Physiological Psychology, 10,* 46–50.

DeVietti, T. L., Wittman, T. K., & Comfort, M. K. 1980. Rapid development of latent inhibition: No effect of implanted site. *Psychological Reports, 47,* 473–474.

DeVietti, T. L., Wittman, T. K., Emmerson, R. Y., & Thatcher, D. O. 1981. Either stimulation of the mesencephalic reticular formation or a flashing light increases latent inhibition to a tone-conditioned stimulus. *Behavioral and Neural Biology, 32,* 308–318.

Dexter, W. R., & Merrill, H. K. 1969. Role of contextual discrimination in fear conditioning. *Journal of Comparative and Physiological Psychology, 69*, 677–681.
Dickinson, A. 1976. Appetitive-aversive interactions: Facilitation of aversive conditioning by prior appetitive training in the rat. *Animal Learning & Behavior, 4*, 416–420.
Dickinson, A., & Mackintosh, N. J. 1979. Reinforcer specificity in the enhancement of conditioning by posttrial surprise. *Journal of Experimental Psychology: Animal Behavior Processes, 5*, 162–177.
Dieter, S. E. 1977. Preexposure to situational cues and shock intensity in two-way avoidance learning. *Animal Learning & Behavior, 5*, 403–406.
Dieter, S. E. 1978. Preexposure, exploration, and similarity of the start and safe boxes in one-way avoidance learning. Unpublished doctoral dissertation, Northern Illinois University.
Djuric, V. J., Markovic, B. M., Lazarevic, M., & Jankovic, B. D. 1987. Conditioned taste aversion in rats subjected to anaphylactic shock. In B. J. Jankovic, B. M. Markovic, & N. H. Spector (Eds.), *Neuroimmune Interactions* (pp. 561–568). New York: New York Academy of Sciences (Vol. 496).
Domjan, M. 1971. The CS preexposure effect without habituation of attentional responses. Presented before the Eastern Psychological Association, New York, January.
Domjan, M. 1972. CS preexposure in taste aversion learning: Effects of deprivation and preexposure duration. *Learning and Motivation, 3*, 389–402.
Domjan, M. 1976. Determinants of the enhancement of flavored water intake by prior exposure. *Journal of Experimental Psychology: Animal Behavior Processes, 2*, 17–27.
Domjan, M. 1978. Effects of proximal unconditioned stimulus preexposure on ingestional aversions learned as a result of taste presentation following drug treatment. *Animal Learning & Behavior, 6*, 113–142.
Domjan, M. 1980. Ingestional aversion learning: Unique and general processes. In J. S. Rosenblatt, R. A. Hinde, C. Beer, & M. C. Busnel (Eds.), *Advances in the Study of Behavior* (Vol. 11, pp. 275–336). New York: Academic Press.
Domjan, M., & Best, M. R. 1977. Paradoxical effects of proximal unconditioned stimulus preexposure: Interference with and conditioning of a taste aversion. *Journal of Experimental Psychology: Animal Behavior Processes, 3*, 310–321.
Domjan, M., & Best, M. R. 1980. Interference with ingestional aversion learning produced by preexposure to the unconditioned stimulus: Associative and nonassociative aspects. *Learning and Motivation, 11*, 522–537.
Domjan, M., & Bowman, T. G. 1974. Learned safety and the CS-US delay gradient in taste aversion learning. *Learning and Motivation, 5*, 409–423.
Domjan, M., & Siegel, S. 1971. Conditioned suppression following CS preexposure. *Psychonomic Science, 25*, 11–12.
Domjan, M., & Wilson, N. E. 1972a. Contribution of ingestive behaviors to taste aversion learning in the rat. *Journal of Comparative and Physiological Psychology, 80*, 403–412.
Domjan, M., & Wilson, N. E. 1972b. Specificity of cue to consequence in aversion learning in the rat. *Psychonomic Science, 26*, 143–145.
Donegan, N. H., Whitlow, J. W., & Wagner, A. R. 1977. Posttrial reinstatement of the CS in Pavlovian conditioning: Facilitation or impairment as a function of individual differences in responsiveness to the CS. *Journal of Experimental Psychology: Animal Behavior Processes, 3*, 357–376.
Dore, F. Y. 1981. l'Effect de la preexposition au SC dans les apprentissages d'evitement bidirectionnel et unidirectionnel. *l'Annee Psychologique, 81*, 23–92.
Dore, F. Y. 1982. Validite du groupe-temoin dans les etudes sur l'inhibition latente. *Revue Canadienne de Psychologie, 36*, 504–507.
Dore, F. Y. 1984. Preexposition a la reponse, effet Lubow et masquage dans un apprentissage d'evitement bidirectionnel. *l'Annee Psychologique, 84*, 21–42.

Douglas, R. J. 1967. The hippocampus and behavior. *Psychological Bulletin, 67*, 416–442.

Douglas, R. J. 1972. Pavlovian conditioning and the brain. In R. Boakes & M. Halliday (Eds.), *Inhibition and Learning* (pp. 529–533). London: Academic Press.

Douglas, R. J., & Pribram, K. H. 1966. Learning and limbic lesions. *Neuropsychologia, 4*, 197–219.

Downey, P., & Harrison, J. M. 1972. Control of responding by location of auditory stimuli: The role of differential and nondifferential reinforcement. *Journal of Experimental Analysis of Behavior, 18*, 453–463.

Elkins, R. L. 1973. Attenuation of a drug-induced bait shyness to a palatable solution as an increasing function of its availability prior to conditioning. *Behavioral Biology, 9*, 221–226.

Elkins, R. L. 1974. Conditioned taste aversions to familiar tap water in rats: An adjustment with implications for aversion therapy treatment of alcoholism and obesity. *Journal of Abnormal Psychology, 83*, 411–417.

Elkins, R. L., & Hobbs, S. H. 1979. Forgetting, preconditioning CS familiarization and taste aversion learning: An animal experiment with implications for alcoholism treatment. *Behaviour, Research and Therapy, 17*, 567–573.

Ellins, S. R., Thompson, L., & Swanson, W. E. 1983. Effects of novelty and familiarity on illness-induced aversions to food and place cues in coyotes (*Canis latrans*). *Journal of Comparative Psychology, 97*, 302–309.

Ellison, G. D., & Eison, M. S. 1983. Continuous amphetamine intoxication: An animal model of acute psychotic episode. *Psychological Medicine, 13*, 751–761.

Epstein, W. 1967. *Varieties of Perceptual Experience*. New York: McGraw-Hill.

Estes, W. K. 1959. The statistical approach to learning theory. In S. Koch (Ed.), *Psychology: A Study of a Science* (Vol. 2, pp. 380–491). New York: McGraw-Hill.

Estes, W. K., & Skinner, B. F. 1941. Some quantitative properties of anxiety. *Journal of Experimental Psychology, 29*, 390–400.

Evans, J. G. M., & Hammond, G. R. 1983. Differential generalization of habituation across context as a function of stimulus significance. *Animal Learning & Behavior, 11*, 431–434.

Eysenck, H. J., & Eysenck, S. B. G. 1975. *Manual of the Eysenck Personality Questionnaire*. London: Hodder and Stoughton.

Eysenck, H. J., & Eysenck, S. B. G. 1976. *Psychoticism as a Dimension of Personality*. London: Hodder and Stoughton.

Farley, J. 1987a. Contingency learning and causal detection in *Hermissenda:* I. Behavior. *Behavioral Neuroscience, 101*, 13–27.

Farley, J. 1987b. Contingency learning and causal detection in *Hermissenda:* II. Cellular mechanisms. *Behavioral Neuroscience, 101*, 28–56.

Farley, J., & Alkon, D. 1982. Long-term associative and neural behavioral change in *Hermissenda:* Consequences of nervous system orientation for light- and pairing-specificity. *Journal of Neurophysiology, 48*, 785–807.

Farley, J. A., McLaurin, W. A., Scarborough, B. B., & Rawlings, T. D. 1964. Pre-irradiation saccharin habituation: A factor in avoidance behavior. *Psychological Reports, 14*, 401–496.

Feldman, M. A. 1977. The effects of preexposure to a warning or a safety signal on the acquisition of a two-way avoidance response in rats. *Animal Learning & Behavior, 5*, 21–24.

Feldman, M. A. 1982. The effects of command preexposure on instruction-following behavior in severely retarded adolescents. *Psychological Record, 32*, 529–535.

Feldon, J. 1974. The role of the serotonergic system in latent inhibition using the conditioned emotional response paradigm. Unpublished master's thesis, Tel Aviv University.

Fenwick, S., Mikulka, P. J., & Klein, S. B. 1975. The effects of different levels of preexposure to sucrose on the acquisition and extinction of a conditioned aversion. *Behavioral Biology, 14,* 231–235.
Fink, H., & Oelssner, W. 1981. LSD, mescaline, and serotonin injected into the medial raphe nucleus potentiate apomorphine hypermotility. *European Journal of Pharmacology, 75,* 289–296.
Fisher, M. A., & Zeaman, D. 1973. An attention-retention theory of retardate discrimination learning. In N. R. Ellis (Ed.), *International Review of Research in Mental Retardation* (pp. 169–256). New York: Academic Press.
Fitzgerald, R. D., & Hoffman, J. 1976. Classically conditioned heart rate in rats following preconditioning exposure to the CS. *Animal Learning & Behavior, 4,* 58–60.
Forbes, D. T., & Holland, P. C. 1980. Positive and negative patterning after CS preexposure in flavor aversion conditioning. *Animal Learning & Behavior, 8,* 595–600.
Forbes, D. T., & Holland, P. C. 1985. Spontaneous configuring in conditioned flavor aversion. *Journal of Experimental Psychology: Animal Behavior Processes, 11,* 224–240.
Foreman, N., & Stevens, R. 1987. Relationships between the superior colliculus and hippocampus: Neural and behavioral considerations. *Behavioral and Brain Sciences, 10,* 101–152.
Foreman, N., & Thinus-Blanc, C. 1987. Weakness of the "missing-stimulus effect" in hooded rats: Gross asymmetry in the Sokolovian orienting response. *Psychobiology, 15,* 265–271.
Forgus, R. H. 1958. The effect of different kinds of form preexposure on form discrimination learning. *Journal of Comparative and Physiological Psychology, 51,* 75–78.
Franchina, J. J., Domato, G. C., Patsiokas, A. T., & Griesemer, H. A. 1980. Effects of number of preexposures on sucrose taste aversion in weanling rats. *Developmental Psychobiology, 13,* 25–31.
Franchina, J. J., Dyer, A. B., Gilley, D. W., Ness, J., & Dodd, M. 1985. Role of ethanol in conditioning aversion to alcoholic beverages in rats. *Behaviour, Research and Therapy, 23,* 521–529.
Franchina, J. J., & Gilley, D. W. 1986. Effects of pretraining on conditioning-enhanced neophobia: Evidence for separable mechanisms of neophobia and aversion conditioning. *Animal Learning & Behavior, 14,* 155–162.
Franchina, J. J., & Horowitz, S. W. 1982. Effects of age and flavor preexposures on taste aversion performance. *Bulletin of the Psychonomic Society, 19,* 41–44.
Franchina, J. J., Silber, S., & May, B. 1981. Novelty and temporal contiguity in taste aversion learning: Within subject conditioning effects. *Bulletin of the Psychonomic Society, 18,* 99–102.
Franken, R. E., & Bray, G. P. 1973. Prolonged duration of the "novelty effect" following prolonged exposure to a single discriminandum. *Animal Learning & Behavior, 1,* 233–236.
Frey, P. W., Maisiak, R., & Dugue, G. 1976. Unconditional stimulus characteristics in rabbit eyelid conditioning. *Journal of Experimental Psychology: Animal Behavior Processes, 2,* 175–190.
Frey, P. W., & Sears, R. J. 1978. Model of conditioning incorporating the Rescorla-Wagner associative axiom, a dynamic attention process, and a catastrophe rule. *Psychological Review, 85,* 321–340.
Furedy, J. J. 1968. Human orienting reaction as a function of electrodermal versus plethysmographic response modes and single versus alternating stimulus series. *Journal of Experimental Psychology, 77,* 70–78.
Galey, D., Durkin, T., Sifakis, S., Kempf, E., & Jaffard, R. 1985. Facilitation of spontaneous and learned spatial behaviours following 6-hydroxydopamine lesions of the lateral septum: A cholinergic hypothesis. *Brain Research, 340,* 171–174.

Gallagher, M. 1985. Re-viewing modulation of learning and memory. In N. M. Weinberger, J. L. McGaugh, & G. Lynch (Eds.), *Memory Systems of the Brain* (pp. 311–334). New York: Guilford Press.

Gallagher, M., Meagher, M. R., & Bostock, E. 1987. Effects of opiate manipulations on latent inhibition in rabbits: Sensitivity of the medial septal region to intracranial treatments. *Behavioral Neuroscience, 3*, 315–324.

Gamzu, E., & Williams, D. R. 1971. Classical conditioning of a complex skeletal response. *Science, 171*, 923–925.

Gamzu, E., & Williams, D. R. 1973. Associative factors underlying the pigeon's key pecking in auto-shaping procedures. *Journal of Experimental Analysis of Behavior, 19*, 225–232.

Garb, J. L., & Stunkard, A. J. 1974. Taste aversion in man. *American Journal of Psychiatry, 131*, 1204–1207.

Garcia, J., Ervin, F. R., & Koelling, R. A. 1966. Learning with prolonged delay of reinforcement. *Psychonomic Science, 5*, 121–122.

Garcia, J. & Kimmeldorf, D. J. 1957. Temporal relationship within the conditioning of a saccharin aversion through radiation exposure. *Journal of Comparative and Physiological Psychology, 50*, 180–183.

Garcia, J., Kimmeldorf, D. J., & Koelling, R. A. 1955. Conditioned aversion to saccharin resulting from exposure to gamma radiation. *Science, 122*, 157–158.

Garcia, J., & Koelling, R. A. 1966. Relation of cue to consequence in avoidance learning. *Psychonomic Science, 4*, 123–124.

Garcia, J., & Koelling, R. A. 1967. A comparison of aversions induced by X-rays, toxins and drugs in the rat. *Radiation Research Suppl., 7*, 439–450.

Garmezy, N. 1977. The psychology and psychopathology of attention. *Schizophrenia Bulletin, 3*, 360–368.

Gately, P. F., Poon, S. L., Segal, D. S., & Geyer, M. 1985. Depletion of brain serotonim by 5,7-dihydroxytryptamine alters the response to amphetamine and the habituation of locomotor activity in rats. *Psychopharmacology, 87*, 400–405.

Geyer, M., Al Puerto, A., Dawsey, W. J. Knapp, S., Bullard, W. P., & Mandell, A. J. 1976. Histologic and enzymatic studies of the mesolimbic and the mesostriatal serotonergic pathways. *Brain Research, 106*, 241–256.

Gibson, E. J. 1969. *Principles of Perceptual Learning and Development.* New York: Appleton-Century-Crofts.

Gibson, E. J., & Walk, R. D. 1956. The effects of prolonged exposure to visually presented patterns on learning to discriminate them. *Journal of Comparative and Physiological Psychology, 49*, 239–242.

Gibson, E. J., Walk, R. D., & Tighe, T. J. 1959. Enhancement and deprivation of visual stimulation during rearing as factors in visual discrimination learning. *Journal of Comparative and Physiological Psychology, 52*, 74–81.

Gillette, K., & Bellingham, W. P. 1982. Loss of within-compound flavour associations: Configural conditioning. *Experimental Animal Behavior, 1*, 1–17.

Ginton, A., Urca, G., & Lubow, R. E. 1975. The effects of preexposure to a nonattended stimulus on subsequent learning: Latent inhibition in adults. *Bulletin of the Psychonomic Society, 5*, 5–8.

Gjerde, P. F. 1983. Attentional capacity dysfunction and arousal in schizophrenia. *Psychological Bulletin, 93*, 57–72.

Glazer, H. I., & Weiss, J. M. 1976a. Long-term and transitory interference effects. *Journal of Experimental Psychology: Animal Behavior Processes, 2*, 191–201.

Glazer, H. I., & Weiss, J. M. 1976b. Long-term interference effect: An alternative to "learned helplessness." *Journal of Experimental Psychology: Animal Behavior Processes, 2*, 202–213.

Goss, A. E., & Nodine, C. F. 1965. *Paired-Associates Learning.* New York: Academic Press.

Graham, F. K. 1979. Distinguishing among orienting, defense, and startle reflexes. In

H. D. Kimmel, E. H. Van Olst, & J. F. Orlebeke (Eds.), *The Orienting Reflex in Humans* (pp. 137–166). Hillsdale, N.J.: Lawrence Erlbaum.

Graham, F. K., & Clifton, R. L. 1966. Heart-rate change as a component of the orienting response. *Psychological Bulletin, 65,* 302–320.

Grant, D. A., Hake, H. W., Riopelle, A. J., & Kostlan, A. 1951. Effects of repeated pretesting with conditioned stimulus upon extinction of the conditioned eyelid response to light. *American Journal of Psychology, 54,* 247–251.

Grant, D. A., Hake, H. W., & Schneider, D. E. 1948. Effects of pretesting with the conditioned stimulus upon extinction of the conditioned eyelid response. *American Journal of Psychology, 61,* 243–248.

Grant, M. J., & Grant, R. M. 1973. The effects of scopalomine on preexposure to a learning apparatus. *Bulletin of the Psychonomic Society, 1,* 238–240.

Grant, M. J., & Young, D. 1971. The effects of preexposure to learning apparatus. *Behavior Research Methods and Instrumentation, 3,* 251–252.

Gray, J. A. 1982. *The Neuropsychology of Anxiety: An Enquiry into the Functions of the Septo-hippocampal Systems.* New York: Oxford University Press.

Gray, J. A., Feldon, J., Rawlins, J. N. P., Owen, S., & McNaughton, N. 1978. The role of the septo-hippocampal system and its noradrenergic afferents in behavioral responses to nonreward. In K. Elliott & J. Whelan (Eds.), *Functions of the Septo-hippocampal System* (pp. 275–300). Ciba Foundation Symposium No. 58 Amsterdam: Elsevier.

Gray, J. A., & McNaughton, N. 1983. Comparison between the behavioral effects of septal and hippocampal lesions: A review. *Neuroscience and Biobehavior Review, 7,* 119–188.

Gray, T. 1978. CS preexposure and acquisition of the CER. Presented before the Canadian Psychological Association, Ottawa.

Green, K. F., & Parker, L. A. 1975. Gustatory memory: Incubation and interference. *Behavioral Biology, 13,* 359–367.

Green, S. E., Nealis, P. M., & Suomi, S. J. 1977. Novelty and familiarity in discrimination learning by rhesus monkeys. *Bulletin of the Psychnomic Society, 10,* 399–401.

Grice, G. R. 1948. The acquisition of a visual discrimination habit following response to a single stimulus. *Journal of Experimental Psychology, 38,* 633–642.

Grice, G. R., & Hunter, J. J. 1964. Stimulus intensity effects depend on the type of experimental design. *Psychological Review, 71,* 247–256.

Griffith, J. D., Cavanaugh, J., Held, J., & Oates, J. A. 1972. Dextroamphetamine evaluation of psychotomimetic properties in man. *Archives of General Psychiatry, 26,* 97–100.

Griffith, J. D., Oates, J., & Cavanaugh, J. 1968. Paranoid episodes induced by drug. *Journal of the American Medical Association, 205,* 39.

Groenewegen, H. J., Vermeulen-Van Der Zee, E., de Kortschot, A., & Witter, M. P. 1987. Organization of the projections from the subiculum to the ventral striatum in the rat. A study using anterograde transport of phaseolus vulgaris leucoagglutinin. *Neuroscience, 23,* 103–120.

Groves, P. M., & Thompson, R. F. 1970. Habituation: A dual-process theory. *Psychological Review, 77,* 419–450.

Groves, P. M., & Rebec, G. V. 1976. Biochemistry and behavior: Some central actions of amphetamines and antipsychotic drugs. *Annual Review of Psychology, 27,* 91–127.

Gustavson, C. R., Kelly, D. J., Sweeney, M., & Garcia, J. 1976. Prey-lithium aversions: I. Coyotes and wolves. *Behavioral Biology, 17,* 61–72.

Halgren, C. R. 1974. Latent inhibition in rats: Associative or nonassociative? *Journal of Comparative and Physiological Psychology, 86,* 74–78.

Hall, G. 1979. Exposure learning in young and adult laboratory rats. *Animal Behaviour, 27,* 586–591.

Hall, G. 1980. Exposure learning in animals. *Psychological Bulletin, 88,* 535–550.

Hall, G., & Channell, S. 1980. A search for perceptual differentiation produced by nondifferential reinforcement. *Quarterly Journal of Experimental Psychology, 32,* 185–195.

Hall, G., & Channell, S. 1983. Stimulus exposure and discrimination in rats: A test of a theory for the role of contextual factors. *Quarterly Journal of Experimental Psychology, 35B,* 135–147.

Hall, G., & Channell, S. 1984. Differential effects of contextual change on latent inhibition and on the habituation of the orienting response. *Journal of Experimental Psychology: Animal Behavior Processes, 11,* 470–481.

Hall, G., & Channell, S. 1985a. Latent inhibition and conditioning after preexposure to the training context. *Learning and Motivation, 16,* 381–397.

Hall, G., & Channell, S. 1985b. Differential effects of contextual change on latent inhibition and on the habituation of an orienting response. *Journal of Experimental Psychology: Animal Behavior Processes, 11,* 470–481.

Hall, G., & Channell, S. 1986. Context specificity of latent inhibition in taste aversion learning. *Quarterly Journal of Experimental Psychology, 38B,* 121–139.

Hall, G., & Minor, H. 1984. A search for context stimulus associations in latent inhibition. *Quarterly Journal of Experimental Psychology, 36B,* 145–169.

Hall, G., & Pearce, J. M. 1979. Latent inhibition of a CS during CS–US pairings. *Journal of Experimental Psychology: Animal Behavior Processes, 5,* 31–42.

Hall, G., & Pearce, J. M. 1982. Restoring the associability of a preexposed CS by a surprising event. *Quarterly Journal of Experimental Psychology, 34B,* 127–140.

Hall, G., & Schachtman, T. R. 1987. Differential effects of a retention interval on latent inhibition and the habituation of an orienting response. *Animal Learning & Behavior, 15,* 76–82.

Hammond, L. J., & Maser, J. 1970. Forgetting and conditioned suppression: Role of a temporal discrimination. *Journal of Experimental Analysis of Behavior, 13,* 333–338.

Har-Evan, D. 1977. The effect of nonreinforced preexposure of a stimulus on discrimination learning. Unpublished master's thesis, Bar-Ilan University.

Harris, J. D. 1943. Habituatory response decrement in the intact organism. *Psychological Bulletin, 40,* 385–422.

Harrison, J. M. 1979. The control of responding by sounds: Unusual effect of reinforcement. *Journal of Experimental Analysis of Behavior, 32,* 167–181.

Hearst, E. 1972. Some persistent problems in the analysis of conditioned inhibition. In R. A. Boakes & M. S. Halliday (Eds.), *Inhibition and Learning* (pp. 5–39). New York: Academic Press.

Hearst, E. 1978. Stimulus relationships and feature selection in learning and behavior. In S. H. Hulse, H. Fowler, & W. K. Honig (Eds.), *Cognitive Processes in Animal Behavior* (pp. 51–88). Hillsdale, N.J.: Lawrence Erlbaum.

Hebb, D. O. 1949. *The Organization of Behavior.* New York: Wiley.

Hebb, D. O. 1955. Drives and C.N.S. (conceptual nervous system). *Psychological Review, 62,* 243–254.

Heisenberg, W. 1958. *Physics and Philosophy.* New York: Harper.

Hellman, P. A., Crider, A., & Solomon, P. R. 1983. Interaction of tail-pressure stress and d-amphetamine in disruption of rat's ability to ignore an irrelevant stimulus. *Behavioral Neuroscience, 97,* 1017–1021.

Herendeen, D. L., & Shapiro, M. M. 1976. Specific and generalized adaptation of salivary conditioning in dogs. *Bulletin of the Psychonomic Society, 8,* 68–71.

Hernandez, L. L., Buchanan, S. L., & Powell, D. A. 1981. CS preexposure: Latent inhibition and Pavlovian conditioning of heart rate and eyeblink responses as a function of sex and CS intensity in rabbits. *Animal Learning & Behavior, 9,* 513–518.

Hill, W. F. 1978. Effects of mere exposure on preferences in nonhuman mammals. *Psychological Bulletin, 85,* 1177–1198.

Hinson, R. E. 1982. Effects of UCS preexposure on excitatory and inhibitory rabbit eyelid conditioning: An associative effect of conditioned context stimuli. *Journal of Experimental Psychology: Animal Behavior Processes, 8*, 49–61.
Hiroto, D. S. 1974. Learned helplessness and locus of control. *Journal of Experimental Psychology, 102*, 187–193.
Hiroto, D. S., & Seligman, M. E. P. 1975. Generality of learned helplessness in man. *Journal of Personality and Social Psychology, 31*, 311–327.
Hoffeld, D. R., Kendall, S. B., Thompson, R. F., & Brogden, W. J. 1960. Effect of amount of preconditioning training upon the magnitude of sensory preconditioning. *Journal of Experimental Psychology, 59*, 198–204.
Holder, M. D., Leon, M., Yirmiya, R., & Garcia, J. 1987. Effect of taste preexposure on taste and odor aversions. *Animal Learning & Behavior, 15*, 55–61.
Holland, P. C. 1980. Influence of visual conditioned stimulus characteristics on the form of Pavlovian appetitive conditioned responding in rats. *Journal of Experimental Psychology: Animal Behavior Processes, 6*, 81–97.
Holland, P. C. 1983. Occasion-setting in Pavlovian feature positive discrimination. In M. L. Commons, R. J. Herrnstein, & A. R. Wagner (Eds.), *Quantitative Analyses of Behavior. Vol. 4: Discrimination Processes* (pp. 183–206). New York: Ballinger.
Holland, P. C. 1985. The nature of conditioned inhibition in serial and simultaneous feature negative discriminations. In R. R. Miller & N. E. Spear (Eds.), *Information Processing in Animals: Conditioned Inhibition* (pp. 267–297). Hillsdale, N.J.: Lawrence Erlbaum.
Holland, P. C., & Forbes, D. T. 1980. Effects of compound or element preexposure on compound flavor aversion conditioning. *Animal Learning & Behavior, 8*, 199–203.
Holman, E. W. 1976. The effect of drug habituation before and after taste aversion learning in rats. *Animal Learning & Behavior, 4*, 329–332.
Honigfeld, G. 1963. The ability of schizophrenics to understand normal, psychotic and pseudo-psychotic speech. *Diseases of the Nervous System, 24*, 692–694.
Horn, G., & Hinde, R. A. 1970. *Short-Term Changes in Neural Activity and Behavior.* Cambridge University Press.
Hull, C. L. 1943. *Principles of Behavior.* New York: Appleton-Century-Crofts.
Hull, C. L. 1952. *A Behavior System.* New Haven: Yale University Press.
Hulstijn, W. 1978. The orienting reaction during human eyelid conditioning following preconditioning exposure to the CS. *Psychological Research, 40*, 77–88.
Hussey, J. S., Vincent, N. D., & Davies, J. A. 1983. The effect of low doses of d-amphetamine on drug-induced hyperactivity in the mouse. *Psychopharmacology, 81*, 327–331.
Iacono, W. G., & Lykken, D. T. 1983. The effect of instructions on electrodermal habituation. *Psychophysiology, 20*, 71–80.
Iacono, W. G., & Lykken, D. T. 1984. The effects of instructions and an engaging visual task on habituation to loud tones: An evaluation of an alternative to the traditional habituation paradigm. *Physiological Psychology, 12*, 23–29.
Isaacson, R. L., Springer, J. E., & Ryan, J. P. 1986. Cholinergic and catecholaminergic modification of the hippocampal lesion syndrome. In R. L. Isaacson & K. H. Pribram (Eds.), *The Hippocampus* (Vol. 4, pp. 127–158). New York: Plenum Press.
Iversen, S. D. 1977. Brain dopamine systems and behavior. In L. L. Iversen & S. H. Snyder (Eds.), *Handbook of Psychopharmacology* (Vol. 8, pp. 333–383). New York: Plenum Press.
Izaks, M. 1981. Latent inhibition in concept formation. Unpublished master's thesis, Bar-Ilan University.
Jackson, R. L., Alexander, J. H., & Maier, S. F. 1980. Learned helplessness, inactivity and associative deficits: Effects of inescapable shock on response choice escape learning. *Journal of Experimental Psychology: Animal Behavior Processes, 6*, 1–20.

Jacobs, B. L., Wise, W. D., & Taylor, K. M. 1974. Differential behavioral and neurochemical effects of the dorsal or median raphe nuclei in the rat. *Brain Research, 79,* 353-361.
James, J. P. 1971. Latent inhibition and the preconditioning-conditioning interval. *Psychonomic Science, 24,* 97-98.
James, W. 1950. *The Principles of Psychology, Vol. 1.* New York: Dover. (Original work published 1890).
Janowsky, D. S., Huey, L., Storms, L., & Judd, L. L. 1977. Methylphenidate hydrochloride effects on psychological tests in acute schizophrenic and nonpsychotic patients. *Archives of General Psychiatry, 34,* 189-194.
Jeeves, M. A., & North, A. J. 1956. Irrelevant or partially correlated stimuli in discrimination learning. *Journal of Experimental Psychology, 52,* 90-94.
Johnson, P. E. 1979. The effects of amount of preexposure and preexposure-to-conditioning interval on the latent inhibition of flavor aversions. Unpublished manuscript, reported in Best (1982).
Johnston, W. A., & Dark, V. J. 1986. Selective attention. *Annual Review of Psychology, 37,* 43-75.
Jones, D. L., Mogenson, G. J., & Wu, M. 1981. Injections of dopaminergic, cholinergic, serotonergic and GABA-ergic drugs into the nucleus accumbens: Effects on locomotor activity in the rat. Unpublished manuscript.
Joyce, E. M., & Iversen, S. D. 1984. Dissociable effects of 6-OHDA-induced lesions of neostriatum on anorexia, locomotor activity and stereotypy: The role of behavioural competition. *Psychopharmacology, 83,* 363-366.
Kahneman, D. 1973. *Attention and Effort.* Englewood Cliffs, N.J.: Prentice-Hall.
Kalat, J. W. 1974. Taste salience depends on novelty, not concentration, in taste aversion learning in the rat. *Journal of Comparative and Physiological Psychology, 86,* 47-50.
Kalat, J. W. 1977. Status of "learned safety" or "learned noncorrelation" as a mechanism in taste aversion learning. In: L. M. Barker, M. R. Best, & M. Domjan (Eds.), *Learning Mechanisms in Food Selection* (pp. 273-293). Waco, Tex.: Baylor University Press.
Kalat, J., & Rozin, P. 1971. Role of interference in taste aversion learning. *Journal of Comparative and Physiological Psychology, 77,* 53-58.
Kalat, J. W., & Rozin, P. 1973. "Learned safety" as a mechanism in long-delay taste aversion learning in rats. *Journal of Comparative and Physiological Psychology, 83,* 198-207.
Kamin, L. J. 1969. Predictability, surprise, attention and conditioning. In B. A. Campbell & R. M. Church (Eds.), *Punishment and Aversive Behavior* (pp. 279-296). New York: Appleton-Century-Crofts.
Kaniel, S., & Lubow, R. E. 1986. Latent inhibition: A developmental study. *British Journal of Developmental Psychology, 4,* 367-375.
Kasprow, W. J., Cacheiro, H., Balaz, M. A., & Miller, R. R. 1982. Reminder-induced recovery of associations to an overshadowed stimulus. *Learning and Motivation, 13,* 155-166.
Kasprow, W. J., Catterson, D., Schachtman, T. R., & Miller, R. R. 1984. Attenuation of latent inhibition by post-acquisition reminder. *Quarterly Journal of Experimental Psychology, 36B,* 53-64.
Kasprow, W. J., Schachtman, T. R., & Miller, R. R. 1985. Associability of a previously conditioned stimulus as a function of qualitative changes in the US. *Quarterly Journal of Experimental Psychology, 37B,* 33-48.
Kay, S. R., & Opler, L. A. 1985. *l*-dopa in the treatment of negative schizophrenic systems: A single-subject experimental study. *International Journal of Psychiatry in Medicine, 15,* 293-298.

Kaye, H., & Pearce, J. M. 1984a. The strength of the orienting response during Pavlovian conditioning. *Journal of Experimental Psychology: Animal Behavior Processes, 10*, 90–109.

Kaye, H., & Pearce, J. M. 1984b. The strength of the orienting response during blocking. *Quarterly Journal of Experimental Psychology, 36B*, 131–144.

Kaye, H., & Pearce, J. M. 1987a. Hippocampal lesions attenuate latent inhibition and the decline of the orienting response in rats. *Quarterly Journal of Experimental Psychology, 39B*, 107–125.

Kaye, H., & Pearce, J. M. 1987b. Hippocampal lesions attenuate latent inhibition of a CS and of a neutral stimulus. *Psychobiology, 15*, 293–299.

Kaye, H., Preston, G. C., Szabo, L., Druiff, H., & Mackintosh, N. J. 1987. Context specificity of conditioning and latent inhibition: Evidence for a dissociation of latent inhibition and associated interferences. *Quarterly Journal of Experimental Psychology, 39B*, 127–145.

Kaye, H., Swietalski, N., & Mackintosh, N. J. 1988. Distractor effects on latent inhibition are a consequence of generalisation decrement. *Quarterly Journal of Experimental Psychology, 40B*, 151–161.

Kelley, A. E., & Domesick, V. B. 1982. The distribution of the projections from the hippocampal formation to the nucleus accumbens in the rat: An anterograde- and retrograde-horseradish peroxidase study. *Neuroscience, 7*, 2321–2335.

Kelly, P. H., Seviour, P. W., & Iversen, S. D. 1975. Amphetamine and apomorphine responses in the rat following 6-OHDA lesions of the nucleus accumbens septi and corpus striatum. *Brain Research, 94*, 507–522.

Kendler, T. S. 1979a. Cross-sectional research, longitudinal theory, and a discriminative transfer ontogeny. *Human Development, 22*, 235–254.

Kendler, T. S. 1979b. The development of discrimination learning. A levels-of-functioning explanation. In H. W. Reese & L. P. Lipsitte (Eds.), *Advances in Child Development and Behavior, Vol. 13* (pp. 83–117). New York: Academic Press.

Kerpelman, L. C. 1965. Preexposure to visually presented forms and nondifferential reinforcement in perceptual learning. *Journal of Experimental Psychology, 69*, 257–262.

Kiefer, S. W., & Braun, J. J. 1977. Absence of differential responses to novel and familiar taste stimuli in rats lacking gustatory neocortex. *Journal of Comparative and Physiological Psychology, 91*, 498–507.

Kiefer, S. W., Phillips, J. A., & Braun, J. J. 1977. Preexposure to conditioned and unconditioned stimuli in taste aversion learning. *Bulletin of the Psychonomic Society, 10*, 226–228.

Killeen, P. R. 1981. Averaging theory. In C. M. Bradshaw, E. Szabadi, & C. F. Lowe (Eds.), *Quantification of Steady-State Behavior* (pp. 21–34). Amsterdam: Elsevier.

Kimble, D. P. 1968. Hippocampus and internal inhibition. *Psychological Bulletin, 70*, 285–295.

Kimble, D. P. 1969. Possible inhibitory functions of the hippocampus. *Neuropsychologia, 7*, 235–244.

Kimble, G. A. 1961. *Hilgard and Marquis' Conditioning and Learning*. New York: Appleton-Century-Crofts.

Kimmel, H. D. 1973. Habituation, habituability, and conditioning. In H. V. S. Peeke & M. J. Herz (Eds.), *Habituation. Vol. 1: Behavioral Studies* (pp. 219–238). New York: Academic Press.

Kimmel, H. D., & Burns, R. A. 1975. Adaptational aspects of conditioning. In W. K. Estes (Ed.), *Handbook of Learning and Cognitive Processes. Vol. 2: Conditioning and Behavior Theory* (pp. 99–142). Hillsdale, N.J.: Lawrence Erlbaum.

Kimmel, H. D., Van Olst, E. H., & Orlebeke, J. F. 1979. *The Orienting Reflex in Humans*. Hillsdale, N.J.: Lawrence Erlbaum.

Kimmel, W. 1967. Judgments of UCS intensity and diminution of the UCR in classical conditioning. *Journal of Experimental Psychology, 73,* 532–543.
Kirk, R. J. 1974. Saliency modification of neocontinuity theory. Unpublished doctoral dissertation, Purdue University.
Klein, D. E., & Davis, J. M. 1969. *Diagnosis and Drug Treatment of Psychiatric Disorders.* Baltimore: Williams & Wilkins.
Klein, S. B., Mikulka, P. J., Domato, G. C., & Hallstead, C. 1977. Retention of internal experiences in juvenile and adult rats. *Physiological Psychology, 5,* 63–66.
Klein, S. B., Mikulka, P. J., & Hamel, L. 1976. Influence of sucrose preexposure on acquisition of a conditioned aversion. *Behavioral Biology, 16,* 99–104.
Klein, S. B., Mikulka, P. J., Rochelle, F. P., & Blair, V. 1978. Postconditioning CS-alone exposures as a source of interference in a taste aversion paradigm. *Physiological Psychology, 6,* 255–260.
Klosterhalfen, S., & Klosterhalfen, W. 1985. Conditioned taste aversion and traditional learning. *Psychological Research, 47,* 71–94.
Kohler, C., & Steinbush, H. 1982 Identification of serotonin and nonserotonin containing neurons of the mid-brain raphe projecting to the entorhinal area and the hippocampal formation: A combined immunohistochemical and fluorescent retrograde tracing study in the rat brain. *Neuroscience, 7,* 951–975.
Kokkinidis, L., & Anisman, H. 1980. Amphetamine models of paranoid schizophrenia: An overview and elaboration of animal experimentation. *Psychological Bulletin, 93,* 551–579.
Konorski, J., & Szwejkowska, G. 1952. Chronic extinction and restoration of conditioned reflexes. IV: The dependence of the course of extinction and restoration of conditioned reflexes on the "history" of the conditioned stimulus (the principle of the primacy of first training). *Acta Biologiae Experimentalis, 16,* 95–113.
Korboot, P. J., & Damiani, N. 1976. Auditory processing speed and signal detection in schizophrenia. *Journal of Abnormal Psychology, 85,* 287–295.
Kornetsky, C. 1972. The use of a simple test of attention as a measure of drug effects in schizophrenic patients. *Psychopharmacologia, 24,* 99–106.
Kostowski, W., Giacalone, E., Garattini, S., & Valzelli, L. 1968. Studies on behavioral and biochemical changes in rats after lesions of the midbrain raphe. *European Journal of Pharmacology, 4,* 371–376.
Kovach, J. K., Fabricius, E., & Fält, L. 1966. Relationships between imprinting and perceptual learning. *Journal of Comparative and Physiological Psychology, 61,* 449–454.
Kraemer, P. J., Hoffman, H., & Spear, N. E. 1988. Attenuation of the CS-preexposure effect after a retention interval in preweanling rats. *Animal Learning & Behavior, 16,* 185–190.
Kraemer, P. J., & Ossenkopp, K. P. 1986. The effects of flavor preexposure and taste interval on conditioned taste aversions in rats. *Bulletin of the Psychonomic Society, 24,* 219–222.
Kraemer, P. J., & Roberts, W. A. 1984. The influence of flavor preexposure and test interval on conditioned taste aversion in rats. *Learning and Motivation, 15,* 259–278.
Kraepelin, E. 1971. *Dementia Praecox and Paraphrenia.* Huntington, N.Y.: Krieger (originally published in 1919).
Kral, P. A. 1971. Electroconvulsive shock during taste-illness interval: Evidence for induced disassociation. *Physiology and Behavior, 7,* 667–670.
Kral, P. A., & Beggerly, H. D. 1973. Electroconvulsive shock impedes association formation: Conditioned taste aversion paradigm. *Physiology and Behavior, 10,* 145–147.
Krane, R. V., & Robertson, D. P. 1982. Trace decay and priming of short-term memory in long-delay taste aversion learning: Disruptive effects of novel exteroceptive stimulation. *Learning and Motivation, 13,* 434–453.

Krane, R. V., Sinnamon, H. M., & Thomas, G. J. 1976. Conditioned taste aversions and neophobia in rats with hippocampal lesions. *Journal of Comparative and Physiological Psychology, 90,* 680–693.

Kraut, A. G. 1976. Effects of familiarization on alertness and encoding in children. *Developmental Psychology, 12,* 491–496.

Kraut, A. G., & Smothergill, D. W. 1978. A two-factor theory of stimulus repetition effects. *Journal of Experimental Psychology: Human Perception and Performance, 4,* 191–197.

Kraut, A. G., & Smothergill, D. W. 1980. New method for studying semantic encoding in children. *Developmental Psychology, 16,* 149–150.

Kraut, A. G., Smothergill, D. W., & Farkas, M. S. 1981. Stimulus repetition effects on attention to words and colors. *Journal of Experimental Psychology: Human Perception and Performance, 7,* 1303–1311.

Krauter, E. E. 1973. Habituation and latent inhibition. Unpublished doctoral dissertation, University of Rochester.

Krech, D. M., Rosenzweig, M. R., & Bennett, E. L. 1966. Environmental impoverishment, social isolation and changes in brain chemistry and anatomy. *Physiology and Behavior, 1,* 99–124.

Krechevsky, D. 1932. "Hypotheses" in rats. *Psychological Review, 39,* 516–532.

Kremer, E. F. 1971. Truly random and traditional control procedures in CER conditioning in the rat. *Journal of Comparative and Physiological Psychology, 76,* 441–448.

Kremer, E. F. 1972. Properties of a preexposed stimulus. *Psychonomic Science, 27,* 45–53.

Kremer, E. F. 1974. The truly random control procedure: Conditioning to the static cues. *Journal of Comparative and Physiological Psychology, 86,* 700–707.

Kremer, E. F. 1979. Effect of post-trial episodes on conditioning in compound conditioned stimuli. *Journal of Experimental Psychology: Animal Behavior Processes, 5,* 130–141.

Kremer, E. F., & Kamin, L. J. 1971. The truly random control procedure: Association or nonassociative effects in rats. *Journal of Comparative and Physiological Psychology, 74,* 203–210.

Kruse, J. M., & LoLordo, V. M. 1986. Pavlovian conditioning with ingested saline solution as the US *Animal Learning & Behavior, 14,* 22–28.

Kubose, S. K. 1970. The stimulus familiarization effect as a function of stimulus similarity. *Psychonomic Science, 18,* 367–368.

Kuczenski, R. 1983. Biochemical actions of amphetamine and other stimulants. In I. Creese (Ed.), *Stimulants: Neurochemical, Behavioral and Clinical Perspectives* (pp. 31–62). New York: Raven.

Kuhn, C. M., & Schanberg, S. M. 1978. Metabolism of amphetamine after acute and chronic administration to the rat. *Journal of Pharmacology and Experimental Therapeutics, 207,* 544–554.

Kuhn, T. S. 1962. *The Structure of Scientific Revolutions.* Chicago: University of Chicago Press.

Kurtz, K. H., & Pearl, J. 1960. The effect of prior fear experiences on acquired-drive learning. *Journal of Comparative and Physiological Psychology, 53,* 201–206.

Kurtz, K. H., & Walters, G. C. 1962. The effects of prior fear experiences on an approach-avoidance conflict. *Journal of Comparative and Physiological Psychology, 55,* 1075–1078.

Kurz, E. M., & Levitsky, D. A. 1982. Novelty of contextual cues in taste aversion learning. *Animal Learning & Behavior, 10,* 229–232.

La Berge, D. 1976. Perceptual learning and memory. In W. K. Estes (Ed.), *Handbook of Learning and Cognitive Processing. Vol. 4: Attention and Memory* (pp. 237–273). Hillsdale, N.J.: Lawrence Erlbaum.

Lantz, A. E. 1973. Effect of number of trials, interstimulus interval, and dishabituation during CS habituation on subsequent conditioning in a CER paradigm. *Animal Learning & Behavior, 1*, 273–277.

Lantz, A. E. 1976. The effects of the intensity of the preexposed stimulus on subsequent conditioning. *Bulletin of the Psychonomic Society, 7*, 381–383.

Lashley, K. S. 1929. *Brain Mechanisms and Intelligence: A Quantitative Study of Injuries to the Brain.* Chicago: University of Chicago Press.

Launey, G., & Slade, P. 1981. The measurement of hallucination predisposition in male and female prisoners. *Personality and Individual Differences, 2*, 221–234.

Lawrence, D. H. 1949. Acquired distinctiveness of cues: 1. Transfer between discriminations on the basis of familiarity with the stimulus. *Journal of Experimental Psychology, 39*, 770–784.

Lawson, J. S., McGhie, A., & Chapman, J. 1964. Perception of speech in schizophrenia. *British Journal of Psychiatry, 110*, 375–380.

Leaf, R. C., Kayser, R. J., Andrews, J. S., Jr., Adkins, J. W., & Leaf, S. R. P. 1968. Block of fear conditioning induced by habituation or extinction. *Psychonomic Science, 10*, 189–190.

Leaton, R. N. 1974. Long-term retention of the habituation of lick suppression in rats. *Journal of Comparative and Physiological Psychology, 87*, 1157–1164.

Lipton, M. A. & Nemeroff, C. B. 1978. In W. E. Fann, I. Karacan, I. Pokorny, R. L. Williams (Eds.), *Phenomenology and treatment of Schizophrenia.* New York: Spectrum.

Locurto, C. M., Terrace, H. A., & Gibbon, J. (Eds.). 1981. *Autoshaping and Conditioning Theory.* New York: Academic Press.

Loeb, J. 1918. *Forced Movements, Tropisms, and Animal Conduct.* Philadelphia: Lippincott.

Logan, M. S., & Schnur, P. 1976. The effects of nonreinforced and randomly reinforced stimulus preexposure on conditioned suppression in rats. *Bulletin of the Psychonomic Society, 8*, 336–338.

Logue, A. W. 1979. Taste aversion and the generality of the laws of learning. *Psychological Bulletin, 86*, 276–296.

Logue, A. W. 1985. Conditioned food aversions in humans. In N. S. Braveman & P. Bronstein (Eds.), *Experimental Assessments and Clinical Appreciations of Conditioned Food Aversions* (pp. 316–329). New York: New York Academy of Sciences.

Logue, A. W., Logue, K. R., & Strauss, K. E. 1983. The acquisition of taste aversions in humans with eating and drinking disorders. *Behavioral Research and Therapy, 21*, 275–289.

Logue, A. W., Ophir, I., & Strauss, K. E. 1981. The acquisition of taste aversions in humans. *Behavioral Research and Therapy, 19*, 319–333.

Lorden, J. F. 1976. Effects of lesions of the gustatory neocortex on taste aversion learning. *Journal of Comparative and Physiological Psychology, 90*, 665–679.

Lorden, J. F., Rickert, E. J., & Berry, D. W. 1983. Forebrain monoamines and associative learning: I. Latent inhibition and conditioned inhibition. *Behavioral Brain Research, 9*, 181–199.

Lorens, S. A., & Guldberg, H. C. 1974. Regional 5-hydroxytryptamine following selective midbrain raphe lesions in the rat. *Brain Research, 78*, 45–56.

Lorens, S. A., Guldberg, H. C., Hole, K., Kohler, C., & Srebro, B. 1976. Activity, avoidance learning and regional 5-hydroxytryptamine following intra-brainstem, 5,7-dihydroxytryptamine and electrolytic midbrain raphe lesions in the rat. *Brain Research, 108*, 97–113.

Lovejoy, E. 1968. *Attention in Discrimination Learning.* San Francisco: Holden-Day.

Lovibond, P. F., Preston, G. C., & Mackintosh, N. J. 1984. Context specificity of conditioning, extinction, and latent inhibition. *Journal of Experimental Psychology: Animal Behavior Processes, 10*, 360–375.

Lubow, R. E. 1965. Latent inhibition: Effect of frequency of nonreinforced preexposure to the CS. *Journal of Comparative and Physiological Psychology, 60,* 454–457.

Lubow, R. E. 1973a. Latent inhibition. *Psychological Bulletin, 79,* 398–407.

Lubow, R. E. 1973b. Latent inhibition as a means of behavior prophylaxis. *Psychological Reports, 32,* 1247–1252.

Lubow, R. E., Alek, M., & Arzy, J. 1975. Behavioral decrement following stimulus preexposure: Effects of number of preexposures, presence of a second stimulus and interstimulus interval in children and adults. *Journal of Experimental Psychology: Animal Behavior Processes, 1,* 178–188.

Lubow, R. E., Caspy, T., & Schnur, P. 1982. Latent inhibition and learned helplessness in children: Similarities and differences. *Journal of Experimental Child Psychology, 34,* 231–256.

Lubow, R. E., Markman, R. E., & Allen, J. 1968. Latent inhibition and classical conditioning of the rabbit pinna response. *Journal of Comparative and Physiological Psychology, 66,* 688–694.

Lubow, R. E., & Moore, A. U. 1959. Latent inhibition: The effect of non-reinforced preexposure to the conditional stimulus. *Journal of Comparative and Physiological Psychology, 52,* 415–419.

Lubow, R. E., Rifkin, B., & Alek, M. 1976. The context effect: The relationship between stimulus preexposure and environmental preexposure determines subsequent learning. *Journal of Experimental Psychology: Animal Behavior Processes, 2,* 38–47.

Lubow, R. E., Rosenblatt, R., & Weiner, I. 1981. Confounding of controllability in the triadic design for demonstrating learned helplessness. *Journal of Personality and Social Psychology, 41,* 458–468.

Lubow, R. E., Schnur, P., & Rifkin, B. 1976. Latent inhibition and conditioned attention theory. *Journal of Experimental Psychology: Animal Behavior Processes, 2,* 163–174.

Lubow, R. E., & Siebert, L. 1969. Latent inhibition within the CER paradigm. *Journal of Comparative and Physiological Psychology, 68,* 136–138.

Lubow, R. E., Wagner, M., & Weiner, I. 1982. The effects of compound stimulus preexposure of two elements differing in salience on the acquisition of conditioned suppression. *Animal Learning & Behavior, 10,* 483–489.

Lubow, R. E., Weiner, I., & Feldon, J. 1982. An animal model of attention. In M. Y. Spiegelstein & A. Levy (Eds.), *Behavioral Models and the Analysis of Drug Action* (pp. 89–107). New York: Elsevier.

Lubow, R. E., Weiner, I., Rosenblatt, R., Lindenbaum, U., & Margolit, H. 1979. The confounding of controllability in the triadic design for demonstrating learned helplessness. Unpublished manuscript.

Lubow, R. E., Weiner, I., Schlossberg, A., & Baruch, I. 1987. Latent inhibition and schizophrenia. *Bulletin of the Psychonomic Society, 25,* 464–467.

Lubow, R. E., Weiner, I., & Schnur, P. 1981. Conditioned attention theory. In G. H. Bower (Ed.), *The Psychology of Learning and Motivation* (Vol. 15, pp. 1–49). New York: Academic Press.

Lucas, R. 1984. Latent inhibition in kindergarten children's paired associate learning. Unpublished master's thesis, Tel-Aviv University.

Lynch, M. R., Porter, J. H., & Rosecrans, J. A. 1984. Latent inhibition in the oral aversion to methadone. *Pharmacology, Biochemistry and Behavior, 20,* 467–472.

Lyness, W. H., & Moore, K. E. 1981. Destruction of 5-hydroxytryptaminergic neurons and the dynamics of dopamine in nucleus accumbens septi and other forebrain regions of the rat. *Neuropharmacology, 20,* 327–334.

Lynn, R. 1966. *Attention, Arousal and the Orientation Reaction.* New York: Pergamon Press.

McAllister, W. R., McAllister, D. E., Dieter, S. E., & James, J. H. 1979. Preexposure to sit-

uational cues produces a direct relationship between two-way avoidance learning and shock intensity. *Animal Learning & Behavior, 7,* 105–173.
McCubbin, R. J., & Katkin, E. S. 1971. Magnitude of the orienting response as a function of the extent and quality of stimulus change. *Journal of Experimental Psychology, 88,* 182–188.
McDaniel, J. W., & White, R. K. 1966. A factorial study of the stimulus conditions of habituation. *Perceptual and Motor Skills, 23,* 259–270.
McFarland, D. J., Kostas, J., & Drew, W. G. 1978. Dorsal hippocampal lesions: Effect of preconditioning CS exposure on flavor aversion. *Behavioral Biology, 22,* 398–404.
McGhie, A. 1970. Attention and perception in schizophrenia. In B. A. Maher (Ed.), *Progress in Experimental Personality Research* (Vol. 5, pp. 1–35). New York: Academic Press.
McGhie, A., & Chapman, J. 1961. Disorders of attention and perception in early schizophrenia. *British Journal of Medical Psychology, 34,* 103–116.
McIntosh, J. M., & Tarpy, R. M. 1977. Retention of latent inhibition in a taste aversion paradigm. *Bulletin of the Psychonomic Society, 9,* 411–412.
Mackintosh, N. J. 1965. Selective attention in animal discrimination learning. *Psychological Bulletin, 64,* 124–150.
Mackintosh, N. J. 1973. Stimulus selection: Learning to ignore stimuli that predict no change in reinforcement. In R. A. Hinde & J. Stevenson-Hinde (Eds.), *Constraints on Learning,* (pp. 75–96). New York: Academic Press.
Mackintosh, N. J. 1974. *The Psychology of Animal Learning.* New York: Academic Press.
Mackintosh, N. J. 1975. A theory of attention: Variations in the associability of stimuli with reinforcement. *Psychological Review, 82,* 276–298.
Mackintosh, N. J. 1978. Cognitive or associative theories of conditioning: Implications of an analysis of blocking. In S. H. Hulse, H. Fowler, & W. K. Honig (Eds.), *Cognitive Processes in Animal Behavior* (pp. 155–176). Hillsdale, N.J.: Lawrence Erlbaum.
Mackintosh, N. J. 1983. *Conditioning and Associative Learning.* Oxford University Press.
Mackintosh, N. J., Bygrave, D. J., & Picton, B. M. B. 1977. Locus of the effect of a surprising reinforcer in the attenuation of blocking. *Quarterly Journal of Experimental Psychology, 29,* 327–336.
McLaurin, W. A., Farley, J. A., & Scarborough, B. B. 1963. Inhibitory effects of pre-irradiation saccharin habituation on conditioned avoidance behavior. *Radiation Research, 18,* 473–478.
Magaro, P., Abrams, L., & Cantrell, P. 1981. The Maine scale of paranoid and nonparanoid schizophrenia: Reliability and validity. *Journal of Consulting and Clinical Psychology, 49,* 438–447.
Maier, S. F., & Jackson, R. L. 1979. Learned helplessness: All of us were right (and wrong): Inescapable shock has multiple effects. In G. H. Bower (Ed.), *Psychology of Learning and Motivation* (Vol. 13, pp. 155–218). New York: Academic Press.
Maier, S. F., & Seligman, M. E. P. 1976. Learned helplessness: Theory and evidence. *Journal of Experimental Psychology: General, 105,* 3–46.
Maier, S. F., Seligman, M. E. P., & Solomon, R. L. 1969. Pavlovian fear conditioning and learned helplessness: Effects on escape and avoidance behavior of (a) the CS–US contingency and (b) the independence of the US and voluntary responding. In B. A. Campbell & R. M. Church (Eds.), *Punishment and Aversive Behavior* (pp. 299–342). New York: Appleton.
Maier, S. F., Zahorik, D. M., & Albin, R. W. 1971. Relative novelty of solid and liquid diet during thiamine deficiency determines development of thiamine-specific hunger. *Journal of Comparative and Physiological Psychology, 74,* 254–262.
Maloney, M. P., Sloane, R. B., Whipple, R., Razani, J., & Eaton, E. M. 1976. Auditory

attention in process and reactive schizophrenics. *Biological Psychiatry, 11* 325–335.

Maltzman, I. 1977. Orienting in classical conditioning and generalization of the galvanic skin response to words: An overview. *Journal of Experimental Psychology: General, 106,* 111–119.

Maltzman, I., Gould, J., Barnett, O. J., Raskin, D. C., & Wolff, C. 1979. Habituation of the GSR and digital vasomotor components of the orienting reflex as a consequence of task instructions and sex differences. *Physiological Psychology, 7,* 213–220.

Maltzman, I., & Raskin, D. C. 1965. Effects of individual differences in the orienting reflex on conditioning and complex processes. *Journal of Experimental Research in Personality, 1,* 1–16.

Maltzman, I., Raskin, D. C., & Wolff, C. 1979. Latent inhibition of the GSR conditioned to words. *Physiological Psychology, 7,* 193–203.

Maltzman, I., Weissbluth, S., & Wolff, C. 1978. Habituation of orienting reflexes in repeated GSR semantic conditioning sessions. *Journal of Experimental Psychology: General, 107,* 309–333.

Mandler, G., Nakamura, Y., & Van Zandt, B. J. S. 1987. Nonspecific effects of exposure on stimuli that cannot be recognized. *Journal of Experimental Psychology: Learning, Memory and Cognition, 13,* 646–648.

Mandler, J. M. 1972. Multiple stimulus discrimination learning. III. What is learned? *Quarterly Journal of Experimental Psychology, 25,* 112–123.

Marlin, N. A., & Miller, R. R. 1981. Associations to contextual stimuli as determinants of long-term habituation. *Journal of Experimental Psychology: Animal Behavior Processes, 7,* 313–333.

Martin, G. M., Bellingham, W. P., & Storlien, L. H. 1977. The effects of varied color experiences on chickens' formation of color and texture aversions. *Physiology and Behavior, 18,* 415–420.

Mason, S. T., & Fibiger, H. C. 1979. Noradrenaline and selective attention. *Life Science, 25,* 1949–1956.

Mason, S. T., & Lin, D. 1980. Dorsal noradrenergic bundle and selective attention in the rat. *Journal of Comparative Physiology and Psychology, 94,* 819–832.

Matthysse, S. 1974. Schizophrenia: Relationships to dopamine transmission, motor control and feature extraction. In F. O. Schmitt & F. G. Worden (Eds.), *Neurosciences: Third Study Program* (pp. 733–737). Cambridge, Mass.: MIT Press.

Matthysse, S. 1978. A theory of the relationship between dopamine and attention. *Journal of Psychiatry Research, 14,* 241–248.

Matthysse, S., Spring B. J., & Sugarman, J. (Eds.). 1979. *Attention and Information Processing in Schizophrenia.* Oxford: Pergamon Press.

Matzel, L. D., Schachtman, T. R., & Miller, R. R. 1988. Learned irrelevance exceeds the sum of CS-preexposure and US-preexposure deficits. *Journal of Experimental Psychology: Animal Behavior Processes, 14,* 311–319.

May, R. B., Tolman, C. W., & Schoenfeldt, M. G. 1967. Effects of pretraining exposure to the CS on conditioned suppression. *Psychonomic Science, 9,* 61–62.

Medin, D. L. 1975. A theory of context in discrimination learning. In G. H. Bower (Ed.), *The Psychology of Learning and Motivation* (pp. 263–314). New York: Academic Press.

Meiri, N. 1984. Latent inhibition with human adults: Visual masking task and its location are critical. Unpublished master's thesis, Tel-Aviv University.

Mellgren, R. L., Hunsicker, J. P., & Dyck, D. G. 1975. Conditions of preexposure and passive avoidance behavior in rats. *Animal Learning & Behavior, 3,* 147–151.

Mellgren, R. L., & Ost, J. W. P. 1971. Discriminative stimulus preexposure and the learning of an operant discrimination. *Journal of Comparative and Physiological Psychology, 77,* 179–187.

Mercier, P., & Baker, A. G. 1985. Latent inhibition, habituation, and sensory preconditioning: A test of priming in short-term memory. *Journal of Experimental Psychology: Animal Behavior Processes, 11*, 485–501.

Meyers, W. J., & Joseph, L. J. 1968. Response speed as related to CS prefamiliarization and GSR responsivity. *Journal of Experimental Psychology, 78*, 375–381.

Midgley, G. C., Wilkie, D. M., & Tees, R. C. 1988. Effects of superior colliculus lesions on rats' orienting and detection of neglected visual cues. *Behavioral Neuroscience, 102*, 93–100.

Mikulka, P. J., & Klein, S. B. 1977. The effect of CS familiarization and extinction procedures on resistance to extinction of a taste aversion. *Behavioral Biology, 19*, 518–522.

Mikulka, P. J., & Klein, S. B. 1980. Resistance to extinction of a taste aversion: Effects of level of training and procedures used in acquisition and extinction. *American Journal of Psychology, 93*, 631–641.

Mikulka, P. J., Leard, B., & Klein, S. B. 1977. Illness-alone exposure as a source of interference with the acquisition and retention of a taste aversion. *Journal of Experimental Psychology: Animal Behavior Processes, 3*, 189–201.

Milgram, N. W., Krames, L., & Alloway, T. M. (Eds.). 1977. *Food Aversion Learning.* New York: Plenum.

Millard, W. J. 1982. Lithium chloride-induced avoidance of a conspecific odor: Effect of prior exposure to the conditional stimulus. *Behavioral and Neural Biology, 34*, 404–410.

Millenson, J. R., & Dent, J. G. 1971. Habituation of conditioned suppression. *Quarterly Journal of Experimental Psychology, 23*, 126–134.

Miller, C. R., Elkins, R. L., & Peacock, L. J. 1971. Disruption of a radiation-induced preference shift by hippocampal lesions. *Physiology and Behavior, 6*, 283–285.

Miller, F. D. 1969. The effects of differential amounts of stimulus familiarization on choice reaction time performance in children. *Journal of Experimental Child Psychology, 8*, 106–117.

Miller, N. E. 1948. Studies of fear as an acquirable drive. *Journal of Experimental Psychology, 38*, 89–101.

Miller, R. R., & Holzman, A. D. 1981. Neophobias and conditioned taste aversions in rats following exposure to novel flavors. *Animal Learning & Behavior, 9*, 89–100.

Miller, R. R., Kasprow, W. J., & Schachtman, T. R. 1986. Retrieval variability: Sources and consequences. *American Journal of Psychology, 99*, 145–218.

Milner, P. M. 1970. *Physiological Psychology.* New York: Holt, Rinehart & Winston.

Mineka, S., & Cook, M. 1986. Immunization against the observational conditioning of snake fear in rhesus monkeys. *Journal of Abnormal Psychology, 95*, 307–318.

Mineka, S., Cook, M., & Miller, S. 1984. Fear conditioned with escapable and inescapable shock: Effects of a feedback stimulus. *Journal of Experimental Psychology: Animal Behavior Processes, 10*, 307–323.

Mirsky, A. F., & Duncan, C. C. 1986. Etiology and expression of schizophrenia: Neurobiological and social factors. *Annual Review of Psychology, 37*, 291–319.

Misanin, J. R., Blatt, L. A., & Hinderliter, C. F. 1985. Age dependency in neophobia: Its influence on taste aversion learning and the flavor preexposure effect in rats. *Animal Learning & Behavior, 13*, 69–76.

Misanin, J. R., Guanowsky, V., & Riccio, D. C. 1983. The effect of CS preexposure on conditioned taste aversion in young and adult rats. *Physiology and Behavior, 30*, 859–862.

Misanin, J. R., Kniss, D. A., Yoder, S. D., Yazujian, D. L., & Hinderliter, C. F. 1984. The effect of electroconvulsive shock on the attenuation of taste aversion conditioning produced by flavor preexposure. *Behavioral and Neural Biology, 41*, 30–40.

Mitchell, D., Kirschbaum, E. H., & Perry, R. C. 1975. Effects of neophobia and habituation on the poison induced avoidance of exteroceptive stimuli in the rat. *Journal of Experimental Psychology: Animal Behavior Processes, 1*, 47–55.

Mitchell, D., Winter, W., & Moffitt, T. 1980. Cross-modality contrast: Exteroceptive context habituation enhances taste neophobia and conditioned taste aversions. *Animal Learning & Behavior, 8,* 524–528.

Moderasi, H. A. 1982. Defensive behavior of the rat in a shock-prod situation: Effects of the subject's location preferences. *Animal Learning & Behavior, 10,* 97–102.

Mohammed, A. K., Callenholm, N. E. B., Järbe, T. U. C., Swedberg, M. D. B., Danysz, W., Robbins, T. W., & Archer, T. 1986. Role of central noradrenaline neurons in the contextual control of latent inhibition in taste aversion learning. *Behavioural Brain Research, 21,* 109–118.

Mohammed, A. K., Jonsson, G., Soderberg, U., & Archer, T. 1986. Impaired selective attention in methylazoxymethanol-induced microencephalic rats. *Pharmacology, Biochemistry, and Behavior, 24,* 975–981.

Moore, D. E., & Kelly, P. H. 1978. Biochemical pharmacology of mesolimbic and mesocortical dopaminergic neurons. In M. A. Lipton, A. DiMascio, & L. F. Killam (Eds.) *Psychopharmacology: A Generation of Progress* (pp. 221–234). New York: Raven.

Moore, J. W. 1979. Brain processes and conditioning. In A. Dickinson & R. A. Boakes (Eds.) 1979. *Mechanisms of Learning and Motivation: A Memorial Volume for Jerzy Konorski* (pp. 111–142). Hillsdale, N.J.: Lawrence Earlbaum.

Moore, J. W., Goodell, N. A., & Solomon, P. R. 1976. Central cholinergic blockade by scopolamine and habituation, classical conditioning, and latent inhibition of the rabbit's nictitating membrane response. *Physiological Psychology, 4,* 395–399.

Moore, J. W., & Stickney, K. J. 1980. Formation of attentional-associative networks in real time: Role of the hippocampus and implications for conditioning. *Physiological Psychology, 8,* 207–217.

Moore, J. W., & Stickney, K. J. 1985. Antiassociations: Conditioned inhibition in attentional-associative networks. In R. R. Miller & N. E. Spear (Eds.), *Information Processing in Animals: Conditioned Inhibition* (pp. 209–232). Hillsdale, N.J.: Lawrence Erlbaum.

Moore, R. Y., & Halaris, A. E. 1975. Hippocampal innervation by serotonin neurons of the midbrain raphe in the rat. *Journal of Comparative Neurology, 164,* 171–184.

Moray, N. 1959. Attention in dichotic listening: Affective cues and the influence of instructions. *Quarterly Journal of Experimental Psychology, 2,* 56–60.

Mostofsky, D. I. 1970. The semantics of attention. In D. I. Mostofsky (Ed.), *Attention: Contemporary Theory and Analysis* (pp. 9–24). New York: Appleton.

Mowrer, O. H. 1947. On the dual nature of learning – a reinterpretation of "conditioning" and "problem solving." *Harvard Educational Review, 17,* 102–148.

Mowrer, R. R. 1987. Latent inhibition of contextual stimuli reduces the US preexposure effect. *Psychological Record, 37,* 239–246.

Nachman, M. 1970. Learned taste and temperature aversions due to lithium chloride sickness after temporal delays. *Journal of Comparative and Physiological Psychology, 73,* 22–30.

Nachman, M., & Ashe, J. H. 1974. Effects of baso-lateral amygdala lesions on neophobia, learned taste aversions, and sodium appetite in rats. *Journal of Comparative and Physiological Psychology, 87,* 622–643.

Nachman, M., & Jones, D. R. 1974. Learned taste aversions over long delays in rats: The role of learned safety. *Journal of Comparative and Physiological Psychology, 86,* 949–956.

Nakaya, T. 1982. Effects of prior exposure of context on aversive classical conditioning. *Hiroshima Forum of Psychology, 9,* 63–72.

Nauta, W. J. H., & Domesick, V. B. 1984. Afferent and efferent relationships of the basal ganglia. In *Functions of the Basal Ganglia,* CIBA Foundation Symposium 107 (pp. 3–29). London: Pitman.

Neale, J. M., & Cromwell, R. L. 1977. Attention and schizophrenia. In B. A. Maher

(Ed.), *Progress in Experimental Personality Research* (Vol. 5, pp. 37-66). New York: Academic Press.

Oberdieck, F., & Tarte, R. D. 1981. The effect of shock prod preexposure on conditioned defensive burying in rats. *Bulletin of the Psychonomic Society, 17*, 111-112.

Obrist, P. A., Sutterer, J. R., & Howard, J. L. 1972. Preparatory cardiac changes: A psychobiological approach. In A. H. Black & W. F. Prokasy (Eds.), *Classical Conditioning. II: Current Research and Theory* (pp. 312-340). New York: Appleton-Century-Crofts.

Ohad, D., Lubow, R. E., Weiner, I., & Feldon, J. 1987. The effects of amphetamine on blocking. *Psychobiology, 17*, 137-143.

Öhman, A. 1979. The orienting response, attention, and learning: An information-processing perspesctive. In H. D. Kimmel, E. H. Van Olst, & J. F. Orlebeke (Eds.), *The Orienting Reflex in Humans* (pp. 443-471). Hillsdale, N.J.: Lawrence Erlbaum.

Öhman, A. 1983. The orienting response during Pavlovian conditioning. In D. Siddle (Ed.), *Orienting and Habituation* (pp. 315-369). New York: Wiley.

O'Keefe, J., & Nadel, L. 1978. *The Hippocampus as a Cognitive Map*. Oxford University Press.

O'Keefe, J., Nadel, L., & Willner, J. 1979. Tuning out irrelevancy? Comments on Solomon's temporal mapping view of the hippocampus. *Psychological Bulletin, 86*, 1280-1289.

Oliverio, A. 1968. Effects of scopolamine on avoidance conditioning and habituation of mice. *Psychopharmacologia, 12*, 214-216.

Oltmanns, T. F., Ohayon, J., & Neale, J. M. 1978. The effects of antipsychotic medication and diagnostic criteria on distractibility in schizophrenia. *Journal of Psychiatric Research, 14*, 81-91.

Olton, D. S., & Isaacson, R. L. 1968. Hippocampal lesions and active avoidance. *Physiology and Behavior, 3*, 719-724.

Oswalt, R. M. 1972. The relationship between level of visual pattern difficulty during rearing and subsequent discrimination in rats. *Journal of Comparative and Physiological Psychology, 81*, 122-125.

Overall, J. E., & Gorham, D. R. 1962. Brief psychiatric rating scale. *Psychological Reports, 10*, 799-812.

Pavlov, I. P. 1927. *Conditioned Reflexes*. Oxford University Press.

Payne, R. W., Hochberg, A. C., & Hawks, D. V. 1970. Dichotic stimulation as a method of assessing disorder of attention in over-inclusive schizophrenic patients. *Journal of Abnormal Psychology, 76*, 185-193.

Payne, R. W., Matussek, P., & George, E. I. 1959. An experimental study of schizophrenic thought disorder. *Journal of Mental Science, 105*, 627-652.

Pearce, J. M., & Hall, G. 1980. A model of Pavlovian learning: Variations in the effectiveness of conditioned but not of unconditioned stimuli. *Psychological Review, 87*, 532-552.

Pearce, J. M., & Kaye, H. 1985. Strength of the orienting response during inhibitory conditioning. *Journal of Experimental Psychology: Animal Behavior Processes, 11*, 405-420.

Pearce, J. M., Kaye, H., & Hall, G. 1983. Predictive accuracy and stimulus associability: Development of a model of Pavlovian learning. In M. L. Commons, R. J. Herrnstein, & A. R. Wagner (Eds.), *Quantitative Analyses of Behavior. Vol. 3: Acquisition Processes* (pp. 241-255). Cambridge, Mass.: Balinger.

Pearce, J. M., Nicholas, D. J., & Dickinson, A. 1982. Loss of associability by a conditioned inhibitor. *Quarterly Journal of Experimental Psychology, 33B*, 149-162.

Pearl, J. D., Walters, G. C., & Anderson, D. C. 1964. Suppressing effects of aversive stimulation on subsequently punished behavior. *Canadian Journal of Psychology, 18*, 343-355.

Peeke, H. V. S., & Herz, M. J. (Eds.). 1973. *Habituation: Behavioral Studies.* New York: Academic Press.
Peeke, H. V. S., & Petrinovitch, L. (Eds.). 1984. *Habituation, Sensitization and Behavior.* New York: Academic Press.
Peeke, H. V. S., & Veno, A. 1973. Stimulus specificity of habituated aggression in the stickleback (*Gasterosteus aculeatus*). *Behavioral Biology, 8,* 427–432.
Perlmuter, L. C. 1966. Effect of CS manipulations on the conditioned eyelid response: Compounding, generalization, the inter-CS-interval, and preexposure. *Psychonomic Science Monographs, 1,* 271–286.
Peterson, C. S., Valliere, W. A., Misanin, J. R., & Hinderliter, C. F. 1985. Age differences in the potentiation of taste aversion by odor cues. *Physiological Psychology, 13,* 103–106.
Petrinovitch, L. 1973. A species-meaningful analysis of habituation. In H. V. S. Peeke & M. J. Herz (Eds.), *Habituation. Vol. 1: Behavioral Studies* (pp. 141–162). New York: Academic Press.
Pfautz, P. L., Donegan, N. H., & Wagner, A. R. 1978. Sensory preconditioning versus protection from habituation. *Journal of Experimental Psychology: Animal Behavior Processes, 4,* 286–295.
Pijnenburg, A. J. J., Honig, W. M. M., & van Rossum, J. M. 1975. Inhibition of d-amphetamine-induced locomotor activity by injection of haloperidol into the nucleus accumbens of the rat. *Psychopharmacologia, 41,* 87–95.
Pinel, J. P. J., & Treit, D. 1979. Conditioned defensive burying in rats: Availability of burying materials. *Animal Learning & Behavior, 7,* 392–396.
Pinel, J. P. J., Treit, D., & Wilkie, D. M. 1980. Stimulus control of defensive burying in the rat. *Learning and Motivation, 11,* 150–163.
Plotkin, H. C., & Oakley, D. A. 1975. Backward conditioning in the rabbit (*Oryctolagus cuniculus*). *Journal of Comparative and Physiological Psychology, 88,* 586–590.
Pogue-Geile, M. F., & Oltmanns, T. F. 1980. Sentence perception and distractability in schizophrenic, manic, and depressed patients. *Journal of Abnormal Psychology, 89,* 115–124.
Poser, E. G. 1970. Toward a theory of "behavioral prophylaxis." *Journal of Behavioral Therapy and Experimental Psychiatry, 1,* 39–43.
Poser, E. G., Baum, M., & Skinner, C. 1970. CS preexposure as a means of "behavioral prophylaxis": An animal paradigm. *Proceedings of the 78th American Psychological Association Meeting, 6,* 521–522.
Poser, E. G., & King, M. S. 1975. Strategies for the prevention of maladaptive fear responses. *Canadian Journal of Behavioral Science, 7,* 279–294.
Posner, M. I. 1978. *Chronometric Explorations of Mind.* Hillsdale, N.J.: Lawrence Erlbaum.
Posner, M. I., & Boies, S. 1971. Components of attention. *Psychological Review, 78,* 391–408.
Prewitt, E. P. 1967. Number of preconditioning trials in sensory preconditioning using CER training. *Journal of Comparative and Physiological Psychology, 64,* 360–362.
Pribram, K. H. 1986. The hippocampal system and recombinant processing. In R. I. Isaacson & K. H. Pribram (Eds.), *The Hippocampus* (Vol. 4, pp. 329–370). New York: Plenum.
Pribram, K. H., & McGuinness, D. 1975. Arousal, activation and effort in the control of attention. *Psychological Review, 82,* 116–149.
Primavera, J., & Wagner, A. R. 1977. The effects of CS priming on performance of the conditioned emotional response. Unpublished manuscript, cited in Wagner (1979).
Pritchard, W. S. 1981. Psychophysiology of P300. *Psychological Bulletin, 90,* 506–540.
Prokasy, W. F. 1972. Developments with the two-phase model applied to human eyelid conditioning. In A. H. Black & W. F. Prokasy (Eds.), *Classical Conditioning. II: Cur-*

rent Theory and Research (pp. 208–225). New York: Appleton.
Prokasy, W. R., Spurr, C. W., & Goodell, N. A. 1978. Preexposure to explicitly unpaired conditioned and unconditioned stimuli retards conditioned response emergence. *Bulletin of the Psychonomic Society, 12,* 155–158.
Raisman, G., Gowan, W. M., & Powell, T. P. S. 1966. Experimental analysis of the efferent projection of the hippocampus. *Brain, 89,* 83–108.
Randich, A., & LoLordo, V. M. 1979. Associative and non-associative theories of the UCS preexposure phenomenon: Implications for Pavlovian conditioning. *Psychological Bulletin, 86,* 523–548.
Rappaport, M., Silverman, J., Hopkins, H. K., & Hall, K. 1971. Phenothiazine effects on auditory signal detection in paranoid and nonparanoid schizophrenics. *Science, 174,* 723–725.
Ray, R. D., & Brener, J. 1973. Classical heart-rate conditioning in the rat: The influence of curare and various setting operations. *Conditional Reflex, 8,* 224–235.
Reilly, S. 1987. Hyperstriatal lesions and attention in the pigeon. *Behavioral Neuroscience, 101,* 74–86.
Reiss, S., & Wagner, A. R. 1972. CS habituation produces "latent inhibition" effect but no active "conditioned inhibition." *Learning and Motivation, 3,* 237–245.
Rescorla, R. A. 1969. Pavlovian conditioned inhibition. *Psychological Bulletin, 72,* 77–94.
Rescorla, R. A. 1971. Summation and retardation tests of latent inhibition. *Journal of Comparative and Physiological Psychology, 75,* 77–81.
Rescorla, R. A. 1985. Facilitation and inhibition. In R. R. Miller & N. E. Spear (Eds.), *Information Processing in Animals: Conditioned Inhibition* (pp. 299–326). Hillsdale, N.J.: Lawrence Erlbaum.
Rescorla, R. A., & Cunningham, C. L. 1978. Within-compound flavor associations. *Journal of Experimental Psychology: Animal Behavior Processes, 4,* 267–275.
Rescorla, R. A., Durlach, P. J., & Grau, J. W. 1985. Contextual learning in Pavlovian conditioning. In P. Balsam & A. Tomie (Eds.), *Context and Learning* (pp. 23–56). Hillsdale, N.J.: Lawrence Erlbaum.
Rescorla, R. A., & Wagner, A. R. 1972. A theory of Pavlovian conditioning: Variations in the effectiveness of reinforcement and nonreinforcement. In A. Black & W. F. Prokasy (Eds.), *Classical Conditioning II* (pp. 64–99). New York: Appleton-Century-Crofts.
Revusky, S. 1971. The role of interference in association over a delay. In W. K. Honig & P. H. R. James (Eds.), *Animal Memory* (pp. 155–213). New York: Academic Press.
Revusky, S. H. 1968. Effects of thirst level during consumption of flavored water on subsequent preference. *Journal of Comparative and Physiological Psychology, 66,* 777–779.
Revusky, S. H., & Bedarf, E. W. 1967. Association of illness with prior ingestion of novel foods. *Science, 155,* 219–220.
Revusky, S., & Garcia, J. 1970. Learned associations over long delays. In G. H. Bower (Ed.), *The Psychology of Learning and Motivation* (Vol. 4, pp. 1–89). New York: Academic Press.
Revusky, S. H., Parker, L., Coombes, J., & Coombes, S. 1976. Rat data suggest alcoholic beverages should be swallowed during chemical aversion therapy, not just tasted. *Behaviour Research and Therapy, 14,* 189–194.
Revusky, S. H., & Takulis, H. K. 1975. Effects of alcohol and lithium habituation on the development of alcohol aversions through contingent lithium injection. *Behaviour Research and Therapy, 13,* 163–166.
Riley, A. L., & Baril, L. L. 1976. Conditioned taste aversions: A bibliography. *Animal Learning & Behavior, 4,* 15–135.

Riley, A. L., & Clarke, C. M. 1977. Conditioned taste aversions: A bibliography. In L. M. Barker, M. R. Best, & M. Domjan (Eds.), *Learning Mechanisms in Food Selection* (pp. 593–616). Waco, Tex.: Baylor University Press.
Riley, A. L., Jacobs, W. J., & LoLordo, V. M. 1976. Drug exposure and the acquisition and retention of a conditioned taste aversion. *Journal of Comparative and Physiological Psychology, 90,* 799–807.
Riley, A. L., Jacobs, W. J., Jr., & Mastropaolo, J. P. 1983. The effects of extensive taste preexposure on the acquisition of taste aversions. *Bulletin of the Psychonomic Society, 21,* 221–224.
Riley, A. L., & Tuck, D. L. 1985. Conditioned taste aversions: A bibliography. *Annals of the New York Academy of Science, 443,* 381–437.
Rizley, R. C., & Rescorla, R. A. 1972. Associations in second-order conditioning and sensory preconditioning. *Journal of Comparative and Phsyiological Psychology, 81,* 1–11.
Robbins, R. J. 1979. The effect of flavor preexposure upon the acquisition and retention of poison-based taste aversions in deer mice: Latent inhibition or partial reinforcement? *Behavioral and Neural Biology, 25,* 387–397.
Robbins, T. W., Everitt, B. J., Cole, B. J., Archer, T., & Mohammed, A. 1985. Functional hypotheses of the coeruleocortical noradrenergic projection: A review of recent experimentation and theory. *Physiological Psychology, 13,* 127–150.
Robbins, T. W., Everitt, B. J., Fray, P. J., Gaskin, M., Carli, M., & de la Riva, C. 1982. The roles of central catecholamines in attention and learning. In M. Y. Spiegelstein & A. Levy (Eds.), *Behavioral Models and the Analysis of Drug Action* (pp. 109–134). Amsterdam: Elsevier.
Robbins, T. W., & Iversen, S. D. 1973. A dissociation of the effects of *d*-amphetamine on locomotor activity and exploration in rats. *Psychopharmacologia, 28,* 155–164.
Robertson, D., & Garrud, P. 1983. Variable processing of flavors in rat STM. *Animal Learning & Behavior, 11,* 474–482.
Robinson, T. E., & Becker, J. B. 1986. Enduring changes in brain and behavior produced by chronic amphetamine administration: A review and evaluation of animal models of amphetamine psychosis. *Brain Research Reviews, 11,* 157–198.
Roelofs, N. 1982. The effects of stimulus preexposure on subsequent learning in adults. Unpublished master's thesis, Tel-Aviv University.
Rommelschpacher, H., & Strauss, S. 1980. Effect of lesions of raphe nuclei on the activity of catecholaminergic and serotonergic neurons in various brain regions of the rat *in vivo*. *Journal of Neural Transmitters, 49,* 51–62.
Rosellini, R. A., DeCola, J. P., & Warren, D. A. 1986. The effect of feedback stimuli on contextual fear depends upon the length of the minimum intertrial interval. *Learning and Motivation, 17,* 229–242.
Ross, R. T. 1983. Relationships between the determinants of performance in serial feature-positive discriminations. *Journal of Experimental Psychology: Animal Behavior Processes, 9,* 349–373.
Ross, R. T., & Holland, P. C. 1981. Conditioning of simultaneous and serial feature-positive discriminations. *Animal Learning & Behavior, 9,* 293–303.
Roth, R. H. 1983. Neuroleptics: Functional neurochemistry. In J. T. Coyle & S. J. Enna (Eds.), *Neuroleptics: Neurochemical, Behavioral, and Clinical Perspectives* (pp. 119–156). New York: Raven Press.
Roth, W. T. 1983. A comparison of P300 and skin conductance response. In A. W. K. Gaillard & W. Ritter (Eds.), *Tutorials in ERP Research: Endogenous Components* (pp. 177–199). Amsterdam: North Holland.
Roth, W. T., Blowers, G. H., Doyle, C. M., & Kopell, B. S. 1982. Auditory stimulus intensity effects on components of the late positive complex. *Electroencephalography and Clinical Neurophysiology, 54,* 132–146.

Royet, J. P. 1983. The behavioral aspects of conditioned aversion and neophobia. *Année Biologique, 22*, 113–167.

Rudy, J. W., & Cheatle, M. D. 1977. Ontogeny of associative learning: Acquisition of odor aversions by neonatal rats. In N. E. Spear & B. A. Campbell (Eds.), *Ontogeny of Learning and Memory* (pp. 157–188). Hillsdale, N.J.: Lawrence Erlbaum.

Rudy, J. W., & Cheatle, M. D. 1978. A role of conditioned stimulus duration in toxiphobia conditioning. *Journal of Experimental Psychology: Animal Behavior Processes, 4*, 399–411.

Rudy, J. W., Krauter, E. E., & Gaffuri, A. 1976. Attenuation of the latent inhibition effect by prior exposure to another stimulus. *Journal of Experimental Psychology: Animal Behavior Processes, 2*, 235–247.

Rudy, J. W., Rosenberg, L., & Sandell, J. H. 1977. Disruption of taste familiarity effect by novel exteroceptive stimulation. *Journal of Experimental Psychology: Animal Behavior Processes, 3*, 26–32.

Russo, J. M., Reiter, L. A., & Ison, J. R. 1975. Repetitive exposure does not attenuate the sensory impact of the habituated stimulus. *Journal of Comparative and Physiological Psychology, 88*, 665–669.

Rutter, D. R., Wishner, J., & Callaghan, B. A. 1975. The prediction of predictability of speech in schizophrenic subjects. *British Journal of Psychiatry, 126*, 571–576.

Salafia, W. R. 1987. Pavlovian conditioning, information processing, and the hippocampus. In I. Gormezano, W. F. Prokasy, & R. F. Thompson (Eds.), *Classical Conditioning III* (pp. 197–216). Hillsdale, N.J.: Lawrence Erlbaum.

Salafia, W. R., & Allan, A. M. 1980a. Conditioning and latent inhibition with electrical stimulation of hippocampus. *Physiological Psychology, 8*, 247–253.

Salafia, W. R., & Allan, A. M. 1980b. Attenuation of latent inhibition by electrical stimulation of hippocampus. *Physiology and Behavior, 24*, 1047–1051.

Salafia, W. R., & Allan, A. M. 1982. Augmentation of latent inhibition by electrical stimulation of hippocampus. *Physiology and Behavior, 29*, 1125–1130.

Samuels, S. J. 1978. Applications of basic research in reading. In H. L. Pick, Jr., H. W. Leibowitz, J. E. Singer, A. Steinschneider, & H. W. Stevenson (Eds.), *Psychology*. New York: Plenum.

Scarborough, B. B., & McLaurin, W. A. 1965. The effect of interperitoneal injection on aversive behavior conditioning with X-irradiation. *Radiation Research, 15*, 829–835.

Scavio, M. J., Ross, R. T., & McLeod, L. M. 1983. Perseveration of associative strength in rabbit nictitating membrane response conditioning. *Animal Learning & Behavior, 11*, 91–94.

Schachtman, T. R., Channell, S., & Hall, G. 1987. Effects of CS preexposure on inhibition of delay. *Animal Learning & Behavior, 15*, 301–311.

Scheele-Kruger, J. 1971. Comparative studies of various amphetamine analogues demonstrating different interactions with the metabolism of the catecholamines in the brain. *European Journal of Pharmacology, 14*, 47–59.

Schmajuk, N. A. 1984. Psychological theories of hippocampal function. *Physiological Psychology, 12*, 166–183.

Schmajuk, N. A. 1987. Animals models for schizophreniz: The hippocampally lesioned animal. *Schizophrenia Bulletin, 32*, 317–327.

Schmajuk, N. A., & Moore, J. W. 1985. Real-time attentional models for classical conditioning and the hippocampus. *Physiological Psychology, 13*, 278–290.

Schneider, S. J. 1976. Selective attention in schizophrenia. *Journal of Abnormal Psychology, 85*, 167–173.

Schneider, W., Dumais, S. T., & Shiffrin, R. M. 1984. Automatic and central processing and attention. In R. Parasuraman & D. R. Davies (Eds.), *Varieties of Attention* (pp. 1–24). New York: Academic Press.

Schneider, W., & Shiffrin, R. M. 1977. Controlled and automatic human information

processing: I. Detection, search, and attention. *Psychological Review, 84*, 1–66.
Schneiderman, N. 1973. *Classical (Pavlovian) Conditioning*. Morristown, N.J.: General Learning Press.
Schnur, P. 1967. Latent inhibition: The effects of nonreinforced preexposure of the CS in differential eyelid conditioning. Unpublished master's thesis, North Carolina State University.
Schnur, P. 1971. Selective attention: Effect of element preexposure on compound conditioning in rats. *Journal of Comparative and Physiological Psychology, 76*, 123–130.
Schnur, P. 1975. Latent inhibition in compound conditioning in rats: The effect of number of conditioning trials. *American Journal of Psychology, 88*, 411–419.
Schnur, P., & Ksir, C. J. 1969. Latent inhibition in human eyelid conditioning. *Journal of Experimental Psychology, 80*, 388–389.
Schnur, P., & Lubow, R. E. 1976. Latent inhibition: The effects of ITI and CS intensity during preexposure. *Learning and Motivation, 7*, 540–550.
Schnur, P., & Lubow, R. E. 1983. Modelling changes in CS effectiveness: Latent inhibition as a case in point. Unpublished manuscript.
Schnur, P., & Lubow, R. E. 1987. The stimulus familiarization effect in adults. *Social and Behavioral Sciences Documents, 17* (No. 2793), 1–24.
Schnur, P., Lubow, R. E., & Ben-Shalom, H. 1979. The stimulus familiarization effect in adults. Presented before the Psychonomic Society, Phoenix.
Segal, D. S. 1976. Differential effects of para-chlorophenylalanine on amphetamine-induced locomotion and stereotypy. *Brain Research, 116*, 267–276.
Segal, D. S., & Janowsky, D. S. 1978. Psychostimulant-induced behavioral effects: Possible models of schizophrenia. In M. A. Lipton, A. DiMascio & K. E. Killam (Eds.), *Psychopharmacology: A Generation of Progress* (pp. 1113–1123). New York: Raven Press.
Segal, D. S., & Schuckit, M. A. 1983. Animal models of stimulant-induced psychosis. In I. Creese (Ed.), *Stimulants: Neurochemical, Behavioral and Clinical Perspectives* (pp. 131–167). New York: Raven Press.
Seligman, M. E. P., & Haber, J. L. 1972. *Biological Boundaries of Learning*. New York: Appleton-Century-Crofts.
Seligman, M. E. P., Maier, S. F., & Solomon, R. L. 1971. Unpredictable and uncontrollable aversive events. In F. R. Brush (Ed.), *Aversive Conditioning and Learning* (pp. 347–400). New York: Academic Press.
Shakow, D. 1962. Segmental set: A theory of the formal psychological deficit in schizophrenia. *Archives of General Psychiatry, 6*, 17–33.
Sharp, P. E., James, J. H., & Wagner, A. R. 1980. Habituation of a "blocked" stimulus during Pavlovian conditioning. *Bulletin of the Psychonomic Society, 15*, 139–142.
Sherrington, C. S. 1906. *The Integrative Activity of the Nervous System*. London: Constable.
Shiffrin, R. M., & Schneider, W. 1977. Controlled and automatic information processing: II. Perceptual learning, automatic attending, and a general theory. *Psychological Review, 84*, 127–190.
Shishimi, A. 1985. Latent inhibition experiments with goldfish. *Journal of Comparative Psychology, 99*, 316–327.
Siddle, D. A. T. 1983. *Orienting and Habituation: Perspectives in Human Research*. New York: Wiley.
Siddle, D. A. T. 1985. Effects of stimulus omission and stimulus change on dishabituation of the skin conductance response. *Journal of Experimental Psychology: Learning, Memory and Cognition, 11*, 206–216.
Siddle, D. A. T., & Heron, P. A. 1975. Stimulus omission and recovery of the electrodermal and digital vasoconstrictive components of the orienting response. *Biological Psychology, 3*, 277–293.

Siddle, D. A. T., Kuiack, M., & Kroese, B. S. 1984. The orienting reflex. In A. Gale & J. Edwards (Eds.), *Physiological Correlates of Human Behavior* (Vol. 2, pp. 149-170). London: Academic Press.

Siddle, D. A. T., Remington, B., & Churchill, M. 1985. Effects of conditioned stimulus preexposure on human electrodermal conditioning. *Biological Psychology, 20,* 113-127.

Siddle, D. A. T., Remington, B., Kuiack, M., & Haines, E. 1983. Stimulus omission and dishabituation of the skin conductance response. *Psychophysiology, 20,* 136-145.

Siebert, L. 1967. The role of the orienting response in latent inhibition using Ivanov-Smolenskii conditioning. Unpublished master's thesis, North Carolina State University.

Siebert, L., Nicholson, L., Carr-Harris, E., & Lubow, R. E. 1969. Conditioning by the method of Ivanov-Smolenskii. *Journal of Experimental Psychology, 79,* 93-96.

Siegel, S. 1969a. Effect of the CS habituation on eyelid conditioning. *Journal of Comparative and Physiological Psychology, 68,* 245-248.

Siegel, S. 1969b. Generalization of latent inhibition. *Journal of Comparative and Physiological Psychology, 69,* 157-159.

Siegel, S. 1970. Retention of latent inhibition. *Psychonomic Science, 20,* 161-162.

Siegel, S. 1971. Latent inhibition and eyelid conditioning. In A. H. Black & W. F. Prokasy (Eds.), *Classical Conditioning II* (pp. 231-247). New York: Appleton-Century-Crofts.

Siegel, S. 1974. Flavor preexposure and "learned safety." *Journal of Comparative and Physiological Psychology, 87,* 1073-1082.

Siegel, S., & Domjan, M. 1971. Backward conditioning as an inhibitory procedure. *Learning and Motivation, 2,* 1-11.

Sigmundi, R. A., & Bolles, R. C. 1982. Transfer of conditioned freezing following a shift in shock intensity. Unpublished manuscript (presented before the Psychonomic Society).

Silver, A. I. 1973. Effects of prior CS presentations on classical conditioning of the GSR. *Psychophysiology, 10,* 583-588.

Sjoden, P. O., Archer, T., & Carter, N. 1979. Conditioned taste aversion induced by 2,4,5-trichlorphenoacetic acid: Dose-response and preexposure effects. *Physiological Psychology, 7,* 93-96.

Smothergill, D. W., & Kaut, A. G. 1980. Functional significance of dimensional dominance hierarchies. *Merrill-Palmer Quarterly, 26,* 197-204.

Smothergill, D. W., & Kraut, A. G. 1981. Stimulus repetition and encoding facilitation: Locus of the effect. *Canadian Journal of Psychology, 35,* 52-57.

Smotherman, W. P., Margolis, A., & Levine, S. 1980. Flavor preexposure in a conditioned taste aversion situation: A dissociation of behavioral and endocrine effects in rats. *Journal of Comparative and Physiological Psychology, 94,* 23-35.

Snyder, S. H. 1973. Amphetamine psychosis: A model schizophrenia mediated by catechloamine. *American Journal of Psychiatry, 130,* 61-67.

Sokolov, E. N. 1960. Neuronal models and the orienting reflex. In M. A. Brazier (Ed.), *The Central Nervous System and Behavior* (pp. 187-276). New York: Macey.

Sokolov, E. N. 1963. *Perception and the Conditioned Reflex.* New York: Pergamon Press.

Sokolov, E. N., & Paramanova, N. P. 1956. Concerning the role of the orientation reflex in the formation of motor conditioned reactions in man. *Journal of Higher Nervous Activity, 6,* 702-709.

Solomon, P. R. 1977. Role of the hippocampus in blocking and conditioned inhibition of the rabbit's nictitating membrane response. *Journal of Comparative and Physiological Psychology, 91,* 407-417.

Solomon, P. R. 1980. A time and place for everything? Temporal processing views of

hippocampal function with special reference to attention. *Physiological Psychology, 8*, 254–261.
Solomon, P. R., Brennan, G., & Moore, J. W. 1974. Latent inhibition of the rabbit's nictitating membrane response as a function of CS intensity. *Bulletin of the Psychonomic Society, 4*, 445–448.
Solomon, P. R., Crider, A., Winkelman, J. W., Turi, A., Kramer, R. M., & Kaplan, L. J. 1981. Disrupted latent inhibition in the rat with chronic amphetamine or haloperidol-induced supersensitivity: Relationship to schizophrenic attention disorder. *Biological Psychiatry, 16*, 519–538.
Solomon, P. R., Kiney, C. A., & Scott, D. S. 1978. Disruption of latent inhibition following systemic administration of parachlorophenylalanine (PCPA). *Physiology and Behavior, 20*, 265–271.
Solomon, P. R., Lohr, C. A., & Moore, J. W. 1974. Latent inhibition of the rabbit's nictitating response: Summation tests for active inhibition as a function of a number of CS preexposures. *Bulletin of the Psychonomic Society, 4*, 557–559.
Solomon, P. R., & Moore, J. W. 1975. Latent inhibition and stimulus generalization of the classically conditioned nictitating membrane response in rabbits (*Oryctolagus cuniculus*) following dorsal hippocampal ablation. *Journal of Comparative and Physiological Psychology, 89*, 1192–1203.
Solomon, P. R., Nichols, G. L., Kiernan, J. M., Kamer, R. S., & Kaplan, L. J. 1980. Differential effects of median and dorsal raphe lesions in the rat: Latent inhibition and septo-hippocampal serotonim levels. *Journal of Comparative and Physiological Psychology, 94*, 145–154.
Solomon, P. R., & Staton, D. M. 1982. Differential effects of microinjections of d-amphetamine into the nucleus accumbens or the caudate putamen on the rat's ability to ignore an irrelevant stimulus. *Biological Psychiatry, 17*, 743–756.
Solomon, P. R., Sullivan, D. J., Nichols, G. L., & Kiernan, J. M. 1979. Cue or place learning in one-way avoidance acquisition. *Bulletin of the Psychonomic Society, 13*, 243–245.
Solso, R. L. 1979. *Cognitive Psychology.* New York: Harcourt, Brace, Jovanovich.
Soubrie, P., Reisine, T. D., & Glowinski, J. 1984. Functional aspects of serotonin transmission in the basal ganglia: A review and an in vivo approach using the push-pull cannula technique. *Neuroscience, 13*, 605–625.
Spear, N. E. 1973. Retrieval of memory in animals. *Psychological Review, 80*, 163–194.
Spear, N. E., & Smith, G. J. 1978. Alleviation of forgetting in preweanling rats. *Developmental Psychology, 11*, 513–529.
Spence, K. W. 1936. The nature of discrimination learning in animals. *Psychological Review, 43*, 427–449.
Spence, K. W. 1966. Cognitive and drive factors in the extinction of the conditioned eyeblink in human subjects. *Psychological Review, 73*, 445–458.
Spence, K. W., Homzie, M. J., & Rutledge, E. F. 1964. Extinction of the human eyelid CR as a function of the discriminability of the change from acquisition to extinction. *Journal of Experimental Psychology, 67*, 545–552.
Spinks, J. A., & Siddle, D. 1983. The functional significance of the orienting response. In D. Siddle (Ed.), *Orienting and Habituation* (pp. 237–314). New York: Wiley.
Spohn, H. E., Lacoursiere, R. B., Thompson, K., & Coyne, L. 1977. Phenothiazine effects on psychological and psychophysiological dysfunction in chronic schizophrenics. *Archives of General Psychiatry, 34*, 633–644.
Spring, B. J., & Zubin, J. 1978. Attention and information processing as indicators of vulnerability to schizophrenic episodes. *Journal of Psychiatric Research, 14*, 289–301.
Spurr, C. W. 1969. Preexposure and stimulus intensity. Unpublished doctoral dissertation, University of Utah.
Squires, K. C., Wickens, C., Squires, N. K., & Donchin, E. 1976. The effect of stimulus

sequence on the waveform of the cortical event-related potential. *Science, 193*, 1142–1145.
Starr, M. D., & Mineka, S. 1977. Determinants of fear over the course of avoidance learning. *Learning and Motivation, 8*, 332–350.
Staton, D. W., & Solomon, P. R. 1984. Microinjections of *d*-amphetamine into the nucleus accumbens and caudate-putamen differentially affect stereotypy and locomotion in the rat. *Physiological Psychology, 12*, 159–162.
Stewart, D. J., Capretta, P. J., Cooper, A. J., & Littlefield, V. M. 1977. Learning in domestic chicks after exposure to both discriminanda. *Journal of Comparative and Physiological Psychology, 91*, 1095–1109.
Suboski, M. D., DiLollo, V., & Gormezano, I. 1964. Effects of unpaired pre-acquisition exposure of CS and UCS on classical conditioning of the nictitating membrane response of the albino rabbit. *Psychological Reports, 15*, 571–576.
Sullivan, L. G. 1984. Long-delay and selective association in food aversion learning: The role of cue location. *Quarterly Journal of Experimental Psychology, 36B*, 65–87.
Surwit, R. S. 1972. The anticipatory modification of the conditioning of a fear response in humans. Unpublished doctoral dissertation, McGill University.
Surwit, R. S., & Poser, E. G. 1974. Latent inhibition in the conditioned electrodermal response. *Journal of Comparative and Physiological Psychology, 86*, 543–548.
Sutherland, N. S. 1964. The learning of discriminations by animals. *Endeavor, 23*, 148–152.
Sutherland, N. S., & Mackintosh, N. J. 1971. *Mechanisms of Animal Discrimination Learning*. New York: Academic Press.
Sutton, S., Braren, M., Zubin, J., & John, E. R. 1963. Evoked potential coarrelates of stimulus uncertainty. *Science, 150*, 1187–1188.
Swanson, L. W., & Cowan, W. M. 1977. An autoradiographic study of the organization of the efferent connections of the hippocampal formation in the rat. *Journal of Comparative Neurology, 172*, 49–84.
Swartzentruber, D., & Bouton, M. E. 1986. Contextual control of negative transfer produced by prior CS–US pairings. *Learning and Motivation, 17*, 366–385.
Swerdlow, N. R., & Koob, G. F. 1987. Dopamine, schizophrenia, mania and depression: Toward a unified hypothesis of cortico-striato-pallido-thalamic function. *Behavioral and Brain Sciences, 10*, 197–245.
Szakmary, G. A. 1977a. Latent inhibition with sequentially preexposed stimuli. Presented at the annual meeting of the Psychonomic Society, Washington, D.C., November.
Szakmary, G. A. 1977b. A note regarding conditioned attention theory. *Bulletin of the Psychonomic Society, 9*, 142–144.
Szakmary, G. A. 1978. Latent inhibition with sequentially preexposed stimuli: More data. Presented at a meeting of the Psychonomic Society, San Antonio, November.
Takaori, S., Sasa, M., Akaike,A., & Fujimoto, S. 1982. Dopamine and neuron activity in the mesotelecephalic system – an electrophysiological study. In M. Kohsaka, T. Shohmori, Y. Tsukada, & G. N. Woodruff (Eds.), *Advances in Dopamine Research* (pp. 341–355). New York: Pergamon Press.
Tarpy, R. M., & McIntosh, J. M. 1977. Generalized latent inhibition in taste aversion learning. *Bulletin of the Psychonomic Society, 10*, 379–381.
Terlecki, L. J., Pinel, J. P. J., & Treit, D. 1979. Conditioned and unconditioned defensive burying in the rat. *Learning and Motivation, 10*, 337–350.
Terry, W. S. 1976. The effects of priming US representation in short-term memory on Pavlovian conditioning. *Journal of Experimental Psychology: Animal Behavior Processes, 2*, 354–370.
Thompson, R. F. 1972. Sensory preconditioning. In R. F. Thompson & M. R. Voss (Eds.), *Topics in Learning and Performance* (pp. 239–271). New York: Academic Press.

Thompson, R. F., Groves, P. M., Teyler, T. J., & Roemer, R. A. 1973. A dual-process theory of habituation: Theory and behavior. In H. V. S. Peeke & M. J. Herz (Eds.), *Habituation. Vol. 1: Behavioral Studies* (pp. 239–271). New York: Academic Press.

Thompson, R. F., & Spencer, W. A. 1966. Habituation: A model phenomenon for the study of neuronal substrates of behavior. *Psychological Review, 73*, 16–43.

Tighe, T. J., & Leaton, R. N. (Eds.). 1976. *Habituation: Perspectives from Child Development, Animal Behavior and Neurophysiology.* Hillsdale, N.J.: Lawrence Erlbaum.

Toga, A. W., & Burton, H. A. 1979. Effects of the lateral septum and latent inhibition on successive discrimination learning. *Journal of Neuroscience Research, 4*, 215–224.

Tolman, E. C. 1932. *Purposive Behavior in Animals and Men.* New York: Appleton-Century.

Tolman, E. C., & Honzik, H. C. 1930. Introduction and removal of reward, and maze performance of rats. *University of California Publications in Psychology, 4*, 257–275.

Tomie, A. 1976a. Retardation of autoshaping: Control by contextual stimuli. *Science, 192*, 1244–1246.

Tomie, A. 1976b. Interference with autoshaping by prior context conditioning. *Journal of Experimental Psychology: Animal Behavior Processes, 2*, 323–334.

Tomie, A., Murphy, A. L., Fath, S., & Jackson, R. L. 1980. Retardation of autoshaping following pretraining with unpredictable food: Effects of changing the context between pretraining and testing. *Learning and Motivation, 11*, 117–134.

Totterdell, S., & Smith, A. D. 1986. Cholecystokinin-immunoreactive boutons in synaptic contact with hippocampal pyramidal neurons that project to the nucleus accumbens. *Neuroscience, 19*, 181–192.

Trabasso, T., & Bower, G. H. 1968. *Attention in Learning: Theory and Research.* New York: Wiley.

Tranberg, D. K., & Rilling, M. 1978. Latent inhibition in the autoshaping paradigm. *Bulletin of the Psychonomic Society, 11*, 273–276.

Tsaltas, E., Preston, G. C., Rawlins, J. N. P., Winocur, G., & Gray, J. A. 1984. Dorsal bundle lesions do not affect latent inhibition of conditioned suppression. *Psychopharmacology, 84*, 549–555.

Van Kammen, D. P., Bunney, W. E., Docherty, J. P., Marder, S. R., Ebert, M. H., Rosenblatt, J. E., & Rayner, J. N. 1982. d-Amphetamine-induced heterogeneous changes in psychotic behavior in schizophrenia. *American Journal of Psychiatry, 139*, 991–997.

Venables, P. H. 1964. Input dysfunction in schizophrenia. In B. A. Maher (Ed.), *Progress in Experimental Personality Research* (Vol. 1, pp. 1–47). New York: Academic Press.

Vogel, J. R., & Clody, D. E. 1972. Habituation and conditioned food aversion. *Psychonomic Science, 28*, 275–276.

Volpicelli, J. R., Ulm, R. R., & Altenor, A. 1984. Feedback during exposure to inescapable shocks and subsequent shock escape. *Learning and Motivation, 15*, 279–286.

Voronin, L. G., Leontiev, A. N., Luria, A. R., Sokolov, E. N., & Vinogradova, O. S. (Eds.). 1965. *Orienting Reflex and Exploratory Behavior.* Washington, D. C.: American Institute of Biological Sciences.

Wagner, A. R. 1976. Priming in STM: An information processing mechanism for self-generated or retrieval-generated depression in performance. In T. J. Tighe & R. N. Leaton (Eds.), *Habituation: Perspectives from Child Development, Animal Behavior, and Neurophysiology* (pp. 95–128). Hillsdale, N.J.: Lawrence Erlbaum.

Wagner, A. R. 1978. Expectancies and the priming of STM. In S. H. Hulse, H. Fowler, & W. K. Honig (Eds.), *Cognitive Processes in Animal Behavior* (pp. 177–209). Hillsdale, N.J.: Lawrence Erlbaum.

Wagner, A. R. 1979. Habituation and memory. In A. Dickinson & R. A. Boakes (Eds.), *Mechanisms of Learning and Motivation: A Memorial to Jerzy Konorski* (pp. 53–82). Hillsdale, N.J.: Lawrence Erlbaum.

Wagner, A. R. 1981. SOP: A model of automatic memory processing in animal behavior. In N. E. Spear & R. R. Miller (Eds.), *Information Processing in Animals: Memory Mechanisms* (pp. 5–47). Hillsdale, N.J.: Lawrence Erlbaum.

Wagner, A. R., Pfautz, P. L., & Donegan, N. 1977. The extinction of stimulus-exposure effects. Unpublished manuscript (reported in Wagner, 1979).

Wagner, A. R., & Rescorla, R. A. 1972. Inhibition in Pavlovian conditioning: Application of a theory. In M. S. Halliday & R. A. Boakes (Eds.), *Inhibition and Learning* (pp. 301–336). London: Academic Press.

Wagner, A. R., Rudy, J. W., & Whitlow, J. W. 1973. Rehearsal in animal conditioning. *Journal of Experimental Psychology, 97*, 407–426.

Walk, R. D., Gibson, E. J., Pick, H. L., & Tighe, T. J. 1959. The effectiveness of prolonged exposure to cutouts vs painted patterns for facilitation of discrimination. *Journal of Comparative and Physiological Psychology, 52*, 519–521.

Waller, T. G. 1970. Effect of irrelevant cues on discrimination acquisition and transfer in rats. *Journal of Comparative and Physiological Psychology, 73*, 477–480.

Wasserman, E. A., Franklin, S. R., & Hearst, E. 1974. Pavlovian appetitive contingencies and approach versus withdrawal to conditioned stimuli in pigeons. *Journal of Comparative and Physiological Psychology, 86*, 616–627.

Wasserman, E. A., & Molina, E. J. 1975. Explicitly unpaired key-light and food presentations: Interference with subsequent autoshaped keypecking in pigeons. *Journal of Experimental Psychology: Animal Behavior Processes, 1*, 30–38.

Weinberger, D. R., Wagner, R. L., & Wyatt, R. J. 1983. Neuropathological studies of schizophrenia: A selective review. *Schizophrenia Bulletin, 9*, 193–212.

Weiner, I. 1983. The effects of amphetamine on latent inhibition: Tests of the animal amphetamine model of schizophrenia using selected learning paradigms. Unpublished doctoral dissertation, Tel-Aviv University.

Weiner, I., & Feldon, J. 1987. Facilitation of latent inhibition by haloperidol. *Psychopharmacology, 91*, 248–253.

Weiner, I., & Feldon, J. 1988. Haloperidol reverses amphetamine-induced abolition of latent inhibition. Unpublished manuscript.

Weiner, I., Feldon, J., & Katz, Y. 1987. Facilitation of the expression but not the acquisition of latent inhibition by haloperidol in rats. *Pharmacology, Biochemistry and Behavior, 26*, 241–246.

Weiner, I., Feldon, J., & Ziv-Harris, D. 1987. Early handling and latent inhibition in the conditioned suppression paradigm. *Developmental Psychobiology, 20*, 233–240.

Weiner, I., Israeli-Telerant, A., & Feldon, J. 1987. Latent inhibition is not affected by acute or chronic administration of 6 mg/kg dl-amphetamine. *Psychopharmacology, 91*, 345–351.

Weiner, I., Lubow, R. E., & Feldon, J. 1981. Chronic amphetamine and latent inhibition. *Behavioral Brain Research, 2*, 285–286.

Weiner, I., Lubow, R. E., & Feldon, J. 1984. Abolition of the expression but not the acquisition of latent inhibition by chronic amphetamine in rats. *Psychopharmacology, 83*, 194–199.

Weiner, I., Lubow, R. E., & Feldon, J. 1988. Disruption of latent inhibition by acute administration of low doses of amphetamine. *Pharmacology, Biochemistry and Behavior, 30*, 871–878.

Weiner, I., Schnabel, I., Lubow, R. E., & Feldon, J. 1985. The effects of early handling on latent inhibition in male and female rats. *Developmental Psychobiology, 18*, 291–297.

Weiner, I., Schnur, P., & Lubow, R. E. 1983. Associability and associative strength: Latent inhibition as a case in point. Unpublished manuscript.

Weiss, K. R., & Brown, B. L. 1974. Latent inhibition: A review and a new hypothesis. *Acta Neurobiologiae Experimentalis, 34*, 301–316.

Weiss, K. R., & Friedman, R. 1975. Stimulus controllability and the latent inhibition effect. *Acta Neurobiologiae Experimentalis, 35*, 241–254.

Weiss, K. R., Friedman, R., & McGregor, S. 1974. Effects of septal lesions on latent inhibition and habituation of the orienting response in rats. *Acta Neurobiologiae Experimentalis, 34*, 491–504.

Wellman, P. J. 1982. Preexposure to flavor and conditioned taste aversion: Amphetamine and lithium reinforcers. *Psychological Reports, 50*, 555–558.

Westbrook, R. F., Bond, N. W., & Feyer, A. M. 1981. Short- and long-term decrements in toxicosis-induced odor-aversion learning: The role of duration and exposure to odor. *Journal of Experimental Psychology: Animal Behavior Processes, 7*, 362–381.

Westbrook, R. F., Provost, S. C., & Homewood, J. 1982. Short-term flavour memory in the rat. *Quarterly Journal of Experimental Psychology, 34*, 235–256.

White, F. J., & Wang, R. Y. 1983. Differential effects of classical and atypical antipsychotic drugs on A9 and A10 dopamine neurons. *Science, 221*, 1054–1056.

Whitlow, J. W. 1975. Short-term memory in habituation and dishabituation. *Journal of Experimental Psychology: Animal Behavior Processes, 1*, 189–206.

Wickens, C., Tuber, D. S., & Wickens, D. D. 1983. Memory for the conditioned response: The proactive effect of preexposure to potential conditioning stimuli and context change. *Journal of Experimental Psychology: General, 112*, 41–57.

Wickens, D. D., Tuber, D. S., Nield, A. F., & Wickens, C. 1977. Memory for the conditioned response: The effects of potential interference induced before and after original conditioning. *Journal of Experimental Psychology: General, 106*, 47–70.

Williams, J. H., & Azmitia, E. C. 1981. Hippocampal serotonin re-uptake and nocturnal locomotor activity after microinjections of 5,7-DHT in the fornix-fimbria. *Brain Research, 207*, 95–107.

Willner, J. A. 1980. Spatial factors in latent inhibition. Presented before the Eastern Psychological Association, Hartford.

Wilson, L. M., Phinney, R. L., & Brennan, J. F. 1974. Age-related differences in avoidance behavior in rats following CS preexposure. *Developmental Psychobiology, 7*, 421–427.

Wilson, L. M., & Riccio, D. C. 1973. CS familiarization and conditioned suppression in weanling rats. *Bulletin of the Psychonomic Society, 1*, 184–186.

Wilson, R. S. 1964. Autonomic changes produced by noxious and innocuous stimulation. *Journal of Comparative and Physiological Psychology, 58*, 290–295.

Wilson, R. S. 1969. Cardiac response: Determinants of conditioning. *Journal of Comparative and Physiological Psychology, 68*, 1–23.

Winefield, A. H. 1978. The effects of prior random reinforcement on brightness discrimination learning in rats. *Quarterly Journal of Experimental Psychology, 30*, 113–119.

Witcher, E. S., & Ayres, J. J. B. 1980. Systematic manipulation of CS–US pairings in negative CS–US correlation procedures in rats. *Animal Learning & Behavior, 8*, 67–74.

Wittlin, W. A., & Brookshire, K. H. 1968. Apomorphine-induced conditioned aversion to a novel food. *Psychonomic Science, 12*, 217–218.

Wittman, T. K., & DeVietti, T. L. 1981. Latent inhibition measured by heart rate suppression in rats. *Bulletin of the Psychonomic Society, 17*, 283–285.

Wolff, C., & Maltzman, I. 1968. Conditioned orienting reflex and amount of preconditioned habituation. *Proceedings of the American Psychological Association, 3*, 129–130.

Woodruff, G., & Williams, D. R. 1976. The associative relations underlying autoshaping in the pigeon. *Journal of Experimental Analysis of Behavior, 11*, 1–8.

Wright, D. C., & Gustavson, K. K. 1986. Preexposure of the conditioning context and latent inhibition from reduced conditioning. *Bulletin of the Psychonomic Society, 24*, 451–452.

Wright, D. C., Skala, K. D., & Peuser, K. A. 1986. Latent inhibition from context-dependent retrieval of conflicting information. *Bulletin of the Psychonomic Society, 24*, 152–154.

Wyers, E. J., Peeke, H. V. S., & Herz, M. J. 1973. Behavioral habituation in invertebrates. In H. V. S. Peeke & M. J. Herz (Eds.), *Habituation* (pp. 1–57). New York: Academic Press.

Wyss, J. M., Swanson, L. W., & Cowan, W. M. 1979. A study of subcortical afferents to the hippocampal formation in the rat. *Neuroscience, 4*, 463–476.

Zahorik, D. M. 1976. The role of dietary history in the effects of novelty on taste aversion. *Bulletin of the Psychonomic Society, 8*, 285–288.

Zajonc, R. B. 1980. Feeling and thinking: Preferences need no inferences. *American Psychologist, 35*, 151–175.

Zeaman, D., & House, B. J. 1963. The role of attention in retardate discrimination learning. In N. R. Ellis (Ed.), *Handbook of Mental Deficiency: Psychological Theory and Research* (pp. 159–223). New York: McGraw-Hill.

Zeiner, A. R. 1970. Orienting response and discrimination conditioning. *Physiology and Behavior, 5*, 641–646.

Author index

Abrams, L., 252, 272, 294
Abramson, C. I., 101, 102–3, 273
Ackil, J. E., 14, 16, 17, 130, 131, 141, 196, 217, 238, 262, 273
Adkins, J. W., 23, 292
Akaika, A., 139, 306
Alek, M., 19, 45–6, 54–5, 64, 76–8, 100, 107–9, 111–12, 209, 216, 220, 227, 270, 293
Alexander, J. H., 154, 287
Alkon, D. L., 102, 280, 282
Allan, A. M., 24, 27, 85–6, 131, 238, 262, 302
Allen, J. A., 26, 113, 156, 273
Alleva, E., 14, 16, 17, 101, 106, 154, 273, 276
Alloway, T. M., 20, 296
Alpert, M., 249, 273
Altenor, A., 242–4, 307
Amorico, L., 14, 16, 17, 101, 106, 273
Ampuero, M. X., 206, 278
Andelman, L., 55, 273
Anderson, D. C., 23, 55, 59, 62–3, 65, 66, 67, 75, 153, 154, 196–7, 241, 266, 273, 280, 298
Andrews, J. S., Jr., 23, 292
Angrist, B., 249, 250, 260, 272, 273
Anisman, H., 134, 153, 249, 250, 273, 274, 290
Anscombe, R., 135, 249, 260
Archer, T., 13, 14, 16, 21, 60, 74, 78, 123, 133, 214, 234, 266, 268, 274, 297, 301
Arwas, S., 128, 129, 266, 274
Arzy, J., 108, 203, 220, 227, 270, 293
Asin, K. E., 14, 134, 274
Ayres, J. J. B., 26, 87, 88, 89, 169, 194, 203, 267, 274, 309
Azmitia, E. C., 139, 274, 309

Babb, H., 56, 274
Badia, P., 207, 275
Baker, A. G., 13, 26, 46–50, 62, 81, 84, 92–3, 178–9, 240, 266, 268, 275, 295
Balaz, A. M., 26, 82, 182, 185, 200, 275, 288
Balsam, P. D., 32, 275
Barber, J. G., 243–4, 275
Baril, L. L., 20, 22, 300
Barker, L. M., 20, 275

Barnett, O. J., 156, 295
Barnett, S., 20, 275
Barrett, O.. V., 62–4, 66, 186–9, 271, 280
Barrett, R. J., 136, 275
Barry, R. J., 271, 275
Baruch, I., 107, 121, 124–5, 127, 130, 219, 223, 252, 254–5, 256, 257–9, 275, 293
Bateson, P. P. G., 55, 275
Batson, J. D., 21, 61, 90, 91, 128, 129, 266, 275, 278
Baum, M., 154, 269, 275, 299
Bauste, R. L., 62, 64, 186–9, 271, 280
Becker, D. E., 157, 275
Becker, J. B., 134, 250, 301
Bedarf, E. W., 21, 22, 300
Bell, D. S., 249, 275
Bell, J. A., 46, 276
Bellingham, W. P., 22, 60, 92, 95, 284, 295
Bennett, E. L., 55, 291
Bennett, T. L., 52–4, 276
Ben-Shahar, O., 138, 203, 276
Berenson, B., 9, 264, 265, 266, 276
Berlyne, D. E., 37, 271, 276
Bernstein, I. L., 129, 276
Berry, D. W., 132–3, 145, 292
Best, M. R., 20–2, 61, 64, 69, 70, 76, 78, 86, 87, 90, 91, 95, 128, 129, 144, 201, 203, 266, 275, 276, 278, 281
Bignami, G., 14, 16, 17, 101, 106, 154, 273, 276
Bitterman, M. E., 30, 101, 102–3, 273, 276
Bjorkstrand, P. A., 268, 276
Blair, V., 60, 290
Blatt, L. A., 21, 97, 98, 100, 296
Bleuler, E., 135, 248, 277
Blodgett, H. C., 2, 277
Blowers, G. H., 37, 157, 277, 301
Boies, S., 108, 231, 299
Bolles, R.. C., 18, 61, 69, 70, 78, 87, 206, 277, 304
Bond, N. W., 21, 22, 61, 64, 69, 70–2, 78–9, 277, 309
Borson, S., 129, 276
Bostock, E., 38–9, 138, 277, 284
Boulton, A. A., 135, 136, 280
Bouton, M. E., 74, 78, 87–8, 267, 268, 271, 277, 306
Bower, G. H., 190, 307

311

Bowman, T. G., 61, 71, 281
Braff, D. L., 255, 277
Brandeis, R., 118, 119, 120, 277
Braren, N., 269, 306
Braud, W. G., 17, 101, 103, 268, 277
Braun, J. J., 22, 289
Braveman, N. S., 59, 277
Bray, G. P., 55, 283
Brener, J., 38–9, 300
Brennan, G., 24, 27, 46–7, 65, 144, 305
Brennan, J. F., 19, 309
Brennan, J. H., 272, 277
Broadbent, D. E., 248–9, 277
Brogden, W. J., 2, 277
Broks, P., 258, 279
Brookshire, K. H., 21–2, 309
Brown, B. L., 7, 308
Brown, E. R., 21, 266, 278
Brown, P. L., 32, 278
Brown-Su, A. M., 87, 278
Bruner, A., 207, 278
Buchanan, S. L., 24, 27, 38–9, 65, 286
Bunney, B. S., 138, 279
Bunney, W. E., 272, 307
Burns, R. A., 207, 208, 289
Burton, H. A., 46–7, 131, 262, 278, 307
Bygrave, D. J., 195, 294

Cacheiro, H., 82, 183, 186, 200, 274, 288
Callaghan, B. A., 248, 302
Callenholm, N. E. B., 78, 297
Campbell, B. A., 206, 278
Cannon, D. S., 21, 128–9, 266, 278
Cantor, G. N., 59, 108, 232, 278
Cantor, J. H., 59, 232, 278
Cantrell, P., 252, 272, 294
Caponigri, V., 23, 75, 273
Capra, S., 26, 275
Capretta, P. J., 55, 306
Carli, M., 46, 48, 51–2, 132, 301
Carlton, P. H., 11–13, 23, 58, 63, 141, 197, 199, 278
Carpenter, D. L., 55, 279
Carr, A. F., 94, 268, 278
Carrell, L. A., 21, 266, 278
Carr-Harris, E., 268, 304
Carter, C. J., 140, 278
Caspy, T., 26, 101, 107–8, 110–12, 115, 127, 219–20, 223, 240, 245, 253, 255, 293
Catanzaro, D. de, 154, 273
Cavanaugh, J., 249, 285
Cecil, L. S., 113, 231–2, 278
Chacto, C., 27, 28, 62, 196, 278
Channell, S., 21, 44–6, 55, 78, 80, 82, 87–8, 158–9, 172, 177, 178, 188, 189, 194, 199–200, 203, 214, 278, 279, 286, 302
Chantrey, D. F., 55, 275, 279
Chapman, L. J., 248, 279

Chapman, J. P., 135, 248, 249, 279, 292, 294
Cheatle, M. D., 22, 61, 71–3, 97, 99–100, 302
Chiodo, L. A., 138, 279
Chronister, R. B., 139, 279
Claridge, G. S., 258, 279
Clarke, C. M., 20, 22, 301
Clarke, M. E., 24, 26, 62, 197, 279
Clifton R. L., 40, 285
Cohen, C. A., 266, 279
Cohen, D. H., 38–9, 106, 279
Cole, B. J., 132, 301
Cole, J., 75, 154, 273
Comfort, M. K., 64, 267, 280
Cone, D. M., 96, 279
Connell, P. H., 249, 279
Conti, L. H., 21, 279
Cook, M., 101, 106, 242, 243, 244, 296
Cools, A. R., 269, 279
Cooper, A. J., 55, 306
Corrigan, J. G., 55, 279
Corriveau, D. P., 242, 279
Corwin, J., 260, 272, 273
Costall, B., 140, 279
Coulombe, D., 26, 280
Cowan, W. M., 139, 306, 310
Coyne, L., 255, 305
Creese, I., 138, 280
Crider, A., 14, 130, 135–6, 137, 251, 255, 280, 285, 305
Crow, T. J., 102, 260, 280
Crowell, C. R., 59, 62–3, 65, 66, 67, 196, 197, 241, 266, 280

Dale, P. S., 112, 280
Daly, K., 62, 64, 187–9, 271, 280
Damiani, N., 255, 290
Danielson, T. J., 135, 136, 280
Danysz, W., 78, 297
Dark, V. J., 181, 288
Davis, J. M., 257, 290
Davis, M., 150, 151, 280
Dawley, J. M., 21, 59, 280
De Acetis, L., 14, 16, 17, 101, 106, 273
Deakin, J. F. W., 139, 280
DeCola, J. P., 272, 301
Defran, R. H., 207, 275
Dent, J. G., 189, 296
DeVietti, T. L., 38–9, 62–4, 66, 131, 186–9, 267, 280, 309
Dexter, W. R., 26, 78, 177, 281
Dickinson, A., 26, 88, 169, 171, 195, 281, 298
Dieter, S. E., 17, 19, 199, 281, 293
DiGiusto, E., 61, 69, 70, 277
DiLollo, V., 24, 26–7, 63, 67, 306
Djuric, V. J., 21, 281
Docherty, J. P., 272, 307
Dodd, M., 22, 283

Author index 313

Domato, G. C., 21, 60, 62, 97, 98, 100, 283, 290
Domesick, V. B., 139, 289, 297
Domjan, M., 6, 11, 13, 20, 23-4, 27, 60, 61, 62, 64, 71, 90, 91, 141, 149, 196, 201, 266, 275, 276, 281, 304
Donchin, E., 269, 305
Donegan, N. H.,178, 195, 281, 308
Doré, F. Y., 16, 19, 281
Douglas, R. J., 130, 282
Downey, P., 50, 282
Doyle, C. M., 157, 301
Drew, W. G., 22, 131, 217, 262
Druiff, H., 44, 87, 177, 289
Dugue, G., 24, 26, 67, 85, 202, 207, 283
Dumais, S. T., 226, 302
Duncan, C. C., 249, 296
Durlach, P. J., 214, 300
Dyck, D. G., 20, 295
Dyer, A. B., 22, 283

Eaton, E. M., 255, 294
Ebert, M. H., 272, 307
Eison, M. S., 250, 282
Elkins, R. L., 60, 61, 62, 67, 68, 282
Ellison, G. D., 250, 282
Emmerson, R. Y., 64, 131, 267, 280
Epstein, W., 53, 282
Estes, W. K., 23, 180, 282
Everitt, B. J., 44, 48, 51-2, 132, 301
Eysenck, H. J., 125, 258, 282
Eysenck, S. B. G., 125, 258, 282

Fabricius, E., 55, 290
Fält, L., 55, 290
Farkas, M. S., 113, 232, 291
Farley, J., 30, 101-2, 282
Farley, J. A., 20, 21, 60-1, 67, 68, 282, 294
Fath, S., 32-5, 307
Feldman, M. A., 14, 128, 129, 278, 282
Feldon, J., 14, 16, 26, 101, 130-1, 135-8, 203, 250-1, 255, 257-8, 276, 282, 285, 308
Fenson, L., 238, 278
Fenwick, S., 60, 62, 283
Feyer, A. M., 22, 61, 69, 72, 78-9, 309
Fibiger, H. C., 269, 295
Fietz, A., 30, 101, 102-3, 276
Fink, H., 140, 283
Fisher, M. A., 167, 283
Fitzgerald, R. D., 38-9, 283
Forbes, D. T., 92, 94, 234-5, 283, 287
Foreman, N., 217, 271, 283
Forgus, R. H., 52, 283
Formica, R., 23, 75, 273
Franchina, J. J., 11, 21, 22, 60, 62, 97, 98, 283
Franken, R. E., 55, 283
Franklin, S. R., 32, 308

Fray, P. J., 46, 48, 51-2, 132, 301
Frey, P. W., 24, 26, 72, 85, 141, 165, 174-5, 190, 193, 202, 207, 283
Friedhoff, A. J., 249, 273
Friedman, R., 14, 131, 202, 238, 262, 270, 309
Frommer, G. P., 14, 16, 17, 130, 131, 141, 217, 238, 262, 273
Furedy, J. J., 156, 283
Fujimoto, S., 139, 306

Gaffuri, A., 92-4, 267, 302
Gallagher, M., 38-9, 138, 277, 284
Gamzu, E., 32, 284
Garattini, S., 139, 290
Garb, J. L., 266, 284
Garcia, J., 6, 20, 21, 284, 287, 300
Garmezy, M., 135, 284
Gaskin, M., 46, 48, 51-2, 132, 301
Gately, P. F., 139, 284
Gemberling, G. A., 22, 61, 64, 69, 70, 86-7, 95, 203, 276
George, E. I., 249, 298
Gershon, S., 249, 250, 273
Geyer, M., 139, 284
Giacalone, E., 139, 290
Giardini, V., 17, 101, 106, 154, 276
Gibbon, J., 31, 292
Gibson, E. J., 45, 46, 52-4, 112, 284, 308
Gillette, K., 60, 92, 95, 284
Gilley, D. W., 11, 21, 22, 283
Ginton, A., 101, 107, 115, 121-4, 125, 127, 219, 223, 236, 252, 253, 255, 284
Gjerde, P. F., 249, 284
Glazer, H. I., 154, 284
Glowinski, J., 139, 305
Goodell, N. A., 24, 26, 67, 133, 297, 300
Gordon, E. L., 82, 278
Gorham, D. R., 258, 298
Gormezano, I., 24, 26-7, 63, 67, 306
Goss, A. E., 112, 284
Gould, J., 157, 295
Gowan, W. M., 139, 300
Graham, F. K., 40, 266, 269, 284-5
Grant, D. A., 26, 113, 285
Grant, M. J., 19, 199, 285
Grant, R. M., 19, 199, 285
Grau, J. W., 214, 300
Gray, J. A., 121, 124-5, 127, 130, 131, 132-3, 219, 223, 256, 257-9, 275, 285, 307
Gray, T., 62, 63-4, 285
Green, S. E., 46, 285
Grice, G. R., 53, 285
Griesemer, H. A., 21, 60, 62, 97, 98, 283
Griffith, J. D., 249, 285
Groenewegen, H. J., 139, 269, 285
Groves, P. M., 137-8, 150-4, 285
Guanowsky, V., 97, 98, 100, 296
Guldberg, H. C., 139, 292

Gustavson, K. K., 62, 267, 309
Gustin, P., 82, 182, 185, 200, 275

Halaris, A. E., 139, 297
Halgren, C. R., 14, 16, 17, 46–7, 52, 94, 130, 131, 141, 144, 217, 238, 262, 273, 285
Hall, G., 9, 21, 26, 44–6, 53, 55, 78, 80–2, 87–9, 131, 141, 156, 158–9, 164, 165, 169–74, 177–9, 180, 188–90, 194, 198–200, 202–5, 214, 216, 235, 267–8, 279, 285–6, 298, 302
Hall, K., 255, 300
Hallstead, C., 97, 98, 100, 290
Hake, H., 26, 113, 285
Hamel, L., 59, 210, 290
Hammond, L. J., 188, 285
Har-Evan, D., 15, 121, 125–7, 219, 223, 285
Harris, J. D., 269, 285
Harrison, J. M., 46, 48–52, 282, 285
Hawks, D. V., 255, 298
Hearst, E., 32, 144, 270, 285, 308
Hebb, D. O., 52, 141, 285
Heisenberg, W., 181, 285
Held, J., 249, 285
Hellman, P. A., 14, 135–6, 251, 285
Hemsley, D. R., 121, 124–5, 127, 130, 219, 223, 256, 257–9, 272, 275, 276
Herendeen, D. L., 31, 101, 106, 285
Hernandez, L. L., 24, 27, 38–9, 65, 285
Heron, P. A., 157, 303
Herz, M. J., 6, 269, 299, 310
Hill, D. F., 55, 285
Hinde, R. A., 269, 287
Hinderliter, C. F., 21, 97–100, 296, 299
Hinson, R. E., 271, 287
Hobbs, S. H., 67, 68, 282
Hochberg, A. C., 255, 298
Hoffman, H., 74, 97, 98, 100, 183–4, 268, 290
Hoffman, J., 38–9, 283
Holder, M. D., 21, 287
Hole, K., 139, 292
Holland, P. C., 41, 92, 94, 214, 234–5, 283, 287, 301
Homewood, J., 61, 69, 70, 86–7, 95, 201, 309
Homzie, M. J., 114, 305
Honig, W. M. M., 138, 299
Honigfeld, G., 248, 287
Honzik, H. C., 2, 307
Hopkins, H. K., 255, 300
Horn, G., 269, 287
Horowitz, S. W., 21, 60, 97, 98, 283
House, B. J., 160, 166, 190, 310
Howard, J. L., 38, 298
Huey, L., 249, 288

Hui, S. C. G., 140, 279
Hull, C. L., 1, 3, 4, 9, 168, 238, 265, 287
Hulstijn, W., 26, 101, 107, 113, 114, 127, 219, 223, 289
Hume, D., 3
Hunsicker, J. P., 20, 295
Hupka, R. B., 24, 26, 62, 197, 279

Iacono, W. G., 221, 287
Isaacson, R. L., 17, 139, 287, 298
Ison, J. R., 149, 269, 302
Israeli-Telerant, A., 135–6, 250, 308
Iversen, S. D., 138, 250, 272, 280, 287, 288, 289, 301

Jackson, R. L., 32–5, 154 287, 307
Jacobs, B. L., 139, 288
Jacobs, W. J., Jr., 22, 61, 301
James, J. H., 17, 19, 26, 72, 199, 288, 303
James, W., 223–6, 228, 288
Jankovic, B. D., 21, 281
Janowsky, D. S., 249, 272, 288, 303
Järbe, T. U. C., 13, 21, 78, 133, 214, 234, 274, 297
Jarvis, P. J., 59, 277
Jeeves, M. A., 56, 288
Jenkins, H. M., 32, 278
John, E. R., 269, 306
Johnson, P. E., 22, 61, 69, 86, 87, 95, 203, 276, 288
Johnston, W. A., 181, 288
Jones, D. L., 140, 288
Jones, D. R., 61, 297
Jonsson, G., 78, 133, 297
Joseph, L. J., 113, 296
Joyce, E. M., 138, 288
Judd, L. L., 249, 288

Kahneman, D., 157, 288
Kalat, J. W., 22, 60–2, 67, 68, 69, 71, 288
Kamer, R. S., 14, 134, 238, 305
Kamin L. J., 89, 164, 205, 288, 291
Kaniel, S., 100, 107–10, 219–20, 229–30, 288
Kaplan, L. J., 14, 130, 134, 135, 137, 238, 251, 255, 305
Kasprow, W. J., 26, 169, 171, 182–4, 185–6, 194, 203, 275, 288, 296
Katkin, E. S., 156, 294
Katz, Y., 130, 137, 251, 257–8, 308
Kay, S. R., 260, 272, 288
Kaye, H., 41–4, 87, 131, 158, 170, 177, 217, 225, 238, 262, 263, 289
Kayser, R. J., 23, 292
Kelley, A. E., 139, 289
Kelly, P. H., 138, 289
Kendler, T. S., 229, 289
Kent, E. W., 14, 134, 274
Kerpelman, L. C., 53, 289

Kiefer, S. W., 22, 289
Kierman, J. M., 14, 19, 134, 238, 305
Killeen, P. R., 170, 289
Kimble, D. P., 130, 289
Kimble, G. A., 148, 289
Kimmel, H. D., 207, 208, 227, 266, 269, 289
Kimmel, W., 207, 290
Kimmeldorf, D. J., 20, 284
Kiney, C. A., 14, 134, 144, 305
King, D. A., 268, 277
King, M. S., 269, 299
Kirschbaum, E. H., 22, 79, 296
Kirk, R. J., 174, 290
Klein, D. E., 257, 290
Klein, S. B., 59, 60, 62, 97, 98, 100, 210, 283, 290
Klosterhalfen, S., ,20, 21, 290
Klosterhalfen, W., 20, 21, 290
Koelling, R. A., 6, 21, 284
Kohler, C., 139, 290, 292
Kokkinidis, L., 134, 249, 250, 290
Konorski, J., 31, 89, 290
Koob, G. F., 250, 261, 306
Kopell, B. S., 157, 301
Korboot, P. J., 255, 290
Kornetsky, C., 255, 290
Kortschot, A. de, 139, 269, 285
Kostas, J., 22, 131, 217, 262, 294
Kostlan, A., 26, 113, 285
Kostowski, W., 139, 290
Kovach, J. K., 55, 290
Kraemer, P. J., 21, 22, 67, 68, 74, 97, 98, 100, 183–5, 268, 290
Kraepelin, E., 135, 248, 263, 290
Kraese, B. S., 38, 266, 304
Kramer, R. M., 14, 130, 135, 137, 251, 255, 305
Krames, L., 20, 296
Krane, R. V., 22, 290, 291
Kraut, A. G., 108, 113, 231–2, 278, 291, 302
Krauter, E. E., 92–4, 149, 267, 291, 302
Krech, D. M., 55, 291
Krechevsky, D., 190, 291
Kremer, E. F., 63, 89, 195, 291
Kruse, J. M., 21, 22 291
Ksir, C. J., 26, 101, 107, 113–14, 127, 196, 219, 223, 303
Kuczenski, R., 135, 291
Kuhn, C. M., 135, 291
Kuhn, T. S., 7, 291
Kuiack, M., 38, 266, 304
Kurtz, K. H., 154, 291
Kurz, E. M., 78, 291

Lacoursiere, R. B., 255, 305
Lantz, A. E., 59, 62, 64, 65, 66, 151, 156, 158, 172, 196–8, 292

Lashley, K. S., 190, 292
Laskowski, B. A., 61, 69, 70, 277
Launey, G., 258–9, 292
Lawrence, D. H., 190, 292
Lawson, J. S., 248, 292
Lazarevic, M., 21, 281
Leaf, R. C., 23, 292
Leaf, S. R. P., 23, 292
Leaton, R. N., 269, 307
Lee, H. K., 250, 273
Leith, N. J., 136, 275
Leon, M., 21, 287
Leontiev, A. N., 155, 307
Levine, S., 60, 304
Levitsky, D. A., 78, 291
Liebig, J. von, 3
Lin, D., 46, 48, 52, 132, 295
Lindenbaum, U., 242, 246, 293
Lipton, M. A., 257, 292
Littlefield, V. M., 55, 306
Livesey, P. J., 46, 276
Locke, J., 3
Locurto, C. M., 31, 292
Loeb, J., 31, 292
Logan, M. S., 63, 292
Logue, A. W., 21, 266, 292
Logue, K. R., 266, 292
Lohr, C. A., 46–7, 144, 266, 305
LoLordo, V. M., 21, 22, 81, 268, 291, 300, 301
Lorden, J. F., 132–3, 145, 292
Lorens, S. A., 139, 292
Lovejoy, E., 160, 166, 190, 292
Lovibond, P. F., 78, 80, 178, 264
Lubow, R. E., 1–3, 7, 13–14, 16–17, 19–21, 23, 26–8, 31, 37, 45–6, 54–5, 58–9, 62–5, 66, 67, 75, 76–8, 83–4, 87, 91–4, 100–1, 106–13, 115, 119, 121–5, 127–30, 135–6, 141–2, 145–6, 148–9, 151, 159–62, 175, 179, 190–2, 196–7, 199–204, 208–10, 219–20, 223, 227, 229–30, 235–46, 250, 252–5, 257, 266–71, 274, 284, 292–3, 303, 304, 308
Lucas, R., 108, 111–12, 118, 293
Luria, A. R., 155, 307
Lykken, D. T., 221, 287
Lynch, M. R., 22, 293
Lyness, W. H., 140, 293
Lynn, R., 37, 155, 266, 269, 293

McAllister, D. E., 17, 19, 54, 199, 276, 293
McCubbin, R. J., 156, 294
McDaniel, J. W., 101, 106, 294
MacDonald, R. L., 38–9, 106, 279
McFarland, D. J., 22, 131, 217, 262, 294
McGhie, A., 135, 248–9, 292, 294
McGregor, S., 14, 131, 238, 262, 270, 309

McIntosh, J. M., 59, 67, 68, 294, 306
Mackintosh, N. J., 9, 32-6, 44, 46-50, 78, 80, 87, 92-4, 131, 141, 159-61, 164-9, 170, 174-5 177, 180, 190, 193, 195, 201, 207, 235, 238, 240, 264, 266-8, 275, 281, 289, 292, 294, 306
McLaurin, W. A., 20, 21, 60-1, 67, 68, 282, 294
McLeod, L. M., 24, 27, 68, 88, 302
McMahon, M. A., 251, 280
McNaughton, N., 131, 285
McVaugh, W., 75, 153, 273
Magaro, P., 252, 272, 294
Maier, S. F., 4, 58, 153-4, 239-40, 245, 267, 269, 287, 294, 303
Maisiak, R., 24, 26, 68, 85, 202, 207, 283
Maloney, M. P., 255, 294
Maltzman, I., 117, 118, 120, 141, 155, 156, 158, 206, 269, 295, 309
Mandler, G., 271, 295
Marder, S. R., 272, 307
Margolis, A., 60, 304
Margolit, H., 242, 246, 293
Markman, R. E., 27-8, 37, 62, 68, 101, 141, 196, 293
Markovic, B. M., 21, 281
Marlin, N. A., 270, 295
Marsden, C. D., 140, 279
Martin, G. M., 22, 295
Maser, J., 188, 285
Mason, S. T., 46, 48, 52, 132, 269, 295
Mastropaola, J. P., 61, 301
Matthysse, S., 135, 260, 272, 295
Matussek, P., 249, 298
Matzel, L.D., 87, 278, 295
May, R. B., 62, 295
Meachum, C. L., 78, 276
Meagher, M. R., 38-9, 138, 284
Medin, D. L., 75, 295
Meiri, N., 127, 220, 295
Mellgren, R. L., 14, 16-17, 20, 46, 48-50, 52, 130, 131, 141, 144, 196, 199, 217, 238, 262, 273
Menzel, R., 30, 101-3, 276
Mercier, P., 13, 26, 81, 84, 92, 93, 178, 179, 275, 295
Merrill, H. K., 26, 78, 177, 281
Messing, R. B., 266, 279
Meyers, W. J., 113, 296
Midgley, G. C., 65, 197, 296
Mikulka, P. J., 59, 60, 62, 97, 98, 100, 210, 283, 290
Milgram, N. W., 20, 296
Millard, W. J., 22, 296
Millenson, J. R., 188, 296
Miller, F. D., 232, 296
Miller, N. E., 18, 296

Miller, R. R., 22, 26, 59, 82, 87-8, 169, 171, 182-4, 185-6, 194, 200, 203, 270, 275, 288, 295, 296
Miller, S., 242, 243, 244, 296
Mineka, S., 101, 106, 242, 243, 244, 296, 306
Minor, H., 26, 78, 80-1, 178, 200, 285
Mirsky, A. F., 249, 296
Misanin, J. R., 21, 97-100, 296, 299
Mitchell, D., 22, 79, 199, 296, 297
Moderasi, H. A., 31, 297
Moffitt, T., 79, 199, 297
Mogenson, G. J., 140, 288
Mohammed, A. K., 13, 21, 78, 132, 133, 214, 234, 274, 297, 301
Molina, E. J., 32-6, 308
Moore, A. U., 1-3, 27-8, 58, 101, 106, 141, 145-6, 161, 196, 266, 293
Moore, D. E., 138, 297
Moore, J. W., 24, 25-7, 46-7, 64-5, 87-8, 130-1, 133, 144, 169, 194, 203, 217, 238, 262, 266-7, 270, 274, 297, 302, 305
Moore, K. E., 140, 293
Moore, R., 139, 297
Moray, N., 224, 297
Mostofsky, D. I., 225, 297
Mowrer, O. H., 18, 297
Mowrer, R. R., 82, 271, 297
Murphy, A. L., 32-5, 307
Musty, R. E., 21, 279

Nachman, M., 21, 61, 297
Nadel, L., 217, 298
Nakamura, Y., 271, 295
Nakaya, T., 271, 297
Nauta, W. J. H., 139, 297
Naylor, R. J., 140, 279
Neale, J. M., 255, 298
Nealis, P. M., 46, 285
Nemeroff, C. B., 257, 292
Ness, J., 22, 283
Nicholas, D. J., 26, 88, 169, 171, 298
Nichols, G. L., 14, 19, 134, 238, 305
Nicholson, L., 268, 304
Nield, A. F., 206, 240, 309
Nilsson, L. G., 74, 266, 268, 274
Nodine, C. F., 112, 284
North, A. J., 56, 288
Nutt, G., 62, 64, 186-9, 271, 280

Oakley, D. A., 24, 26, 63, 67, 90, 299
Oates, J., 249, 285
Oberdieck, F., 30, 298
Obrist, P. A., 38, 298
Oelssner, W., 140, 283
O'Farrell, T., 23, 75, 273
Ohayon, J., 255, 298

Öhman, A., 38, 157, 206, 270, 298
O'Keefe, J., 217, 298
Oliverio, A., 17, 133, 298
Oltmanns, T. F., 255, 298, 299
Olton, D. S., 17, 298
Ophir, I., 266, 292
Opler, L. A., 260, 272, 288
Orlebeke, J. F., 227, 266, 269, 289
Ossenkopp, K. P., 21, 74, 184–5, 290
Ost, J. W. P., 46, 48–50, 52, 144, 196, 199, 295
Oswalt, R. M., 53, 298
Overall, J. E., 258, 298
Owen, S., 131, 285

Paramanova, N. P., 115, 304
Patsiokas, A. T., 21, 60, 62, 97, 98, 283
Pavlov, I. P., 37, 40, 146, 188, 298
Payne, R. W., 249, 255, 298
Pearce, J. M., 9, 26, 41–4, 87–91, 131, 141, 156, 158–9, 164, 165, 169–75, 179, 180, 190, 194, 198, 202–3, 205, 216–17, 225, 235, 238, 262–3, 267–8, 285, 289, 298
Pearle, J. D., 154, 291, 298
Peeke, H. V. S., 6, 269, 299, 310
Perlmuter, L. C., 26, 113, 299
Perry, R. C., 22, 79, 296
Peselow, E., 260, 272, 273
Peterson, C. S., 97, 99, 100, 299
Petrali, E. H., 135, 136, 280
Petree, A. D., 62, 64, 186–9, 271, 280
Peuser, K. A., 267, 270, 310
Pfautz, P. L., 178, 308
Phillips, J. A, 22, 289
Phinney, R. L., 19, 309
Picton, B. M. B., 195, 294
Pijnenburg, A. J. J., 138, 299
Pinel, J. P. J., 30, 299, 306
Plotkin, H. C., 24, 26, 63, 67, 90, 299
Pogue-Geile, M. F., 255, 299
Poon, S. L., 139, 284
Porter, J. H., 22, 293
Poser, D. G., 117–18, 120, 269, 299, 306
Posner, M. I., 108, 231, 299
Powell, D. A., 24, 27, 38–9, 65, 285
Powell, T. P. S., 139, 300
Preston, G. C., 44, 78, 80, 87, 132–3, 178, 264, 289, 292, 307
Pribram, K. H., 130, 131, 282, 299
Priedite, G., 21, 78, 133, 274
Primavera, J., 178, 299
Pritchard, W. S., 37, 299
Prokasy, W. F., 24, 26, 72, 205, 299–300
Provost, S. C., 61, 70, 73, 86–7, 95, 201, 309
Pycock, C. J., 140, 278

Raisman, G., 139, 300

Randich, A., 81, 268, 300
Rappaport, M., 255, 300
Raskin, D. C., ,117, 118, 120, 141, 156, 295
Ray, O. S., 136, 275
Ray, R. D., 38–9, 300
Rawlings, T. D., 20, 60–1, 282
Rawlins, J. N. P., 131–3, 285, 307
Razani, J., 255, 294
Rebec, G., 137–8, 285
Reilly, S., 32–4, 36, 101, 300
Reisine, T. D., 139, 305
Reiss, S., 24, 27, 36, 46–7, 58, 64, 67, 141, 143–4, 162, 268, 300
Reiter, L. A., 149, 269, 302
Remington, B., 107, 118–21, 304
Remington, G., 154, 273
Rescorla, R. A., 36, 46–7, 75, 81, 141–3, 145, 162–9, 174–6, 214, 238, 268, 270, 300, 308
Revusky, S. H., 21, 22, 300
Riccio, D. C., 96–8, 100, 296, 309
Rickert, E. J., 54, 132–3, 145, 276, 292
Rifkin, B., 19, 45–6, 54–5, 64, 76–8, 100, 107–9, 111–12, 209, 216, 293
Riley, A. L., 20, 22, 61, 69, 70, 277, 300–1
Rilling, M., 32–4, 36, 101, 307
Riopelle, A. J., 26, 113, 285
Riva, C., de la, 46, 48, 51–2, 132, 301
Robbins, R. J., 60, 67, 68, 301
Robbins, T. W., 46, 48, 51–2, 78, 132, 250, 297, 301
Roberts, W. A., 22, 67, 68, 74, 183–5, 290
Robertson, D. P., 22, 290
Robinson, T. E., 134, 250, 301
Rochelle, F. P., 60, 290
Roelofs, N., 113, 301
Rolnick, A., 128, 129, 266, 274
Rommelschpacher, H., 139, 301
Rosecrans, J. A., 22, 293
Rosellini, R. A., 272, 301
Rosenberg, L., 86–7, 198, 201, 302
Rosenblatt, R., 192, 242, 243–4, 246, 293
Rosenzweig, M. R., 55, 291
Ross, R. T., 24, 27, 67, 88, 214, 301, 302
Rossum, J. M. van, 138, 269, 279, 299
Roth, W. T., 157, 301
Rotrosen, J., 260, 272, 273
Royet, J. P., 20, 202
Rozin, P., 60–62, 67, 68, 69, 71, 288
Rubenstein, J. A., 21, 266, 278
Rubinstein, M., 260, 272, 273
Rudy, J. W., 22, 61, 71–3, 86–7, 92–4, 97, 99–100, 178, 198, 201, 267, 302, 308
Russo, J. M., 149, 269, 302
Rutledge, E. F., 114, 305

Author index

Rutter, D. R., 248, 302
Ryan, J. P., 139, 287

Saccuzzo, D. P., 255, 277
Salafia, W. R., 24, 27, 85–6, 131, 238, 262, 302
Sandell, J. H., 86–7, 198, 201, 302
Sasa, M., 139, 306
Scarborough, B. B., 20, 21, 60–1, 67, 68, 282, 294
Scavio, M. J., 24, 27, 67, 88, 302
Schachtman, T. R., 44, 87–8, 171, 183, 188–9, 194, 203, 285, 295, 296, 302
Schäfer, S., 30, 101, 102–3, 276
Schanberg, S. M., 135, 291
Scheele-Kruger, J., 250, 302
Schlossberg, A., 107, 121, 124, 127, 130, 219, 223, 252, 254–5, 257, 293
Schmajuk, N. A., 131, 262, 302
Schnabel, I., 16, 101, 308
Schneider, D. E., 26, 113, 285
Schneider, S. J., 255, 302
Schneider, W., 226, 228, 302, 303
Schneiderman, N., 206–7, 303
Schnur, P., 9, 17, 26, 63, 65, 66, 83–4, 87–8, 91, 101, 106–8, 113–15, 121, 127, 141–2, 151, 156–8, 159–61, 175, 179, 190, 196–7, 200–2, 204, 208, 219–20, 223, 235, 237, 239, 240–1, 244–5, 253, 255, 267–8, 270–1, 292, 293, 303
Schoenfeldt, M. G., 62, 295
Schuckit, M. A., 250, 303
Scott, D. S., 14, 134, 144, 305
Sears, R. J., 141, 166, 174–5, 190, 193, 283
Segal, D. S., 139, 249, 250, 272, 284, 303
Segal, M., 139, 274
Seligman, M. E. P., 4, 58, 153, 239–40, 267, 269, 294, 303
Seviour, P. W., 138, 289
Shakow, D., 135, 303
Shapiro, D., 157, 275
Shapiro, M. M., 31, 101, 106, 285
Sharp, P. E., 26, 303
Shek, D. T. L., 37, 157, 277
Sherrington, C. S., 146, 303
Shiffrin, R. M., 226, 228, 302, 303
Shishimi, A., 17, 27–8, 101, 103–5, 268, 303
Siddle, D. A. T., 38, 107, 118–21, 156, 227, 266, 269, 303–4
Siebert, L., 13, 23, 63, 115–16, 149, 196, 218–19, 268, 293, 304
Siegel, S., 13, 23–4, 27, 40–4, 58, 61–2, 64, 69, 71, 74, 90, 141, 149, 196, 210, 266, 281, 304
Sigmundi, R. A., 87, 304

Sikes, R. W., 139, 279
Silver, A. I., 117–20, 304
Silverman, J., 255, 300
Sjöden, P. O., 74, 266, 268, 274
Skala, K. D., 267, 270, 310
Skinner, B. F., 23, 282
Skinner, C., 269, 299
Slade, P., 258–9, 292
Sloane, R. B., 255, 294
Smith, A. D., 139, 269, 307
Smith, N. F., 242, 279
Smothergill, D. W., 108, 113, 231, 232, 278, 291, 304
Smotherman, W. P., 60, 304
Snyder, S. H., 250, 304
Soderberg, U., 78, 133, 297
Sokolov, E. N., 37, 117, 155–7, 176, 205, 226, 266, 271, 304, 307
Solomon, P. R., 14, 19, 24, 26, 27, 46–7, 64–5, 130, 131, 133–8, 144, 217 238, 251, 255, 262–3, 266, 280, 285, 297 304, 305, 306
Solomon, R. L., 4, 239–40, 245, 269, 294, 303
Solso, R. L., 224, 305
Soubrie, P., 139, 305
Spear, N. E., 74, 75, 97, 98, 100, 183–4, 268, 290, 305
Spence, K. W., 113–14, 127, 223, 265, 305
Spencer, W. A., 6, 150–1, 269, 307
Sperber, S. B., 266, 279
Spinks, J. A., 37, 157, 277
Spohn, H. E., 255, 305
Spring, B. J., 135, 258, 295, 305
Springer, J. E., 139, 287
Spurr, C. W., 24, 26, 67, 151, 300, 305
Squires K. C., 269, 305
Squires M. K., 269, 305
Srebro, B., 139, 292
Starr, M. D., 242, 244, 306
Staton, D. M., 14, 130, 137, 138, 263, 305, 306
Steinbush, H., 139, 290
Stevens, R., 217, 283
Steward, D. J., 55, 306
Stickney, K. J., 131, 270, 297
Storlien, L. H., 22, 295
Storms, L., 249, 288
Strauss, K. E., 266, 292
Strauss, S., 139, 301
Stunkard, A. J., 266, 284
Suboski, M. D., 24, 26–7, 63, 67, 306
Sugarman, J., 135, 295
Sullivan, D. J., 19, 305
Sullivan, L. G., 267, 306
Sullivan, P., 23, 75, 273
Suomi, S. J., 46, 285

Author index 319

Surwit, R. S., 117–18, 120, 269, 306
Sutherland, N. S., 55, 160-1, 165, 166, 190, 273, 306
Sutterer, J. R., 38, 298
Sutton, S., 269, 306
Swanson, L. W., 139, 306, 310
Swartzentruber, D., 87-8, 267, 271, 277, 306
Swedberg, M. D. B., 78, 297
Swerdlow, N. R., 250, 261, 306
Swietalski, N., 87, 201, 289
Szabo, L., 44, 87, 177, 289
Szakmary, G. A., 83-4, 88-9, 179, 197, 201, 203, 244, 268, 306
Szwejkowska, G., 31, 89, 290

Takaori, S., 139, 306
Takulis, H. K., 21, 300
Tarpy, R. M., 59, 67, 68, 294, 306
Tarte, R. D., 30, 298
Taylor, K. M., 139, 288
Tees, R. C., 65, 197, 296
Terlecki, L. J., 30, 306
Terrace, H. A., 31, 292
Terry, W. S., 178, 306
Thatcher, D. O., 64, 267, 280
Thinus-Blanc, C., 271, 283
Thompson, K., 255, 305
Thompson, R. F., 6, 89, 150-4, 237, 269, 285, 306-7
Tighe, T. J., 53-4, 269, 284, 307, 308
Titchner, E. B., 224
Toga, A. W., 46-7, 131, 262, 278, 307
Tolman, C. W., 62, 295
Tolman, E. C., 1, 2, 3, 9, 307
Tomie, A., 32-5, 268, 275, 307
Totterdell, S., 139, 269, 307
Trabasso, T., 190, 307
Tranberg, D. K., 32-4, 36, 101, 307
Treit, D., 30, 299, 306
Tsaltas, E., 132-3, 307
Tuber, D. S., 27-8, 58, 69, 74, 89, 101, 106, 199, 206, 240, 309
Tuck, D. L., 20, 22, 301
Turi, A., 14, 130, 135, 137, 251, 255, 305
Tyson, H. W., 153, 273

Ulm, R. R., 242-4, 307
Urca, G., 101, 107, 115, 121-4, 125, 127, 219, 223, 236, 252, 253, 255, 284

Valliere, W. A., 97, 99, 100, 299
Valzelli, L, 139, 290
Van Kammen, D. P., 272, 307
Van Olst, E. H., 227, 266, 269, 289
Van Zandt, B. J. S., 271, 295
Venables, P. H., 135, 307
Vermeulen-Van Der Zee, E., 139, 269, 285

Vigorito, M., 26, 87, 88, 89, 169, 194, 203, 274
Vinogradova, O. S., 155, 307
Vogel, J. R., 11-13, 23, 58, 63, 141, 197, 199, 278
Volpicelli, J. R., 242-4, 307
Voronin, L. G., 155, 307

Wagner, A. R., 24, 26-7, 36, 46-7, 58, 64, 67, 75, 81, 141, 143-5, 150, 162-76, 177-9, 195, 235-6, 238, 268, 270, 280-1, 299, 300, 303, 308
Wagner, M., 26, 62, 92-4, 161, 197, 199, 293
Wagner, R. L., 262, 308
Walk, R. D., 45, 46, 52-4, 284, 308
Waller, T. G., 55, 154, 274, 308
Walters, G. C., 154, 291, 298
Wang, R. Y., 138, 309
Warren, D. A., 272, 301
Wasserman, E. A., 32-6, 308
Weinberger, D. R., 262, 308
Weiner, I., 7, 9, 14, 16, 26, 62, 87, 92-4, 101, 107, 121, 124, 127, 130, 135-8, 141-2, 159-60, 161, 190, 192, 197, 199, 203-4, 208, 219, 223, 235, 237, 239, 242-4, 246, 250-2, 254-5, 257-8, 270, 276, 293, 308
Weiss, J. M., 154, 284
Weiss, K. R., 7, 14, 131, 202, 238, 262, 270, 308-9
Weissbluth, S., 155-6, 206, 269, 295
Westbrook, R. F., 21, 22, 61, 64, 69, 71, 72, 78-9, 86-7, 95, 201, 277, 309
Whipple, R., 255, 294
White, F. J., 138, 309
White, L. E., Jr., 139, 279
White, N., 26, 280
White, R. K., 101, 106, 294
Whitlow, J. W., 178, 195, 281, 308, 309
Wickens, C., 27-8, 58, 69, 74, 89, 101, 106, 199, 206, 240, 269, 305, 309
Wickens, D. D., 27-8, 58, 69, 74, 89, 101, 106, 199, 206, 240, 309
Wilkie, D. M., 30, 65, 197, 296, 299
Williams, D. R., 32, 34, 284, 309
Williams, F., 153, 273
Williams, J. H., 139, 309
Willner, J. A., 31, 309
Wilson, L. M., 19, 96, 309
Wilson, N. E., 6, 281
Wilson, R. S., 38-40, 309
Winefield, A. H., 55, 243-4, 275, 309
Winkelman, J. W., 14, 130, 135, 137, 251, 255, 305
Winocur, G., 132-3, 307
Winter, W., 79, 199, 297
Wirschafter, D., 14, 134, 274

Wise, W. D., 139, 288
Wishart, T. B., 135, 136, 280
Wishner, J., 248, 302
Witcher, E. S., 26, 309
Witter, M. P., 139, 269, 285
Wittlin, W. A., 21–2, 309
Wittman, T. K., 38–9, 64, 131, 267, 280, 309
Wolf, D., 23, 75, 273
Wolff, C., 117, 118, 120, 141, 155–6, 206, 269, 295, 309
Woodruff, G., 34, 309
Wright, D. C., 62, 267, 270, 309, 310
Wu, M., 140, 288

Wundt, W., 224
Wyatt, R. J., 262, 308
Wyers, E. J., 6, 310
Wyss, J. M., 139, 310

Yirmiya, R., 21, 287
Young, D., 19, 199, 285

Zahorik, D. M., 59, 310
Zajonc, R. B., 55, 271, 310
Zeaman, D., 160, 166, 190, 283, 310
Zeiner, A. R., 117–18, 120, 310
Ziv-Harris, D., 101, 308
Zubin, J., 258, 269, 305, 306

Subject index

acquired distinctiveness of cues, 238
age, 96–9, 109–10
alpha response, 148–50
amphetamine, 130, 134–8, 249–51, 255, 260–3, 269, 272
amphetamine psychosis, 249–50
antipsychotic drugs
  normalization of attention with, 255–7
  see also chlorpromazine, haloperidol, sulpiride
apparatus preexposure effects, 199–200
appetitive conditioning, 41, 47–55, 158, 199, 200
associative biases, 4–7
associative learning theory, 1–3
attention, 7, 9
  active–passive, 229–30
  adult vs. children, 218, 230
  animal model, 225,
  automatic-controlled, 226
  course of, 221–3
  definitions, 223–5, 237–8
  internal–external overrides, 226–30, 247–8
  voluntary–nonvoluntary, 225–6
  voluntary override, 260
attentional decrement overrides
  external, 226–8, 247–8
  internal, 228–30, 247–8
autoshaping, 32–7, 266, 268
avoidance conditioning, 13–20
  amphetamine, 250–1
  conditioned inhibition, 144
  one-way, 18–19
  passive, 19–20
  septal lesions, 131
  two-way, 14–17, 199, 202

backward conditioning, 90
blocking, 4, 5, 41, 74, 164, 182–4, 195, 270
blocking of LI, 198
Brief Psychiatric Rating Scale, 258
British empiricists, 3
Broadbent's information procesing model, 249

cat, 89, 106, 107
catastrophe rule, 270
chemotherapy, 129

children, 76, 77, 108–12
  age, 109–10
  context, 109
  masking, 110–11
  paired associative learning, 111
  reaction time, 108
  social class, 109–10
  stimulus specificity, 110–11
chlorpromazine, 225
cholinergic manipulations, 130, 133–4, 138–40
classical conditioning
  stimulus preexposure parallels, 215
  two-stage theories, 205–8
competing responses, 141, 145–7
complementary responses, 141
compound conditioning, 74
compound stimulus preexposure, 82, 92–5, 199
conditioned appetitive responding, 131
conditioned attention theory, 142, 152, 190–238
  applied to humans, 218–38
  assumptions, 190–1
  conventional conditioning and, 205–8
  modifications and elaborations, 208–17, 230–1
  predictions, 195–205, 208
conditioned defensive burying, 30–1
conditioned leg flexion, 146
conditioned suppression, 11–13, 23, 26, 47, 83–4, 89, 149, 196–201, 266, 268
  age, and, 96
  amphetamine, and, 250–1
  compound stimulus preexposure, 92–4
  context, and, 75, 80, 81
  preexposed stimulus parameters, 62–8, 81, 83, 91
  septal lesions, and, 131
  second stimulus effects, 86–7, 90–1
conditioned taste aversion, 10–11, 20–3, 196, 198, 199, 201, 203, 210, 218, 234, 267
  adult human, 129, 265–6
  age, 97–9
  compound stimulus preexposure, 92, 94–5
  conditioned inhibition, 144
  context, 74, 79–81
  preexposed stimulus 21–2

321

conditioned taste aversion (*cont.*)
  preexposed stimulus parameters, 60–1, 64, 70–3
  second stimulus effects, 86–7, 90–1
conditioning of inattention to S_1
  S_1 predictions, 196–205
  S_1–S_2 predictions, 200–5
conditioning of orienting responses, 37–44
context, 53–5, 74–82, 109, 176–8, 208, 213–15, 217, 235–7, 267–8, 270–1
  change, 76–9, 178
  extinction, 80–1, 178, 235, 270
  encoding, 233–4
  familiarization, independent of S_1, 79
  preexposure, 81–2
  –S_1 associations, 179–80, 234, 236
  specificity, 159, 177
  tonic vs. phasic, 74, 78–9, 81
  –US associations, 81–2
contiguity, 1, 3, 5

dark rearing, 55
defensive conditioning preparations, 27–30
defensive reactions, 40, 155, 266, 269
discrimination learning, 44–56, 159–60, 266
  appetitive test, 47–55
  classical conditioning test, 45–7
  compound test stimulus, 45–7
  conditioned inhibition, 144
  context change, 78
  continuous stimulus preexposure, 52–5
  DNAB lesions, 132
  discrete stimulus preexposure, 45–52
  instrumental conditioning test, 47
  noxious US, 45–7
  septal lesions, 131
dishabituation, 38
distractor stimuli, 177
dog, 31, 38–9, 106
dopamine agonists–antagonists, 260–2
dopaminergic activity, 272
dopaminergic manipulations, 134–8, 138–40
dorsal noradrenergic bundle, 132
dual process theory of habituation, 152–4

encoding
  property, 247
  relationship, 247
  see also stimulus encoding
enriched environment, 55
excitation, conditioned, 162, 165
extinction, 113
  LI, 199
eyeblink conditioning, 24–5, 26–7, 40, 47, 90, 113–15, 196

conditioned inhibition, 144
eyemovement, 222
Eysenck's Personality Questionnaire, 258–9

fear relevant, irrelevant stimuli, 268
feature positive effect, 171–2, 270
Frey-Sears model, 174–5

galvanic skin response (GSR), 13, 38, 155–6, 157–8, 206, 218, 221, 227, 269
  conditioning, 116–21
generalization gradient for LI, 196, 210
goat, 27–9, 106, 107
goldfish, 27–9, 103, 107

habituation, 6, 9, 38, 221, 226, 269, 271
  alpha response, 148–50
  dual process theory, 152–4
  long-term, 150–1
  orienting response, of, 65, 154–9, 161, 227
  theories, 147–60
Hall-Pearce effect, 87–9, 171–4, 179, 194, 198, 202–3, 205, 216, 267–8
haloperidol, 137–8, 249–51, 255, 258, 260–2
handling, 101
heart-rate conditioning, 38–40, 138
hippocampal lesions, 131, 158–9, 217, 225, 238, 262–3, 269
hippocampus, 86, 130–2, 262
honeybee, 102, 107
human adults
  age, 100–1, 107
  conditioned taste aversion, 129, 265–6
  eyeblink conditioning, 113–15
  GSR conditioning, 116–21
  instrumental learning, 121–8
  Ivanov-Smolenskii conditioning, 115–16
  masking, 101–2, 107, 113–15, 116, 121–8, 129
  reaction time, 111–12
  retarded, adolescents, 128

imprinting, 6, 55
inattention, 9, 152
information processing theories of LI, 162–82, 249
inhibition
  conditioned, 141, 142–5, 158, 162, 165, 265, 266, 268
  external, 197
  Hullian, 2
  Pavlovian, 2
inhibition of delay hypothesis of LI, 183, 188–9

intercranial stimulation, 85, 202
nterreflex inhibition, 149
Ivanov-Smolenskii conditioning, 115–16, 218
investigatory reflex, 37, 146
  see also orienting response (OR)

latent inhibition (LI)
  definition 1, 6
  first study, 2, 3–4
  significance, 3
  theories, 9, 141–238
  variables, see stimulus variables and specific entries
latent learning, 1–2, 265
Launey-Slade hallucination scale, 258–9
learned helplessness, 4, 9, 237, 239–46, 267, 272
  context specificity, absence of, 244–6
  frustration-instructions, 244
  relationship to LI, 239–40, 244–6
  S_1–S_2 effects, 242–6, 272
long-term memory (LTM), 175–6, 180, 211–213

Mackintosh's theory of attention, 165–8, 170, 174, 180
Maine Scale, 252, 254
masking task, 110–11, 113–16, 121–9, 218–23, 228, 248, 253, 267, 272
  ontogenetic considerations, 229
  phylogenetic considerations, 229–30
  spatial relationships, 219–22
"mere" exposure effect, 55
missing stimulus effect, 271
mollusc, 101, 107
monkey, 106
mouse, 106

nalaxone, 138
neophobia, 11, 74
nictitating membrane conditioning, 24–5, 26–7, 47, 89–90, 131, 133, 197, 268
noradrenergic manipulations, 130, 132–3, 138–40
nucleus accumbens, 262–3, 269
nucleus locus coeruleus, 132

observing response, conditioning of, 40–4
occasion setting, 214–15, 217, 271
one-trial learning, 5
overshadowing, 4, 5, 6, 74, 195
overshadowing of LI, 198–9
opiate manipuations, 130, 138
orienting response (OR), 40–4, 65, 116, 121, 155–60, 162, 188, 205–6, 211, 221, 225–7, 266, 269–71
  conditioning, 37–44

P300, 37, 157, 269
paired-associate learning, 111
Pearce-Hall model, 169–74, 180
perceptual learning, 52–5, 76–8, 209
  context change, and, 53–5, 76–8
phobia, 269
pigeon, 32–7, 38–9, 106, 107
Posner's two-process theory, 231
post-trial episode, (PTE), 195
prediction theories, evaluation of, 180–2
primacy effect, 89, 268
priming tests, 178–80
prophylactic treatment, LI as, 269
psychoticism, 258–9
  see Eysenck's Personality Questionnaire

rabbit, 38–40, 89, 106, 107, 144, 149, 279
Raven matrices, 111
reaction time, 108, 111–12, 227, 231–3, 271
reflex figures, 146
reinforcement schedule effects, 55
Rescorla-Wagner model, 75, 81, 145, 162–6, 168, 170, 174–6, 238, 270
retardation test of conditioned inhbition, 142–4, 162–3, 266
retarded adolescents, 128
retention of LI, 66, 68–74, 196, 199
retrieval failure hypothesis of LI, 182–7
  acquisition test delay, 183–4
  reminder treatments, 184–6
reversal shift, 229
rotation US, 129

S_1–S_2, 82–95, 179, 239, 242, 267, 270–1
  number of pairings, 87, 191–5, 200–5, 208, 216–17, 227–8
  interstimulus interval, 203
  S_2 change, 205
  S_2 intensity, 205
  S_2 omission, 202–3
  S_2 prior exposure, 203
S_1-weak shock effect, see Hall-Pearce effect
S_2–S_1, 89–92, 267, 271
salivary conditioning, 31
schizophrenia, 7, 9, 130, 134–5, 237, 239, 247–64, 272
  animal amphetamine model of, 130, 134–8, 249–50
  attentional processes in, 247–50
  chronic and acute, 256, 257–9
  dopaminergic theory of, 249–51, 260–2
  latent inhibition tests with
    Lubow et al. (1987) Study, 252–7
    Maudsley Studies, 257–9
  paranoid and nonparanoid, 249, 251, 252–7, 272

schizophrenia (*cont.*)
  positive and negative symptoms, 260–1, 271
  relation to latent inhbition, 249, 263–4
  selective attention deficits, 255
schizotypal personality, 258–9
  questionnaire, 258–9
scopolamine, 133
second stimulus effects, 82–95
  see Hall-Pearce effect, S_1–S_2, S_2–S_1
selective attention, 160–1, 167, 238
  inverse hypothesis in, 160–1
selective filters, 141
sensitization, 74
sensory adaptation, 196
sensory preconditioning, 1–2, 195, 207, 265
septo-hippocampal manipulations, 130–2, 138–40
serotoneric manipulations, 130, 134, 138–40, 145
sex, 101
sheep, 27–9, 106, 107
short-term memory (STM), 152, 172, 175–6, 178–80, 211–13, 235–7
socioeconomic class and LI, 109–10, 229–30
species differences, 101–17
  habituation vs. LI, 6
spontaneous labeling, 112
spontaneous recovery of LI, 199
state conditioning, 153
stereotypy, 272
stimulus encoding, 209–11, 215–17, 229, 230–5
  property extractions, 232–3
  stimulus relationship, 233–5

two processes, 230–2
stimulus differentiation, 112
stimulus equipotentiality, 4, 6
stimulus salience, 141, 145, 197
stimulus specificity, 58–9, 110–11, 114, 196, 209, 247, 267, 271
stimulus variables in PE affecting LI
  duration, 63–4, 70–2, 188, 207, 208, 209–13
  intensity, 64–5, 197, 204–5, 210, 241, 243, 244, 269
  interval, 65, 67, 197
  number, 59–63, 117–19, 188, 196, 207
successive vs. simultaneous stimulus preexposures, 112, 118
sulpiride, 138
summation test of conditioned inhibition, 142–4, 163, 266, 268
superior colliculus, 65

temporal conditioning, 153
trace conditioning hypothesis of LI, 182, 186–8
triadic design, 242
tropism, 6

unconditioned suppression, 11–13, 83, 149, 270
unconditioned responding, *see also* alpha response, 11–13
US preexposure effect, 82, 90, 268, 271

Wagner's priming theory, 172, 175–80, 235–6, 271
  retrieval generated priming, 175
  self-generated priming, 176